THE SOUND OF FREEDOM

1918 - 2006

THE SOUND OF
FREEDOM
Naval Weapons Technology at Dahlgren, Virginia

1918-2006

For sale by the Superintendent of Documents, U.S. Government Printing Office
Internet: bookstore.gpo.gov
Phone: toll free (866) 512-1800
DC area (202) 512-1800
Fax: (202) 512-2104
Mail: Stop IDCC, Washington, DC 20402-0001

ISBN 0-16-077712-7

NSWCDD/MP-06/46
Approved for public release; distribution is unlimited.

About the Authors

Rodney P. Carlisle is vice president and senior associate of History Associates Incorporated (HAI), a historical services firm located in Rockville, Maryland. He is professor emeritus of history at Rutgers University, The State University of New Jersey, at the Camden Campus. He holds an A.B. in History from Harvard College and an M.A. and Ph.D. in History from the University of California at Berkeley.

Dr. Carlisle is the author of several prior works in the fields of military and naval history. Recent works include *Where the Fleet Begins: A History of the David Taylor Research Center* (Naval Historical Center, 1998); *Supplying the Nuclear Arsenal: American Production Reactors, 1942-1992* (Johns Hopkins University Press, 1996); and *Powder and Propellants: Energetic Materials at Indian Head, Maryland, 1890-2001* (2nd ed., University of North Texas Press, 2002). He has also written a number of shorter studies for the Navy Laboratory/Center Coordinating Group, published by the Naval Historical Center, which provide detailed examinations of Research, Development, Testing, and Evaluation (RDT&E) in the Navy. Among these are *Management of the U.S. Navy Research and Development Centers During the Cold War: A Survey Guide to Reports* (Naval Historical Center, 1996) and *Navy RDT&E Planning in an Age of Transition: A Survey Guide to Contemporary Literature* (Naval Historical Center, 1997). With Dr. James Lide of HAI, he coauthored *The Complete Idiot's Guide© to Communism* (Alpha, 2002). Dr. Carlisle has also edited the *Encyclopedia of the Atomic Age* (Facts on Files, 2001) and the *Encyclopedia of Intelligence and Counterintelligence* (Facts on File, 2004). Additionally, he has recently published *The Persian Gulf War* (Facts on File, 2003) and *The Iraq War* (Facts on File, 2004). He and his wife, Loretta, make their home in Cherry Hill, New Jersey.

James P. Rife is a historian with HAI and a colleague of Dr. Carlisle. He holds a B.S. in Electrical Engineering Technology from Bluefield State College, Bluefield, West Virginia, a B.A. in History from King College, Bristol, Tennessee, and an M.A. in history from the Virginia Polytechnic Institute and State University, Blacksburg, Virginia. Mr. Rife has broad research and writing experience in the fields of American military and naval history, and has taught Modern European History and Western Civilization as a graduate assistant at Virginia Tech and the University of Tennessee–Knoxville. This is his first published book. He and his wife, Samantha, reside in Gettysburg, Pennsylvania.

Table of Contents

Table of Contents (Continued)

Foreword

The Sound of Freedom: Naval Weapons Technology at Dahlgren, Virginia, 1918-2006

When the United States Navy sails into harm's way, our warships and weapons must be as good as we can make them. Since the 19th century, officers like Lieutenant John Dahlgren have worked to bring scientific advancement into the fleet. Through his leadership and that of others like him, much of the Navy's shore establishment has been devoted to making better ships, more powerful and more accurate guns, stronger armor, and in more recent years, improved aircraft, bombs, missiles, and electronics to support the mission of the sailor as both a warfighter and peacekeeper. We pride ourselves on the character and training of the men and women who serve in uniform; we also take great pride in the quality of equipment that we provide them.

This book tells the story of one part of the Navy's research and development effort. Rooted in tradition and heritage traced directly back to the first efforts of Lieutenant Dahlgren to improve the scientific study of ordnance and naval weapons technology, our facility on the shores of the Potomac River started life at the end of the First World War, when the Navy needed a longer testing range than the existing facility at Indian Head, Maryland. Every major naval gun and every lot of ammunition had to be tested, not only to guarantee safety, but also to calculate the ballistic data necessary to ensure accuracy in fire control.

The Naval Proving Ground at Dahlgren, Virginia, first known as the "Lower Station" of the Indian Head facility, soon expanded its mission area into other technologies such as aerial bombing and formally separated from Indian Head in 1932. Many of the types of research here were far ahead of their time. We see this in the previously little-known story of the attempt to develop automatic and remote-controlled aircraft that could serve as weapons, forerunners of modern missiles and unmanned aerial vehicles (UAVs). Often the work at Dahlgren was highly secret, and only in later years could the details be released, as with the development of the Norden bombsight. Other once-classified stories, such as the development and testing of proximity fuzes, the ballistic experiments conducted on early scale models of the "Little Boy" atomic bomb, the ballistic experiments conducted on its later derivative, the "Light-Case" ground penetrator bomb, and Dahlgren's movement into the fields of computing technology and systems engineering, spell out the important role in the nation's defense that the station has played over the years.

As James Rife and Rodney Carlisle point out, Dahlgren continued to take on new missions, building on established reputations and achievements. There was a logical progression from the ballistic computation of gun projectiles, to calculating high-altitude bomb trajectories, to guiding long-range ballistic missiles to their designated aim points. Some of the first large computers built immediately after World War II were installed at Dahlgren, and naturally, our people were ready to act whenever new demands for computer knowledge appeared.

As a center of innovation devoted to the Navy's needs, and more broadly to national defense needs, Dahlgren has constantly adapted to change. Beyond the Cold War, sailors and Marines needed new technologies in Vietnam, in the Middle East, and elsewhere around the globe. And the nation has faced the need for fresh technological innovation to deal with the War on Terror. Taking a lead in sensors technology and a "joint" approach to defense and security needs, including "naval operations other than war," at Dahlgren we have worked on literally thousands of technical advances, many still classified, that serve to strengthen the nation against new and ever shifting threats.

At Dahlgren, we were fortunate to recruit the services of Dr. Carlisle and Mr. Rife to assist us in bringing together the many fascinating aspects of naval technological history presented in this volume. Both experienced professional writers and historians, the two authors brought their experience from History Associates Incorporated to the task. Dr. Carlisle's previous works include studies of other parts of the naval shore establishment, including *Where the Fleet Begins: A History of the David Taylor Research Center*, a study of the Carderock Division of the Naval Surface Warfare Center, and *Powder and Propellants: Energetic Materials at Indian Head, Maryland, 1890-2001*, which chronicles the story of Dahlgren's parent facility at Indian Head, Maryland. Mr. Rife and Dr. Carlisle interviewed and corresponded with dozens of key people who had worked on many of the weapon and sensor systems directly. Mr. Rife worked particularly closely with many of Dahlgren's current scientists and engineers, checking and rechecking the facts and the phrasing, touching base again and again to make sure that the information was fully documented and clearly presented.

Even for those of us directly involved in one or another specialized piece of work, much of the story we find here is fresh. Technical work requires that specialists know their individual research and development areas well, and consequently, they may not be familiar with the tasks or challenges faced by others outside their respective internal organizations. Because such specialization can sometimes generate a narrow perspective, for more than a generation at Dahlgren, civilian and naval managers have worked to overcome the natural compartmentalization, or "stove piping," of technical work, by rotating managers within departments, divisions, branches, and even sections, so that a broader view of the tasks and capabilities can inform their decisions and broaden their outlooks.

In addition to exploring the history of the technologies, this volume explains the evolution of these management styles, what many called "The Dahlgren Way." That part of our heritage lives on, and this volume will help newcomers to our institution better understand the roots of our broader outlook and to learn of the great range of tasks that our researchers have explored. For others inside the defense establishment, the history of the Dahlgren Way can serve to explain a technique of R&D management that may find application elsewhere. For readers outside the Navy and beyond Dahlgren who simply want to understand the Navy and its equipment better, we are sure this book will provide a rich and readable reference.

Captain Joseph McGettigan, USN
Commander, Naval Surface Warfare Center, Dahlgren Division

Acknowledgments

A book project of this magnitude could not be completed without the assistance of a great many people, at History Associates Incorporated (HAI), at Dahlgren, and at the various archives and repositories in which we worked. We therefore wish to express our sincere gratitude to the following individuals for their assistance, input, and support through the course of this two-year endeavor.

First and foremost, HAI History Division director Kenneth Durr was a ruthless editor, combating technical jargon and paring down cumbersome, overly wordy passages in early drafts to make the text more presentable for readers. His editorial surgery improved the work, and for that we are grateful. Likewise, manuscript specialist Gail Mathews patiently read every word of every draft, addressing stylistic problems and suggesting revisions for clarity. Her diligence and seemingly endless patience through all phases of the project have contributed significantly to the book's overall improvement. We also wish to thank our colleagues Paul Veneziano, Amber Moulton-Wiseman, Jennifer Rogers, and Janet Holsinger for repeatedly pitching in at critical moments to help with short-notice archival and library research as deadlines loomed. Likewise, Garry Adelman not only helped conduct research at the National Archives on Dahlgren's early history, but he also provided significant assistance in the formatting and preparation of the electronically scanned photographs used to illustrate this work. Special thanks also goes out to HAI's support staff, including Carol Spielman, Mary Ann FitzGerald, Jenny Bradfield, and Camille Regis, whose assistance in administrating the project's correspondence, managing our research files, and photocopying documents made our task much easier. Darlene Wilt provided invaluable assistance by managing both invoicing and progress report procedures and, along with Gene Jurasinski, helped us overcome computer difficulties whenever they threatened to overwhelm us. Ruth Dudgeon and Barbara Hunt also helped us keep the project's contract and finances in order. We would also like to extend our thanks to our colleagues Philip Cantelon, Richard Hewlett, Brian Martin, Mike Reis, James Lide, and Adrian Kinnane for allowing us to tap their considerable knowledge of repositories and sources. Their support and guidance provided new perspectives on the context of Dahlgren's history and opened inroads into potential new avenues of research.

We also wish to thank a large number of individuals at Dahlgren who assisted us over the course of this work, beginning with Naval Surface Warfare Center, Dahlgren Division's former Executive Director, Tom Pendergraft and Captain Lyal Davidson (Dahlgren's commanding officer from March 2001 to April 2004), both of whom launched the project in 2002 to preserve Dahlgren's history and corporate memory in the face of continuing uncertainty about the future of the Navy's RDT&E establishment. We were fortunate to capture much of Dahlgren's memory by interviewing thirteen people either formerly or currently affiliated with Dahlgren who kindly agreed to share their vast knowledge of different time periods and aspects of the station's organization and history. These include Captains Davidson, Paul Anderson, and Norm Scott, along with Lemmuel Hill, Tom Clare, Armido DiDonato, Rob Gates, Gene Gallaher, Sheila Young, Tommy Tschirn, Barry Dillon, Joe Francis, Dave Colby, and Charles Roble. They not only shared many previously obscure facets of Dahlgren's history but also added much needed color to otherwise dense discussions of Navy weapons technologies and capabilities.

The complex technical language associated with Navy RDT&E presented us with a real challenge. Consequently, we are grateful to the panel of Dahlgren readers who reviewed our chapter drafts for accuracy and offered constructive criticism and written recommendations for improvement and revision. Included among these readers were Rob Gates, Gene Gallaher, Bob Hudson, Wayne Harman, Walter Hoye, Robin Staton, and Tommy Tschirn. During the revision process, they readily answered technical questions and thereby ensured that the book is as accurate as possible, particularly in covering the most recent years. Rob Gates took a particularly active interest in the book and provided guidance on a wide variety of matters based on his professional experience and academic interest in Dahlgren's history, SLBMs, and Navy RDT&E. Bob Hudson and Gene Gallaher likewise were invaluable editors of several of the book's later chapters by helping us tell the story of Dahlgren's Joint Warfare Applications (or "J") Department and its vital national defense work. Additionally, Wayne Harman contributed a great deal of insight into Dahlgren's role in the TOMAHAWK program and helped us identify, collect, and caption many of the photographs used in this book.

We also owe a debt of thanks to several members of Dahlgren's support staff who assisted us during our research. An always cheerful Karen Melichar not only handled administrative issues on Dahlgren's end during the project, but she also got us access to the station's historical files and unarchived records in both the Headquarters Administrative Vault and the so-called Wine Cellar Collection. She likewise opened her own collection of unclassified files for our review. Whenever Karen was unavailable, command staff secretaries Diane Ovadia and Joan Fridell did not hesitate to step in and help us access the headquarters' records and to maintain our formal communications with Dahlgren's leadership. At the Technical Library, Wes Pryce helped identify and photocopy vintage technical reports, photographs, and other historic materials using Dahlgren's internal electronic catalog. He also directed us to Dahlgren's Museum Collection, under the care of Naval District Washington (NDW)–NEPA Program Manager Patricia Albert. Even though she was very busy with her primary job in Dahlgren's environmental office, Patricia often went above and beyond the call of duty in responding to our numerous research requests and repeatedly scheduling, with some difficulty, a high-demand conference room for our use. For that, we give her our sincere thanks.

Following Karen Melichar's transition to the NDW staff, Corporate Communications Director Russ Coons helped move the project to completion by coordinating with us, the Dahlgren review panel, and contracting officer Connie Salisbury, who worked out the publication arrangements with the Government Printing Office and the book's publisher. Additionally, Russ's staff members Janice Miller, Stacia Courtney, Kathy Rector, Lucia Sanchez, John Joyce, and Shedona Chisley all worked extremely hard to coordinate the transmission of chapter drafts to the review panel members. Jan was especially helpful and patient as we worked through a number of manuscript formatting issues, as well as acting as liaison with Bob Hudson, during a final round of revisions conducted in September and October 2005. Russ's staff also scanned many of the historic photographs used in this book before sending them to NSWCDD graphic designer Clement Bryant, who expertly supervised the final formatting and publication process. No book of this kind can come to fruition without the assistance of the professional archivists and curators who manage external repository collections and records. We received tremendous help in locating Dahlgren-related files from naval archivists Barry Zerby and Charles Johnson at the National Archives facilities in College Park, Maryland, and Washington, D.C., respectively. Mike Walker at the Operational Archives Branch and Eric Hazell of the Navy Laboratory/ Center Coordinating Group archives, both located at the Naval Historical Center (NHC) at the Washington Navy Yard, also provided key archival support for their collections. Likewise, the staff of the Navy Department Library proved equally friendly and helpful during our research there. At the NHC's Photographic Section, Rob Sanshaw and Ed Finney helped us locate and collect additional photographs for illustrative purposes, as did Kay Peterson at the Archives Center of the Smithsonian Institution's National Museum of American History in Washington, D.C., and Sam Bono, Tom Salazar, and Myra O'Canna of the National Atomic Museum in Albuquerque, New Mexico. Their assistance is greatly appreciated.

Although we consulted a large number of secondary works during the preparation of this book, as noted in the bibliography, it is largely based on primary sources located at Dahlgren and in Washington, D.C., area repositories. We have benefitted from the assistance of Dahlgren's review panel and the oral history participants in our attempt to place those sources within their proper context, but any mistakes, omissions, or erroneous conclusions remain the responsibility of the authors.

Finally, we have reserved our most heartfelt thanks for our wives, Loretta Carlisle and Samantha Rife, who somehow tolerated the intrusion of naval weapons and warfare systems into their homes by virtue of our research and writing. Without their love, patience, and support, we could not have completed the book, and we thank Loretta and Samantha for so graciously allowing us to impose upon them our fascination with military technology.

THE SOUND OF
FREEDOM
Naval Weapons Technology at Dahlgren, Virginia

1918-2006

Chapter

Introduction:
Proving Ground to Warfare Center

From the earliest days of the American Republic, the United States Navy has had a huge stake in mastering the changing technology of warfare. No less than the naval technology of warships and weapons, the knowledge of how to build, operate, maintain, and use them in battle has always been crucial for the country's seagoing warfighters. Beginning with the establishment of the original procurement bureaus in 1842, naval ordnance officers worked alongside a dedicated corps of civilians to apply new advances in science and technology to the design and construction of ships, armor, guns, projectiles, and propellants. After World War I ended in 1918, much of this work was done on an isolated point of land in Virginia that overlooked the lower Potomac River, originally known as the Lower Station of the Indian Head Naval Proving Ground but later called the Dahlgren Naval Proving Ground.

Since its establishment in 1918, Dahlgren has been repeatedly transformed and restructured, and enters the twenty-first century as the Naval

Surface Warfare Center, Dahlgren Division. Despite these changes and its diversification into new fields of naval warfighting technology, Dahlgren has retained its core missions of gun and ammunition testing and fire control computation, building upon the scientific and mathematical methodologies established in the mid-nineteenth century by the station's namesake, Rear Admiral John Dahlgren. Even as the Dahlgren Naval Proving Ground evolved to keep up with swift advances in twentieth century science and technology that came ever more quickly after World War II and during the Cold War, it did so in a way consonant with Rear Admiral Dahlgren's original vision. His principles of ordnance, science, mathematics, and engineering have proven even more important to the modern Navy than they were to the Navy of the mid-nineteenth century.

Accurately and safely striking a target with a projectile through ballistic trajectory computation was the key challenge taken up by then Lieutenant John A. B. Dahlgren in the 1850s. The problem was not only complex but also never ending, since naval weapons, propellants, and fire control technology improved steadily over time. By the early 1900s, a heritage in ballistics had already been well established at the Indian Head proving ground, with naval ordnance officers, chemists, industrial artisans, and enlisted gunners having worked together to assure the quality of propellants and explosives and the reliability of naval guns and armor. After the Bureau of Ordnance opened the "Lower Station" at Dahlgren, Virginia, in late 1918, military and civilian personnel from Indian Head brought that heritage with them and embedded it within the new proving ground's organizational culture.

The addition of a chief physicist, Dr. L. T. E. Thompson, to the station's civilian staff in 1923 strengthened Dahlgren's commitment to furthering science and technological change. Under Thompson's direction, naval proof and experimental officers applied the science of physics not only to the ballistics problems concerning shipboard guns, projectiles, and armor, but also to the fields of offensive and defensive aircraft ordnance. Dahlgren's movement into aerial warfare technology resulted in the development of Carl Norden's Mk 15 bombsight and the VT "proximity" fuze, both of which were critical to the Allied victory in World War II. Similarly, Elmer Sperry's early "flying bomb" experiments and Norden's work in automatic pilot technology were ahead of their time, and through the examples of their early research, Dahlgren established a beachhead in the missile revolution that followed the war.

Engagement in difficult ballistics problems led the Dahlgren Naval Proving Ground into ever more sophisticated research and development activities. Expertise in the mathematics of trajectory computation and fire control led Dahlgren's scientists to play important roles in the development of early analog computers, such as the Mk 15 bombsight and the Aiken Relay Calculator. This post-World War II emphasis on computing technology converged with Dahlgren's early aviation ordnance experiments to take the proving ground into the exciting new fields of rocketry and ballistic missiles just as the Cold War heated up in the 1950s. In parallel fashion, Dahlgren's involvement in satellite geodesy and space surveillance, in conjunction with its missile fire control work during the POLARIS, POSEIDON, and TRIDENT programs, positioned it to assume a leading role in AEGIS and higher order systems engineering during the last two decades of the twentieth century. AEGIS, in turn, transformed the intellectual underpinnings not only of military hardware design and engineering but also the philosophy of modern warfighting.

This record of technological achievement notwithstanding, Dahlgren's heritage encompasses more than an ability to handle the complex mathematics and engineering associated with ballistic weapons and projectiles. Part of it stems from a concept that came to be known as the "Dahlgren Way," which Thompson and his colleagues first formulated during the 1930s. Although different individuals have expressed the phrase differently, all shared a common understanding. Thompson's research, development, testing, and evaluation (RDT&E) philosophy revolved around an entirely self-sufficient laboratory in which concepts were quickly researched, developed, analyzed, designed, built, tested, and evaluated all in one place without involving outside institutions or contractors. The idea was to work fast, to make mistakes fast, and to learn fast in order to develop the best possible weapons and ordnance for the Navy without bureaucratic meddling or burdensome contract negotiations. Technical knowledge was at the heart of the Dahlgren Way. Thompson believed that Navy laboratories were most qualified to ensure that the Navy got the best product for its money; that they owed it to the service to know more about weapon and ordnance engineering than defense contractors. He warned that laboratories like Dahlgren had to resist pressure to focus on the Navy's short-term needs and insisted that doing responsible, long-term science and engineering meant doing right by the Navy. This was a persistent refrain in later years as Thompson's successors struggled to hold the line against the Navy's short-term technical fads in the greater interest of keeping the Navy ready for the warfare needs of the distant future.

As it developed, the Dahlgren Way not only encompassed RDT&E but also laboratory management. Thompson's system imparted upon Dahlgren a distinctive approach that was in many ways far ahead of the times. That approach was founded upon the recognition that, very often, talent and expertise resided far down in the chain of command. At Dahlgren, technical leaders were encouraged to channel responsibility to talented individuals at the bench level. Their efforts were very often rewarded with initiative and brilliance. Although management schools began to recognize the virtues of participatory management and the "power of the individual" widely by the 1980s, Dahlgren had already established just such a pattern, under the leadership of key individuals like Thompson, Bernard Smith, and James Colvard. When in a later period Tom Clare and Tom Pendergraft became Dahlgren's top civilian leaders, these patterns of responsibility and recognition for talent were deeply rooted and still bearing fruit.

Organizationally, the Dahlgren Way perpetuated the dual laboratory leadership system pioneered by Thompson, in which a Navy captain oversaw the station's administration and security, while its senior scientists managed its technical RDT&E operations. Although the system was subject to stress depending upon the personalities and leadership styles of Dahlgren's various military commanders and technical directors, it directly linked the station's scientists to the fleet and proved a successful mechanism for Dahlgren's scientists to quickly react to fleet problems whenever they arose. For the most part, Dahlgren's skippers and their superiors recognized the strength in the heritage they encountered—they quickly understood that the civilians in the structure knew how to hire good people and to give the resources, challenges, and rewards it took to succeed. Most of the officers reveled in this flexible and responsive atmosphere and helped ensure that the focus of research and development groups at Dahlgren responded to newly arising defense needs.

This intertwining relationship between scientists and military personnel helped Dahlgren develop, maintain, and update the technical knowledge of its staff and enabled it to react to fleet technical problems, often under combat conditions, on a moment's notice. To some extent that process was made easier by the close proximity of Dahlgren to Washington, D.C., and Navy headquarters and defense procurement offices. More important, though, was the laboratory's location within the Navy's procurement structure. Indeed, to fully understand the management aspect of Dahlgren's heritage, one has to carefully follow its shifting position within the Navy Department and within the Department of Defense hierarchy. As the station rose from humble beginnings as an isolated "auxiliary" proving ground under the old Bureau of Ordnance to become the Navy's premier R&D laboratory under the Naval Sea Systems Command, the Dahlgren organization gravitated upward within the Navy RDT&E establishment, ultimately reaching "center" status in the 1970s, thanks to its stature as a crack technical troubleshooter and research facility that had been heightened in the post-World War II world. This progression through the Navy's procurement structure over the years stands as a testament to a legacy of forward-thinking and capable management that began with Thompson and continued through the years.

As the Navy and the Defense Department struggled through cycles of downsizing during the post-Cold War era, Dahlgren's structure and heritage gave it enormous resiliency, and so it survived a series of congressionally mandated defense drawdowns, reductions-in-force, and budget cuts. Geography, of course, helped. Not only was it close to Washington, but more importantly it remained isolated enough to continue serving as an excellent river range for the testing of guns and ordnance, with fixed observation points ashore for monitoring and measuring shotfall. It remained, in fact, the only such test installation in the world. With some 4,500 acres of real estate (much of it too swampy for profitable private development), long water frontage, and its own airfield, Dahlgren represented a solid physical and technical asset that the Navy and the nation decided to preserve. Minimal suburban encroachment helped shield the station from some of the pressures that closed the White Oak Naval Ordnance Laboratory and threatened other naval shore establishments in the 1990s.

Over and above these factors though, Dahlgren's heritage of technical expertise enabled it to not only survive but also flourish, moving beyond serving the Navy to meeting national defense needs. The laboratory's 1980s emphasis on systems engineering spurred several innovative managers to help shape new warfighting philosophies such as the doctrines of "theater warfare" and "jointness." As the Defense Department gradually adopted the integrated approach called for in the 1986 Goldwater-Nichols reforms, this special work at Dahlgren was legitimized and institutionalized in two wholly new technical departments. The technologies arising from these departments would become vitally important to the country by the early twenty-first century, as government policy makers and American troops began facing regional threats from rogue nations, armed with modern strategic missiles and "weapons of mass destruction," and even more dangerous asymmetrical threats from terrorists, armed drug smugglers, and guerilla forces operating in the Third World.

Naval weapons and ordnance technology has drastically changed over time, but Rear Admiral Dahlgren's original four principles remain as valid today as they were in the mid-nineteenth century. Nowhere are they more in evidence than at the laboratory complex at Dahlgren, Virginia, that has served both the Navy and the nation well for over eighty years.

2

Chapter

Finding the Range, 1841-1932

The origins of the Naval Surface Warfare Center, Dahlgren Division, can be traced to the 1840s, a period in which the U.S. Navy began seriously exploring technological innovation in ordnance and gunnery. These efforts were led by Secretary of the Navy Abel Parker Upshur, a champion of naval expansion and modernization. Beginning in 1841, Upshur lobbied Congress hard for meaningful naval reforms and also urged the frustratingly hidebound Board of Navy Commissioners to allow him to start a concerted program of experimentation in the physical sciences to improve naval ordnance. Upshur's modernization program gained steam in 1842 when Congress finally accepted his recommendations and authorized a major reorganization of the Navy, which established a bureau system of management and placed emphasis on scientific applications in naval design and engineering. The Navy's new organization included a Bureau of Ordnance and Hydrography that was charged with developing and constructing shipboard weapons and armor.[1]

Upshur's ideas were reflected in the Navy steam sloop USS *Princeton*, which was launched in 1843. The Princeton, the first screw-powered warship in naval history, was designed and built by Swedish engineer John Ericsson, under the direction of railroad financier and Navy Captain Robert Stockton. The *Princeton's* technological improvements were not limited to her propulsion system. During an 1839 visit to England, Stockton had seen British and French experiments with lighter-weight wrought iron guns and was suitably impressed. After consulting experts in wrought iron technology and considering the relative strengths of different materials, he finally commissioned in 1842, on his own volition, an experimental 12-inch, 225-pound wrought iron gun from the Mersey iron works in England.[2]

The new gun, called the "Oregon," was forged according to his specifications and shipped to New York in the summer of 1843. Stockton had his special assistant, Lieutenant William E. Hunt, pick up the gun and transport it to the U.S. Army's Sandy Hook, New Jersey proving ground, where the captain planned to test it himself. At Sandy Hook, Hunt and several of the *Princeton's* crewmen set the Oregon up on a sand emplacement, and once Stockton arrived they loaded it with 35 pounds of black powder and a 212-pound solid shot for the first test. Stockton then fired it. The test appeared entirely successful, but when Stockton ordered Hunt to mount the Oregon on a carriage for additional firing, the crewmen discovered a longitudinal crack underneath its breech. Stockton had them complete the mounting, and he fired the damaged gun three more times, using only a reduced charge of 14 pounds of powder and solid shot to see if the crack opened further. When it did not, he ordered the breech reinforced with 3½ inch-thick iron bands, which covered the crack but did not seal it. Over the next several days, he fired the Oregon approximately 150 more times at full charge, sometimes 15 to 20 times a day in rapid succession. Under the repeated stress, the crack finally did expand inward, so much that water poured into its chamber when his crew washed the gun. Despite the obvious warning sign, Stockton convinced himself that the technology was safe since overstressed wrought iron appeared to simply split open rather than fly apart like cast iron.[3]

As Hunt subsequently recalled, Stockton believed that American iron was superior to English iron, and that a domestically manufactured wrought iron gun could withstand "any number of pounds of powder that could be burnt in it." Stockton therefore commissioned a second gun modeled on the Oregon, this time from the New York foundry of Hogg & Delamater. The new gun was finished in late 1843 and Stockton named it the "Peacemaker." The Peacemaker, weighing some 27,000 pounds, had the same chamber size as its parent design but was constructed with twelve inches of additional metal around the breech, which Stockton, Hunt, and Ericsson all agreed was far stronger than the Oregon's reinforcing bands. The gun's strength appeared to be confirmed by an especially severe test conducted in New York, when the manufacturers proof-fired it using over 49 pounds of powder. A close examination revealed no cracks, and after four more test firings at Sandy Hook, with incrementally greater charges, Stockton pronounced the Peacemaker safe and fit for service.[4]

In January 1844 Stockton installed both the Oregon and the Peacemaker aboard the *Princeton* and then steamed to Washington, D.C., arriving on 13 February. Three days later, he began taking passengers down the Potomac to showcase the ship's capabilities. During these public relations excursions, he occasionally fired the Peacemaker to impress his passengers with the size of its shot and its muzzle blast. He also hoped to convince Congress of the benefits of outfitting more warships with more heavy guns like his, a measure supported by President John Tyler. On 28 February, he once again demonstrated the ship and the Peacemaker, this time for some of the government's highest officials. Among the 350 dignitaries and guests in attendance were President Tyler, various senators and congressmen, and several cabinet members, including Upshur, Secretary of State since the previous July, and new Navy Secretary Thomas W. Gilmer (only nine days in office). Stockman sailed the *Princeton* fifteen miles down the Potomac and fired the Peacemaker twice for the delighted crowd. On the return trip, as the ship passed Fort Washington, Gilmer asked Stockman to fire his gun once more. Stockman complied. This time the Peacemaker exploded, killing Upshur, Gilmer, New York State Senator David Gardiner, Chief of the Bureau of Construction, Equipment, and Repairs Commodore Beverly Kennon, the U.S. Charge d'Affaires to Belgium Virgil Maxcy, two sailors, and President Tyler's valet. Stockton and Lieutenant Hunt were wounded in the blast, as were Missouri Senator Thomas Hart Benton and nine sailors.[5]

During the ensuing investigation into the "awful and distressing catastrophe," a naval Court of Inquiry and the House Naval Affairs Committee learned that the purchase of both the ship and its guns had been carried out entirely under the supervision of Stockton, without naval approval. After reviewing the Court of Inquiry proceedings, the Naval Affairs Committee concluded that "everything seems to have been left to Captain Stockton, to enable him to carry out his peculiar views" regarding wrought iron gun technology. It was a tragedy, but one that strengthened the hand of the Bureau of Ordnance in arranging future gun procurement and testing in a more scientific and organized fashion. Specifically, the committee stated that while it had "no disposition to advise an interference with the duties of the Executive by undertaking to prescribe the exact mode of arming our public ships," it felt "bound to express the opinion that an unusual species of armament, attended with danger, should not be introduced into the public service until it receives the full approbation of the ordnance officers as to its efficiency and safety."[6]

Although both the Navy and Congress ultimately absolved Stockton of blame for the accident, the reaction to the Peacemaker explosion led immediately to the Navy's adoption of a new policy of proper testing for all future naval ordnance work. In early 1845, John Y. Mason, Upshur's successor as Secretary of the Navy, appointed an ordnance board comprised of the chiefs of the bureaus. The board recommended tightening quality control in the manufacture of new guns as well as testing the range and power of naval weapons more systematically. The board also specifically called for the establishment of an onshore practice battery to test and range guns before their installation aboard ship. This "practice battery" became the forerunner of the Navy's proving grounds.

ENTER DAHLGREN

In 1847 Lieutenant John A. B. Dahlgren was assigned to the Washington Navy Yard. Although the yard had produced anchors, blocks, ammunition, cables, and gun carriages since the early 1800s, it had only recently become the Navy's center for metalworking, housing a rolling mill, a foundry, and extensive metal shops. Commodore Lewis Warrington, the head of the Bureau of Ordnance and Hydrography, had assigned Dahlgren the tasks of transforming the Navy Yard into an ordnance establishment and working on the development of war rockets. Additionally, Dahlgren did double-duty as Professor of Gunnery at the Naval School (renamed the U.S. Naval Academy in 1850) in Annapolis, teaching the subject of ordnance there twice a week. Dahlgren's scientific outlook and openness to new ideas in naval technology were unusual for an American naval officer of his time. Not only was he an experienced oceanographer and surveyor, but he displayed an eagerness to learn the scientific practices employed by the other navies of the world.[7]

One of Dahlgren's top concerns was gun ranging, and he found that Britain and France had already developed scientific means of doing this. Gun ranging was critical, since each production run had slightly different characteristics. Different minute alterations in angle were required for each of several guns aboard a ship to hit the same target, angles that could only be determined by consulting detailed range tables. Range tables for each gun, therefore, had to be worked out, showing the range with a specific charge and weight of projectile at different angles. Specific range tables for each weapon allowed gun crews to fire accurately under battle conditions, and the crucial work in establishing the tables could only be conducted by a rigorous program of test firing each new gun, under scientifically controlled conditions, by experienced officers keeping meticulous records. Thus, testing of guns before installation aboard ship would not only contribute to safety but would also allow greater accuracy through the use of the scientifically established range tables.[8]

To the casual observer, naval gunnery appeared to be a simple matter of pointing a weapon and shooting it. But the high-flying arcs of long-range ballistic shots made accurate naval gunfire a complicated affair since the length of the gun, the wear on the rifling of the barrel (called erosion), its charge, and the weight and shape of the projectile all interacted in mathematically complex ways. Moreover, the motions of both an attacking ship and its target, relative to one another, further complicated gunnery calculations to the extent that a successful hit often seemed a product of luck rather than ballistic science.

After further studying existing literature concerning ordnance and the determination of range tables, Lieutenant Dahlgren established a regular gunnery regimen at the Navy Yard, firing down the Anacostia River from what came to be known as the Experimental Battery, the first such test battery established for the Navy. Mounted on a "gun deck" platform overlooking the river, the guns, with a range of nearly five miles, had a clear line of sight down the Anacostia, across the Potomac past Buzzard's Point, intersecting the Virginia shore about where the modern Reagan National Airport occupies filled land just upriver from the city of Alexandria. For the new installation, Dahlgren himself designed special instruments including a gunner's quadrant, a micrometer for measuring distances, and an alidade for recording the impact of the shots. In a detailed report submitted in 1849, Dahlgren spelled out his methods and gave a description of his systematic procedures for testing and recording results. In subsequent years Dahlgren refined his methods and steered the Navy toward instituting a more defined research and development establishment, especially after his promotion to Rear Admiral and appointment as Chief of the reorganized Bureau of Ordnance (BUORD) in 1862, which had just transferred its former Hydrography function to the newly created Bureau of Navigation.[9]

Dahlgren died in 1870, but not before his legacy for using scientific methods had been well established at the Washington Navy Yard, and his Anacostia battery became the prototype for shore facilities later used to test naval guns.

FROM ANNAPOLIS TO MACHODOC CREEK

By 1872 gun ranges had increased to the point that BUORD was compelled to shift the Experimental Battery from its cramped quarters on the Anacostia to Annapolis, across the Severn River from the Naval Academy. Then, in 1890, as the Navy began developing all-steel ships and even longer-range weapons for the so-called "New Navy," the battery was moved again to Indian Head, Maryland, where a new 13,000-yard testing facility was built under the guidance of Navy Ensign Robert Brooke Dashiell. For the next twenty-seven years, Indian Head continued the gun ranging and testing program begun by Lieutenant Dahlgren in the late 1840s.[10]

After only a decade in operation, however, the Indian Head facility began to show its limitations, particularly in geography. The gun emplacements were installed next to the river in a small valley that was about 100 yards wide and drained by a small stream. Guns hauled by barge from the Navy Yard could be off-loaded from a scow at a dock right at the shore of the valley, wheeled on rails to the emplacement, tested, and hauled back to the scow. Shells could be fired directly across the valley into butt emplacements that held 10- and 12-inch armor plates as a means to test both shells and armor. In down range testing, however, the guns had to be fired blindly over the south embankment and the intervening land before passing over the open Potomac. On the riverfront, a spotter posted under a primitive lean-to phoned in reports of river traffic and plotted the guns' shotfall.[11]

Since housing at Indian Head was built on the high ground above the river, stray shots, flying pieces of armor, and the rotating bands from shells would occasionally fall into the civilian and military residences. In 1900 Lieutenant Joseph Strauss, the Officer in Charge at Indian Head, had grown concerned after witnessing shells passing over a civilian neighborhood at Stump Neck, immediately to the south of the station. In 1901 the Navy purchased more than a thousand additional acres at Indian Head to reduce the hazards and annoyance to the residents. This was only a short-term solution to a long-term problem, as the Marines later billeted on the acquired land discovered when they were forced to evacuate their barracks whenever firing was under way. In 1902 Chief of BUORD Rear Admiral Charles O' Neil officially informed Secretary of the Navy William H. Moody that "the great increase in the power of guns in recent years, and their greatly extended range, renders a more isolated location necessary for proving and ranging them," and that "the time is not far distant when the matter will have to be seriously considered." However, the Navy Department found it difficult to convince Congress of the situation's urgency, and therefore the matter was not "seriously considered" for another fourteen years.[12]

As O'Neil and others at BUORD recognized, Indian Head had become even more dangerous after a smokeless powder factory was built there in 1900. Lieutenant Strauss, who had surmounted numerous technological and administrative problems in order to get the factory into operation, reported the manufacture of 250,000 pounds of powder in the first year of operation alone. But keeping such a volume of smokeless powder in the same vicinity as gun testing presented significantly increased risks to both civilian and military personnel at Indian Head. The dangerous mix of powder factory, housing, and gun testing put local residents on edge. Since a fire or detonation in the powder factory itself could be fatal to workers, few became inured to the blasts from the guns when an explosion could signal a catastrophe to friends and relatives. Furthermore, as river traffic increased, the blind firing, lookout or no, remained an uncomfortable arrangement.[13]

As the Navy Department waited, no real long-term solutions appeared, and the number of incidents continued to mount. In 1908 a shell struck the water about forty feet from a Standard Oil tug pushing a barge, and the following year Congressman John Hull reported that a fisherman had complained that a shell fragment had damaged his nets. In 1911 at least two 12-inch projectiles fell on a residential area at Indian Head, fortunately without injury to the personnel living there.[14]

During 1910 and 1911, the Navy attempted to alleviate at least part of the problem at Indian Head by using the old monitor *Tallahassee* as an experimental gun platform and firing high-powered guns mounted on her at the condemned ram *Katahdin*, which served as a floating target. The results proved unsatisfactory, however. In reporting on the *Katahdin* operation, BUORD Chief Rear Admiral Newton E. Mason noted of Indian Head, "This station, while very conveniently situated for the work of a proving ground in the most restricted sense of the term—the actual proving of guns, powders, armor plates, projectiles, etc.—is altogether unfit for an 'experimental station.'" He pointed out that the range down the Potomac crossed the Virginia side of the river, prohibiting the use of explosive projectiles, and the result was to "tie the hands of the bureau in the matter of nearly all experimental work." In 1912 BUORD again reminded Congress of the problem, noting that "owing to the very limited facilities of the proving ground as an experimental station and to the danger to life and property," badly needed experiments were not carried out. It soon became a standard refrain among successive BUORD chiefs that "a new proving ground was worth more to the Navy than the price of a battleship."[15]

Then, in summer of 1913, Lieutenant Garret L. "Mike" Schuyler, testing a 14-inch gun, fired his second shot of the day, just as a yacht had cleared the range. Unfortunately, it was the presidential yacht *Mayflower*, with President Woodrow Wilson aboard, along with his personal physician and friend, Dr. Cary Grayson. Both Wilson and Grayson watched as a shell component, probably the rotating band, struck the water a few hundred feet away. The press made quite a story out of the fact that the Navy had fired on the Democratic President, then considered a bit of a pacifist and not too keen on a large defense budget. Assistant Secretary of the Navy (ASN) Franklin D. Roosevelt conducted an investigation into the incident but exonerated Lieutenant Schuyler of any wrongdoing, leaving the *Mayflower's* captain grumbling that Schuyler's reckless firing near the yacht had been a manifest "impropriety." The episode certainly impressed upon the President the limitations and dangers of Indian Head.[16]

Between 1912 and 1918, the Navy took a few limited steps to improve the safety of the range, purchasing small lots to round out the holdings on Stump Neck and also another 1,270 acres to bring the Indian Head holdings up to more than 3,200 acres. But Indian Head as a proving ground finally reached the breaking point when the demands of World War I swamped the Navy's range-testing program. During 1916-17, Indian Head tested 494 guns; during 1917-18, the facility tested more than 1,100 guns; and in 1918, the total number tested exceeded 3,400.[17]

Despite the increasing workload at Indian Head, Congress still hesitated, and the Navy, at the beginning of the war, deemed it "impossible . . . to entertain any idea of immediate transfer of proof activities to another site." The cramped valley became even more confining immediately after America's entry into World War I in April 1917, with a new lot of batteries, stringent traffic rules, twenty-four-hour testing schedules, and overlapping firing ranges that made operations a continual exercise in frustration.[18]

One pressing wartime need was a proper facility in which to test the big 16-inch, 45-caliber battleship gun, which had been developed and proved in 1914. Because of its enormous power and range, full elevation testing and accurate ranging of the 16-inch gun simply could not be achieved at the Indian Head site, and even horizontal, low-angle proving was a hazardous proposition. Despite extra precautions during an August 1916 test, a 16-inch gun blasted its projectile completely through a 13-5/8-inch belt armor plate, a braced butt built from 16-inch thick oak timbers, and twenty-seven feet of sand reinforced by 5/8-inch skin plates. After passing through the armor and butt, the projectile angled up, tumbled a mile below the station, and wrecked a house owned by a farmer named William Swann. No one was injured in the incident, and the Navy agreed to repair the family's house completely.[19]

The hazards only promised to heighten when the Navy planned to upgrade the 16-inch gun from 45 to 50 calibers in length, thereby increasing its power and lethality even further. The first of these monsters was tested at Indian Head, without incident, in April 1918, and Navy Secretary Josephus Daniels pronounced the new 16-inch, 50-caliber guns as "the last word in American naval design." Naval ordnance officers estimated that a full broadside from these guns would "produce energy equal to that required to lift a battleship to the height of the Washington Monument." Some 104 of these guns were planned for construction as part of the wartime construction program, and it was clear that Indian Head was not equal to the task of proving them without an inevitable accident.[20]

The idea of establishing a new proving ground away from Indian Head found a champion in the person of new BUORD Chief Rear Admiral Ralph Earle. Earle had entered the Naval Academy in 1892 and, having earned his commission as an ensign in 1898, served as a line officer in the fleet before coming to BUORD. His penchant for science led to assignments as a powder inspector and then, in 1908, as the Officer in Charge of the Chemical Laboratory at Naval Station, Puerto Rico. In August 1916, BUORD named him Inspector of Ordnance in Charge at Indian Head, and he served in that capacity until he became BUORD Chief in December. Earle was a creative, forward-thinking officer who strongly supported the adoption of 16-inch guns for the Navy's new battleships and, after America entered the war, was instrumental in organizing the North Sea "mine barrage." Additionally, he had conceived the novel idea of mounting reserve 14-inch naval guns on railway mounts and putting them into land service with the American Expeditionary Forces in France. Earle's railway guns were designed, built, and proof-fired only four months after the Chief of Naval Operations authorized them on 26 November 1917. The five-gun battery, manned by sailors trained at Indian Head and commanded by Rear Admiral Charles Peshall Plunkett, saw action in France late in the war, pummeling German railroads and supply depots behind the front lines near Verdun. A grateful Navy Secretary Daniels declared that "it was more than good fortune that in these testing times the Navy had Admiral Earle, one of the ablest and fittest officers, in direction of great ordnance plans and operations."[21]

The anticipated 14-inch railway guns brought the proving ground issue to a head for Earle and BUORD, since the Navy had no firing range that was capable of testing them at full elevation under battlefield conditions and would have to use the Army's Sandy Hook, New Jersey, proving ground for that purpose. If the Navy was going to send its guns and personnel into combat on the Western Front, then an alternative site to Indian Head would have to be acquired, and quickly.[22]

Although BUORD's first and preferred plan was the construction of a wholly new proving ground, Earle initially ordered Indian Head Inspector of Ordnance in Charge Commander Henry E. Lackey to confer with Army ordnance officers in Washington, D.C., to see whether or not BUORD could share the new Aberdeen Proving Ground, which had just opened near Baltimore, Maryland, in December 1917. Lackey quickly found that Aberdeen was so restricted, crossing the Baltimore Channel as it does, that firing longer-ranged naval guns there would be "highly unsafe." Additionally, Aberdeen's grounds were so laid out and its emplacements were so different from what BUORD needed for its gun mounts that Lackey determined that too much money would have to be spent to modify them for naval use. Since building a new proving ground would likely be much more cost effective than trying to operate at Aberdeen, Earle accordingly decided to pursue that option. Secretary Daniels agreed to support Earle but told him that BUORD could only purchase approximately one thousand acres and spend no more that $1,000,000 on the endeavor.[23]

With Daniels' blessing, Earle ordered Lackey to locate a one-thousand-acre site with a long range for the new proving ground. Ideally, it would be similar to the British range at Shoeburyness in Essex, located on the north mouth of the Thames River, where His Majesty's ordnance officers could receive Woolwich-manufactured, barge-transported naval guns and then "fire over the water at high tide and recover shells on the sand at low water." A special board appointed several years before had already combed the East Coast of the United States for geographically similar sites but had found none. However, Lackey was confident that he could find a suitable one based on his intimate knowledge of the Lower Potomac. Knowing that Earle wished to keep the new proving ground as close to BUORD, the gun factory, and the powder factory as possible, he promptly identified a spit of land lying along Machodoc Creek on the Virginia side of the Potomac, located about twenty-two miles downriver from Indian Head, as the best prospect. Geographically, the site was far superior to Indian Head and somewhat comparable to Shoeburyness. It provided a straight, unimpeded, over-water range of nearly 90,000 yards toward Chesapeake Bay (more than fifty miles away), and guns could still be shipped by barge from the Washington Navy Yard foundry. In addition, its isolation guaranteed that accidents such as those that had hampered Indian Head would never happen there. In short, the Machodoc Creek site was ideal for transporting and safely testing long-range, major-caliber guns, and so he recommended its acquisition to Rear Admiral Earle.[24]

Earle acted immediately. On 18 January 1918, he asked Congress for a $1 million appropriation for what he characterized as an "auxiliary" proving ground "for such guns as can not be safely tested at the present grounds." He had to explain to the House Naval Affairs Committee, rather delicately, a number of points that to him must have seemed self-evident. No, it would not be possible to combine Navy and Army gun testing at Aberdeen. No, the 90,000-yard range was not over land, and the million dollars was not intended to purchase a piece of land that large. Yes, there would be money left over for building the facility. Yes, he had reduced the amount from $2 million at the request of the Secretary of the Navy. The admiral kept his composure throughout the tedious budget hearings, and Congress subsequently approved the appropriation in Public Law 140 on 26 April 1918.[25]

Throughout Earle's testimony, the Battle of Jutland, fought in the evening and night of 31 May-1 June 1916, was very much on his mind. Although it is often offered as truism that admirals and generals are engaged in "fighting the last war" rather than preparing for the next, Admiral Earle and his staff in the bureau were a bit more current in their use of "lessons learned." Indeed, they had studied the battle closely and had carefully analyzed its ramifications in terms of science and technology, and how advances in both had affected its outcome. Although usually considered a "draw," in many ways the battle did reveal superior German gunnery technology while exposing inadequacies in British ordnance operations. The Germans had superior fire control and greater range and had also used illuminating shells—spotting shells with dye to distinguish the blasts from their ship's guns (during the night action)—and superior armor-penetrating projectiles. On the other hand, poor British design of the elevators to the magazines had resulted in the loss of at least one British capital ship and possibly a few others in the action. American naval officers took these lessons to heart, as did Navy Secretary Daniels when he explicitly tied the new proving ground to the battle in his Annual Report: "In order to keep pace with the rapidly increasing ranges of battles as shown by the action of the Dogger Bank and the Battle of Jutland, the Navy Department [acquired a] tract of land on the Potomac near Machodoc Creek, Virginia. . . . The creation of this new proving ground makes it possible for the Navy to test its biggest guns at their longest ranges, which heretofore could not be done."[26]

As authorized by Congress, President Wilson commandeered for the Navy 1,366 acres at Machodoc Creek on the Virginia side of the Potomac through two presidential proclamations. He signed the first of these on 10 June 1918 and acquired the initial 994 acres between Machodoc Creek and Lower Cedar Point Light. Within a few months, it became evident that more land beyond the stipulated one thousand acres was needed, and so he signed another proclamation on 4 November adding the adjoining 372-acre Arnold farm to the new reservation. Later, on 4 March 1919, Wilson also took control of Blackistone Island. The marshy, 70-acre island was situated on the Maryland side of the Potomac, some 30,500 yards (about eighteen miles) downriver from the new site. Its lighthouse would make an ideal observation station, and it also could serve as an excellent target for major-caliber projectiles, which could be recovered for examination even more easily than from the river. BUORD also wanted to use the island as an airfield, a seaplane and boat refuge, a range supply station, and a center for the range's communication service.[27]

CONSTRUCTING THE LOWER STATION

BUORD lost no time in making preparations for its new proving ground. In January 1918, just as Earle was requesting the initial $1 million appropriation and before Congress passed Public Law 140, Commander Lackey at Indian Head directed Lieutenant Commander S. A. Clement to begin making construction arrangements and to manage the project. Lackey assigned the task of actually laying out the new proving ground to Naval Reserve Force Lieutenant Swepson Earle, a hydrographic engineer who would later become a noted Maryland Conservation Commissioner and an expert on the Chesapeake Bay's ecology. Accordingly, through the spring of 1918, Clement made his administrative and logistical preparations while Earle (no relation to the BUORD chief) surveyed the site and drew up a topographic map. Earle was particularly sensitive to the Potomac's ecosystem and carefully planned the range so that the large shells would fall into deep water a safe distance from the main oyster bars and rocks in the Lower Potomac.[28]

On 28 May 1918, almost two full weeks before Wilson issued his first proclamation, Clement, Earle, and all of the administrative personnel slated for the new "Lower Station" of the Indian Head Naval Proving Ground moved there permanently and began construction of the facility. Among those who accompanied Clement and Earle were Lieutenant W. H. Caldwell, Ensign L. A. Rehfuss, civil engineer John W. Russell, and draftsman Charles Isbell. To form the nucleus of the new station, buildings from Stump Neck vacated by Indian Head's Marine detachment were disassembled, floated down the Potomac on barges, unloaded, and reassembled at Machodoc Creek. Foundations for temporary buildings were completed on 30 June, while bulkheads for fill along the shoreline, laying sewer pipe, and grading a railroad from the wharf site to the designated main battery location were finished soon after. Clement likewise leased a suction dredge from the Corps of Engineers to deepen a nearby basin and a channel between Upper Machodoc Creek and the Potomac and to fill the site's various marshes with the discharge. By October, Clement and Earle had made enough progress that Rear Admiral Earle proudly reported to Secretary Daniels that "the Bureau will soon be in possession of ample proving ground facilities which will be utilized to the utmost in performing the experiments and tests desired for many, many years, toward the improvement of both ordnance and guns."[29]

Rear Admiral Earle's confidence must have been bolstered further by the news from the Lower Station that reached BUORD on 16 October 1918. Under the supervision of Navy Lieutenant Commander H. K. Lewis, a detachment of Marines hoisted the colors at the new proving ground and officially opened it on that date. With representatives of the U.S. Army's Ordnance Department watching, they then fired the Lower Station's first shot, a 153-pound projectile, some 24,000 yards down the Potomac from a 7-inch, 45-caliber naval gun mounted on a special caterpillar-propelled tractor carriage. The gun and mount, which had originally been requested from the Navy by the artillery-strapped U.S. Army, was one of twenty that BUORD had earmarked for the new 10th Marine Artillery Regiment, then in training at Quantico, Virginia, for service on the Western Front. That weapon, which BUORD later described as "the heaviest and hardest hitting gun for which a mobile field mount of this kind had ever been requested by any nation or army," represented the ancestor of self-propelled artillery that was to play a major role in later wars.[30]

While the 7-inch, 45-caliber tractor gun test signaled that the Lower Station was open for business, the facility was far from complete. Chronic labor shortages and difficulty obtaining materials stalled construction in late 1918. From Indian Head in 1919, Lackey reported to BUORD that "the work at the Station has been materially handicapped by lack of drafting and clerical force," and that the station's Administration Building and the elegant Commandant's (or Inspector's) house were only 15 percent complete. Despite the news, Lackey did indicate some progress, including 100 percent completion of a warehouse and gun emplacements for 3-, 4-, and 7-inch guns, an artesian well, an oil storehouse, and sixteen complete bungalows for civilian employees. Fortunately for Lackey, the Armistice solved his labor supply problem. From the approximately 250 civilian laborers reported for the period December 1918 through April 1919, his work force increased to around 500 between April and July, up significantly from the 75 to 125 civilians reported working the previous summer. The work pace quickened.[31]

By July 1919, BUORD Chief Earle reported that a new fuze battery had been completed and that the Lower Station was now in active operation. All fuze testing was now being conducted there, and the experimental ranging work of the new 6-inch, 53-caliber guns, along with the testing and ranging of major caliber ordnance, accompanied the erection of new buildings, butts, and magazines. Additionally, and quite contrary to what he had told Congress the year before about the Lower Station's supposed "auxiliary" status, Earle announced that all routine proof work would be transferred from Indian Head to the new proving ground as soon as possible to eliminate conflicts with the parent facility's experimental work.[32]

As BUORD completed more of the Lower Station's facilities, larger scale testing began at the site. The Navy's Mk II 14-inch railway gun, an improved version of the design used in France, was the first "big" gun to be tested there. The Mk II was capable of firing at a maximum elevation of forty-three degrees directly from the rails. BUORD originally contracted for five of the new railway gun's support carriages but, after the Armistice, canceled three of them. The contractor, Baldwin Locomotive Works, completed the first carriage on 17 July 1919 and sent it to the Washington Navy Yard, where sailors and civilian laborers mounted a 14-inch, 50-caliber gun on it. From there, BUORD transported the complete, 305-ton weapon down the Potomac to the Lower Station by barge.[33]

On the morning of 16 August, the Mk II railway gun was successfully tested at the Lower Station before an audience of Army and Navy officials and prominent engineers. Lieutenant Swepson Earle, who had become the station's first range officer, witnessed the test. He later recalled that the big gun, fixed at a 30-degree elevation, fired a 1,400-pound projectile 31,680 yards (18 miles) down the Potomac. The Navy had grand plans for the Mk II, which it envisioned running from coast to coast on America's rail system to defend the country's shores from enemy attack. Unfortunately for the Navy, the Joint Army and Navy Board recommended later in the year that BUORD turn over its five Mk I and two Mk II railway mounts, without their naval guns, to the War Department. Both the Secretaries of the Navy and War approved the recommendation on 27 December, and BUORD complied, sending the railway mounts to the Aberdeen Proving Ground at the request of the Army's Chief of Ordnance.[34]

By August 1920, the new facility was reaching an advanced stage of construction. BUORD Chief Earle's successor, Rear Admiral Charles B. McVay Jr., reported to Secretary Daniels that the Lower Station's main and broadside batteries were finally in commission, complete with velocity instrumentation and other physical laboratory equipment. Moreover, magazines, the shell-house establishment, main bombproof butts, and other proofing structures were being pushed rapidly to completion, and a 200-ton gantry crane had been transferred from Indian Head to the Lower Station. Although McVay felt that the whole outlay was "still largely a construction problem," he felt that he could safely predict that "by next summer the new proving ground will be a smoothly operating reality" since proof facilities for major caliber powder, as well as guns, were expected to be completed in September. In view of the Lower Station's near completion, McVay noted that BUORD had stopped all major caliber powder proof at Indian Head and would transfer it to the new proving ground once those specific facilities became operational.[35]

NAMING THE LOWER STATION

Late in 1918, as the infant Lower Station grew, it became obvious to Commander Logan Cresap in BUORD that it needed a separate identity from Indian Head. Cresap's job in BUORD's Armor and Projectile Section included handling the bureau's correspondence concerning the new proving ground's construction. He found that routing all correspondence, materials, and ordnance to the Lower Station through Indian Head was unnecessarily cumbersome. A long-established post office designated as "Dido" existed on the reservation, but BUORD planned to remove it in the near future and establish a new post office, tentatively called "Machodoc Creek." However, postal officials quickly told BUORD that a post office with that designation already existed in Virginia and that another name would be necessary.[36]

Thus informed, Cresap sensed an opportunity to memorialize the achievements of "some Naval officer who had been eminent in the development of Naval ordnance" by lending his name to the new station. He therefore recommended to Rear Admiral Earle that BUORD abandon the Navy's practice of naming shore establishments after geographic locations, as in Indian Head's case, and name the Lower Station after one of these individuals. Cresap suggested Robert Stockton, John Dahlgren, Robert Dashiell, and ordnance expert and Naval Academy professor Philip Alger as the most likely candidates for the honor. Earle liked the idea and added former BUORD Chiefs William Sampson and George Converse to the list.

After some discussion within BUORD, Earle ultimately chose to name the new proving ground after Dahlgren, who he considered to be "the father of modern ordnance and gunnery" and a hero who had pulled the Navy out of an ordnance rut in which it had been stuck since the War of 1812.[37]

As Cresap had observed, it was against Navy tradition to name a new shore installation for a person rather than a place, but Earle finessed the issue by working with the Postal Service to create a local post office named "Dahlgren" at the site. In January 1919 Earle persuaded Navy Secretary Daniels to request that the Postal Service change the name of the existing Dido post office to "Dahlgren." Daniels obliged, and on 15 January he proposed the new name to the Postmaster-General, who accepted the recommendation and directed the name change on 24 January. BUORD was then able to call the new facility the United States Naval Proving Ground at Dahlgren, Virginia. Shortly thereafter, though, the distinction between the post office address and the name for the new proving ground was blurred, even in official correspondence. The Superintendent of Naval Records, for example, noted that relatives of the late Rear Admiral Dahlgren would be glad to learn that Earle had "decided to name the Proving Ground on Machodoc Creek for that officer." In time, the proving ground simply became known as "Dahlgren."[38]

A FIGHT IN CONGRESS

Following the 1919 Versailles Treaty, Dahlgren very nearly became a victim of changing postwar politics, as Earle began having trouble with a suddenly stingy Congress over the cost and necessity of BUORD's new proving ground. He repeatedly testified before the House Naval Affairs Committee to defend, among other things, his request from the House Appropriations Committee in October 1918 for an additional $980,000 for the site, which was rapidly becoming much more than an "auxiliary" proving ground. Objections and tough questioning came from two sources. First, a new Republican congressman and House Naval Affairs Committee member, Ambrose Everett Burnside Stephens of Ohio, believed that BUORD should not have authority to spend money for improvements or replacement of damaged buildings without explicit congressional approval. He therefore appointed himself as a watchdog over BUORD expenditures. According to a Democratic colleague on the committee, "Buzz" Stephens was "always stern and unbending in anything having the least suspicion of waste, extravagance or wrongdoing." Therefore, he became Dahlgren's most vociferous critic since he doubted both the wisdom and necessity of a new proving ground.[39]

Next, Republican Congressman Sydney E. Mudd, the representative from Charles County, Maryland, whose constituency included civilian employees at Indian Head, objected to the new proving ground, calling it redundant. Like his friend Stephens, he was not at all sure that the new proving ground was needed and was opposed to shifting facilities and manpower away from Indian Head. As Earle confronted the skeptical congressmen, he was forced to sharply disagree with their assertions against Dahlgren's necessity, thereby keeping a delicate balance between respect for the elected representatives and advocacy of what he knew was right and required.[40]

During the 1919 House Naval Affairs Committee hearings on the Secretary of the Navy's budget estimates, Republican Congressman Patrick Henry Kelley of Michigan was particularly annoyed that the budgets for Indian Head and for the "new ground" were not clearly separate. Earle explained that the two facilities were under one command, that the books were maintained as one unit, and that as work gradually shifted from Indian Head to Dahlgren, the mix would change. The budget of the two facilities, he noted, would depend on "how much work we will drop from Indianhead [sic] and put there. That is changing as we put that in commission. We take work from Indianhead [sic]." Although he meant to indicate that the bottom line did not change and that the total proving budget would be better spent, the concept of "taking work" from Indian Head did not sit well with Congressman Mudd or with the large local community that had provided the old proving ground's workforce.[41]

Earle thought that the $1,980,000 that BUORD spent on Dahlgren was a real value considering that the Army had spent $12,000,000 on the Aberdeen Proving Ground. Accordingly, when asked to cut the expense estimates further, Earle stood his ground, piqued at what must have seemed a particularly obtuse Naval Affairs Committee. He insisted that "I could not reduce it and say that I was carrying out the work of the Navy. Anything that goes wrong on board ship comes right back to me, and the first thing that happens is the statement that I did not carry out the proving of a gun, that I did not fire the necessary number of rounds. Why? Because I did not have the money. It all comes right back to me, and that is all. Anything that goes wrong on board ship comes back to us." The explosion of the "Peacemaker," although not explicit in Earle's testimony, certainly continued to echo throughout BUORD, and the grim prospect of a catastrophic failure arising from an improperly tested gun weighed heavily upon him and his subordinates.[42]

In 1920, before the congressional battle for Dahlgren was over, Earle left the Bureau to take command of the battleship USS *Connecticut* (BB-18). As Earle steamed away from Washington, D.C., Dahlgren's defense fell to McVay, who quickly discovered that the battle was only heating up. Congressman Stephens, suspicious that BUORD was wasting money, had successfully inserted an amendment into the 1921 Naval Appropriation Act barring the expansion of any naval ordnance station. Shortly thereafter, he began hearing rumors that BUORD was spending $100,000 on a commandant's home at Dahlgren, and the new proving ground immediately fell under his scrutiny.[43]

In late May 1921, Stephens prodded the Naval Affairs Committee chairman, Republican Thomas S. Butler of Pennsylvania, to submit six specific questions to Navy Secretary Edwin Denby concerning all expenditures made at Dahlgren since 1 July 1918. Denby's written answers were startling. The Secretary said that BUORD had spent over $2,200,000 on Dahlgren, and that the two-story commandant's home, comprising twenty-three rooms, two sleeping porches, five bathrooms, and a large 40- by 10-foot attic, would not cost $100,000 as rumored but an estimated total of $52,000, and would be completely furnished for another $8,400. Further, BUORD had spent over $180,000 on officers' quarters, with plans to furnish them for an additional $20,000. All of these expenditures had been funded out of the original "ordnance and ordnance stores" appropriation obtained by Earle in 1918.[44]

Although the commandant's house cost considerably less than what Stephens had originally believed, its price tag was still hefty for that time, as was the cost of the other officers' housing. Needless to say, Stephens was apoplectic. He, along with Mudd and Butler, launched a formal investigation into how Earle had procured Dahlgren's funding and how the money was being spent. Heading up a special committee of the House Naval Affairs Committee, Stephens chaired a series of hearings in late July, in which he not only pored over extensive itemized lists of expenses incurred during Dahlgren's construction but also meticulously reviewed Earle's previous testimony. He also grilled a number of BUORD officers, including Dahlgren and Indian Head's second Inspector of Ordnance in Charge, Captain John W. Greenslade, over the numbers, types, and costs of Dahlgren's facilities.[45]

During these initial hearings, Stephens and his Republican colleagues learned exactly how sly Earle had been. The former BUORD chief, who had initially wanted $2,000,000 for the new proving ground but had been compelled by Secretary Daniels to ask for only $1,000,000, had received that much from the House Naval Affairs Committee, but he later bypassed that committee and approached the House Appropriations Committee for an additional $980,000 deficiency appropriation before the original $1,000,000 was exhausted. The Appropriations Committee had obliged him, and he ultimately had walked away with nearly all of the $2,000,000 that he needed to build Dahlgren. As Stephens pieced together the facts, he became increasingly galled at Earle's perceived deception of both committees and the thwarting of Congress in its oversight role. Moreover, the congressman found that rather than constructing a mere "auxiliary" proving ground, as Earle had originally told Congress, BUORD was building a full-scale installation with approximately seventy buildings, landplane and seaplane hangars, a radio station, four and a half miles of railroad track, and other fixtures typical of the Navy's other shore establishments.[46]

The hearings continued into early August, and Stephens' special committee heard testimony from a number of Indian Head's current employees and residents who would be most affected by the transfer of work to Dahlgren. Among those who testified were ballistician Roger Dement and the chief chemist and powder expert George W. Patterson. The group argued on behalf of their community that the old Indian Head proving ground was perfectly safe and that its location and facilities were unsurpassed for gun proofing. Conversely, Dahlgren was uneconomical and redundant in their view. Also, several of the men claimed that proof work at Dahlgren would unnecessarily damage the Potomac's fish and oyster industry and interfere with transportation and navigation. During Patterson's extended testimony, Congressman Mudd entered into the record a petition signed by 456 Indian Head residents and employees, including Patterson, asking Congress to prevent the transfer of the United States Proving Ground from Indian Head to Dahlgren. Repeating in detail the same points made by their representatives, the petitioners trusted that their arguments "will be deemed of sufficient and good reasoning to convince you of the rightful cause and justness of our claims."[47]

Stephens became convinced that he had uncovered a major scandal. In August 1921 he introduced a resolution that, if enacted, would forbid the use of any existing appropriations for Dahlgren except for the operation and maintenance of its existing facilities, thereby preventing further construction and expansion. When Denby submitted the resolution to BUORD for comment, McVay responded that it served no useful purpose and benefitted only Indian Head's inhabitants, especially those with commercial interests. Denby agreed with McVay's blunt assessment and wrote Butler back on

26 September, telling the chairman that "such a law would not advantage the government in any respect" and urging the Naval Affairs Committee to reject it.[48]

The Naval Affairs Committee began considering Stephens' resolution on 19 October. During the hearing, Denby, McVay, and Inspector of Ordnance Greenslade all testified in Dahlgren's defense. Predictably, it was an uncomfortable experience for them. Citing Earle's earlier testimony that the new proving ground was only an "auxiliary" facility, Butler, Mudd, and Stephens all complained that they had understood that Dahlgren would only be used for the ranging of guns of large caliber, above 8 inches, that were too large for Indian Head. When the Navy began to shift all testing from Indian Head to the Lower Station, they felt they had been deceived. Butler remarked, "I thought that this place would be used only on rare occasions where we had a great gun which was to be ranged; I had no idea that it was proposed at Dahlgren to establish another and distinct station." Mudd added, "I think it was the understanding of the committee that it was to be used exclusively for long-range guns. It was so represented to this committee. . . . We had no intimation that all the guns would be moved [from] Indianhead [sic] and tested on the other side." Stephens concurred with both of his colleagues, stating that "there was no idea or intention, so far as I have been able to learn, of establishing a separate station with new officers' homes and quarters for the men and a large civilian establishment."[49]

The Navy men dodged and parried the congressmen's probes and thrusts. Denby was particularly effective in withstanding their chain-fire questioning and pushed back, arguing that "Dahlgren must be held . . . as a testing ground," regardless of cost or the wisdom of how the money was spent, since "as long as we have big guns we must have a testing ground of that character." Moreover, if Stephens' resolution was intended to halt Dahlgren's operations, then he would vigorously oppose it since he would not stand by and see the Navy crippled without protest.[50]

When his turn came, McVay testified that BUORD had nothing to hide and suggested that Stephens' resolution was really moot since the past year's naval appropriation law had halted all ordnance station expansion, including at Dahlgren. Furthermore, it really made no difference since he was not planning on spending any more money on shore station development beyond that already appropriated. Concerning the allegation that BUORD had built a wholly separate station without authorization, McVay pointed out that he had no intention of operating Dahlgren and Indian Head separately, but that they would be under the "same person." The only question would be where that officer should be located.[51]

The hearing adjourned before McVay finished his testimony but reconvened on 27 October. During that final session, a re-energized Stephens hammered the admiral on the questions of Earle's two appropriation requests and especially the commandant's and officers' quarters. He specifically wanted to know whether Earle or Lackey had authorized their construction, and if McVay thought that it was proper to use the "Ordnance and Ordnance Stores" appropriation for such purposes. Stephens also demanded to know where BUORD obtained the balance between the $2,200,000 actually spent and the $1,980,000 appropriated in 1918. McVay was evasive on who was responsible for the quarters, noting that he was not at BUORD when the decisions were made, but he admitted that BUORD had to authorize both building plans and expenditures suggested by the inspectors. Since the commandant's house was between 80 percent and 90 percent finished when he became BUORD chief, in his estimation it would have been a waste of money to halt work on it, and so he allowed its completion. As for the funding discrepancy, McVay also admitted that the extra funds had come from subsequent "Ordnance and Ordnance Stores Appropriations," which BUORD customarily tapped for the improvement of ordnance stations until Stephens had stopped the practice in 1920. Stephens was wholly unsatisfied with McVay's answers and, along with Butler and Mudd, badgered him for additional facts and figures for the remainder of the hearing. The hassled admiral responded as best he could, ultimately insisting that the $2,200,000 spent at Dahlgren was a "very reasonable" sum for this type of proving ground.[52]

The Naval Affairs Committee tabled Stephens' resolution, but at a 7 December meeting, it instructed Butler to inform Denby that it would frown upon any further spending at Dahlgren. The chairman did so, telling Denby that the committee did not want to send Stephens' resolution to the full House of Representatives since most of its members felt that an open discussion on the House floor would greatly embarrass the Navy. However, Butler made it perfectly clear that the committee would be induced to take that drastic step if the Navy Department further ignored its wishes. "For the good of the Navy," Butler suggested, Denby should comply immediately.[53]

Faced with this bit of congressional blackmail, Denby quickly ordered McVay to stop all new construction work at Dahlgren under the current appropriation, which only had about $1,350 left anyway. Although the Naval Affairs Committee stopped expansion until further notice, it allowed Dahlgren to continue its proving ground work using its existing facilities.[54]

Despite this limited punitive action, Stephens still bristled at BUORD's apparent misbehavior, especially after the Naval Affairs Committee visited

Dahlgren and saw the commandant's and officers' homes for itself. On 27 February 1922, he finally hauled Lackey before the Naval Affairs Committee to answer the question of who had authorized their construction. Beset by Stephens, Butler, and other hostile committee members, Lackey finally admitted that he was solely responsible for planning the homes and submitting their designs to the Chief of BUORD and the Secretary of the Navy for approval. Concerning the commandant's house, he explained that he had primarily planned it so that "a fair-sized committee or commission could be entertained in proper manner by the Government or its representatives when called upon." Lackey recalled that this had been impossible at Indian Head when he had to entertain various foreign naval delegations as well as some of the committee's members since the commandant's house there was "exceedingly embarrassing; people were tripping over each other and there was not sufficient room to move around comfortably." Moreover, he needed the new house at Dahlgren to include a special bedroom with connecting bath for the Navy Secretary's use whenever he came down for inspection, because "being a remote station he could not come down and go back on the same day."[55]

Lackey's explanation hardly appeased Stephens, who was no longer satisfied in just halting Dahlgren's expansion but spoke of closing the station outright. On 18 April he introduced an amendment to the 1923 Naval Appropriation Bill which, if enacted, would strip Dahlgren of all funding other than what was necessary to maintain it on a "closed-down basis." The Lower Station was built "for the purpose of ranging large guns, an absolute war activity," he declared. "The war is over . . . the necessity for this proving ground has disappeared." Dahlgren, therefore, must be closed.[56]

In the fierce floor fight that followed, Stephens, Mudd, and their Republican allies squared off against a bloc of largely southern Democrats, including Lemuel P. Padgett of Tennessee, Robert W. Moore of Virginia, William B. Oliver of Alabama, and William F. Stevenson of South Carolina. By far, the venerable Padgett was Dahlgren's greatest defender in the House. A longtime member of the Naval Affairs Committee, he was familiar with Indian Head's prior troubles and refuted Stephens and Mudd point by point. The Republicans ultimately overpowered the Democrats in the debate, though, and Stephens' amendment passed by a 106 to 67 vote.[57]

The legislative battle over Dahlgren's future then shifted to the Senate. In May 1922 a subcommittee of the Senate Committee on Appropriations convened to consider Stephens' amendment. During the hearing, Virginia's powerful Democratic Senators Claude A. Swanson (a future Navy Secretary under Franklin Roosevelt) and Carter Glass (formerly Treasury Secretary in Wilson's Cabinet), along with Republican Senators Miles Poindexter of Washington (born in Tennessee but educated in Virginia, and a Virginian in mind and heart) and Truman H. Newberry of Michigan (a former Navy Secretary in Theodore Roosevelt's administration), elicited statements from ASN Theodore Roosevelt Jr., Admiral McVay, and the second Chief of Naval Operations, Admiral Robert E. Coontz, supporting not only Dahlgren's retention but also its expansion.[58]

Republican Senator Joseph I. France of Maryland backed Mudd and Stephens and arranged for Stephens to testify before the Senate subcommittee. Armed with his lengthy, itemized list of Dahlgren's construction expenses, Stephens recited for the subcommittee excerpts of Earle's original testimony as well as that of Captain Lackey. Once again, the expensive commandant's "mansion" figured prominently in his condemnation of Earle, BUORD, and Dahlgren.[59]

Glass was unimpressed, to say the least, and scolded Stephens throughout the hearing. After listening to the congressman's seemingly endless complaints about Dahlgren, the Virginian incredulously asked Stephens if his remedy against building expensive naval plants was to abandon them after they had been built. Stephens sidestepped the question, but Glass was not finished. In his ensuing onslaught, he suggested that the Ohioan's opposition to Dahlgren sprang not from any concern with economy but from personal pique at Earle. Moreover, Glass pointed out that Stephens' argument concerning economy made no sense, as he was proposing Dahlgren's closure instead of the more expensive Indian Head facility. Should it not be the other way around, the Virginian asked? "No," said the irrepressible Stephens, "I propose to abandon Dahlgren in order to stop any further wastefulness. . . . "[60]

The confrontation ended without any love being lost between Stephens and Glass, and the Senate subsequently struck Stephens' amendment from the Navy's appropriation bill. France tried to save it by offering a compromise amendment that would restrict Dahlgren to only testing those long-range guns that could not be tested elsewhere. Swanson, Glass, and Poindexter stopped France cold and spared Dahlgren from both total and partial closure. However, Stephens, who remained a staunch "defender of the public purse" until his death in February 1927, managed to keep his original restriction on ordnance facilities expansion in place. As a result, BUORD could not build or transport any more facilities there until the late 1920s.[61]

The clashes within Congress underscored several political trends that worked against Dahlgren in the early 1920s. First, Americans, disillusioned by World War I and Wilsonian internationalism, began exhibiting a backlash against "foreign entanglements." After the war, it became apparent to many Americans that the United States had committed its young men to fight, not for Wilson's ideals but for British and French hegemony in Europe and in the League of Nations. This mix of disillusionment and resentment became known as "isolationism," a position that numerous Republican and anti-Wilson Democratic congressmen came to share. Furthermore, many congressmen from both parties sought to restrict government spending across the board without necessarily taking an anti-military viewpoint, a stance taken by Chairman Butler and Senator France.[62]

This mood of postwar government frugality, in conjunction with isolationism, was especially manifested in the Washington Conference of 1921-22, which Republican President Warren G. Harding had called to avoid the expense and danger of a naval arms race. The resulting international Naval Arms Limitation Treaty of 1922 was devastating for the Navy. Among other things, it proclaimed a ten-year "holiday" in capital ship construction and also required the United States to scrap fifteen battleships and to cancel eleven of the fifteen capital ships then under construction. Once the ten-year holiday was in force and the battleships were scrapped, it became difficult to justify the Lower Station's existence since no new major-caliber guns were needed for at least a decade.[63]

At the local level, the politics of government budget and government employment had a very different character. Congressman Mudd, representing the petitioners at Indian Head, fought not so much to restrict the Navy's expenditure, as his friend Stephens had, but to ensure that employment at the Maryland facility would not be shifted to Virginia. On the other hand, Senators Swanson and Glass, with the help of their Democratic and pro-Navy colleagues, acted decisively to secure and protect Dahlgren for their own Virginia constituents. As a result of their intervention, Dahlgren survived the political and legislative turmoil of its formative early years, but no major construction took place there between 1921 and 1927, and the total complement of employees remained at roughly 200 to 230 personnel through the period.

TESTING ORDNANCE

On 1 August 1921, BUORD transferred "practically the entire volume of ordnance work" from Indian Head to Dahlgren, leaving only a small force at Indian Head to care for remaining ordnance material and the powder factory, and to supervise a few smaller tests and special work in underwater high explosives and mines. The move had really started on 10 March, when Dahlgren's ordnance officers submitted their first powder test report to BUORD, and accelerated on 25 July, following the completion of the Plate Battery, when they conducted the first armor plate firing test, using a 9-inch Class A plate for the USS *Indiana* (BB-50). Just as the final transfer occurred, the Inspector of Ordnance in Charge for both Indian Head and Dahlgren, Captain John W. Greenslade, reported to BUORD that, excepting the shell house, the Lower Station was now fully equipped to conduct all proof and experimental work. Consequently, he was organizing a new, more centralized Proof Department to manage the ordnance-testing program at the Lower Station, and accordingly expected that Dahlgren would turn out considerably more work than Indian Head in the future.[64]

When the Naval Affairs Committee ordered all construction and improvements on the proving ground stopped in December 1921, BUORD's engineers and architects had completed construction of most of its physical plant, including the Commandant's (or Inspector's) Quarters, the elegant Administration Building, a Recreation Hall, a machine shop, and shell storage and loading buildings. In keeping with racial segregation customs in federal facilities at the time, separate mess halls and dormitories for white and African-American employees were also built. After this first group of buildings was completed, additional housing was obtained simply by transporting a number of small bungalows by barge from Indian Head down to Dahlgren.[65]

As soon as the ordnance proving and testing work had fully shifted to Dahlgren, the senior leadership relocated as well. In August 1922 Captain Greenslade reported that he personally was in residence at Dahlgren. Joining him there were a Senior Assistant to the Inspector and an Executive Officer, and a total complement of fifteen officers and petty officers. Ten Navy enlisted men remained at Indian Head, while some sixty-six were stationed at Dahlgren. By 1923, therefore, the subtle movement toward the future formal independence of Dahlgren from its mother institution at Indian Head was already under way.[66]

As the period of cutbacks set in at the Lower Station, and under continued assault by Stephens, Greenslade was called upon to provide yet another justification for Dahlgren. Responding to the "repeated [congressional] assertions that with the adoption of the Limitation of Armament Treaty there would be little further use in maintaining the Proving Ground, Lower Station, Dahlgren, Virginia," Greenslade submitted to McVay a detailed compilation of the work in the first two years, spelling out routine and experimental work, and reported on complimentary remarks made by various visiting experts and officials.[67]

First of all, Greenslade noted that by canceling and delaying contracts and orders, the routine work had diminished but the volume of "work along developmental and experimental lines has increased rather than decreased." In the light of the history of the Navy's laboratories and stations, Greenslade's early emphasis on what a later generation would categorize as exploratory development was prescient. In a rather formal fashion, he expected the ratio of experimental work would increase over routine, particularly as battleships were regunned. Greenslade understood the need to look ahead to future developments to justify and explain the function of the new facility, stating that "the Proving Ground at Dahlgren, Virginia, was carefully and wisely selected as to location and its development has been such that it can efficiently and economically take care of all ordnance work for some time to come, including new developments of the ordnance features of air warfare."[68]

Further, Greenslade proudly noted the opinion of Dr. John Curtis of the Bureau of Standards, who had remarked after touring Dahlgren that he had visited nearly all the European proving grounds and that none compared with Dahlgren in "efficiency of operation, judicious concentration of facilities, and ability to obtain the results sought." Greenslade also mentioned that a representative from the Krupp Gun Works in Germany had informed him that two of the seven proving grounds in Germany were in populated areas and that when firing occurred, the inhabitants had to be removed from the range and compensated for their time. In short, Greenslade intimated, Dahlgren's safe over-water range was the best in the world.[69]

With the ten-year naval holiday in force, Greenslade remarked that the only way to "keep abreast of the time in Naval Ordnance matters and be prepared to take up active building of improved guns and armament for future construction in ten years time" would be to take advantage of the proving ground. He then described seventeen projects that were ordered, planned, or "in contemplation." The list, which was compiled, so to speak, under the gun of the disarmament mood, is instructive. Among the forward-looking projects he listed were: major-caliber fuzes; moments of inertia on projectiles; development of 6-inch twin mounts; developmental work on 5-inch anti-aircraft guns and 6- and 9-inch guns; and special projects with new types of projectiles and star shells. New oscillographs would allow close timing tests to determine the exact cause, either human or mechanical, of the timing interval between a directoscope operator's spotting of a target and the actual firing of the gun. By calculating these extremely short intervals, firing accuracy could be improved.[70]

As Greenslade detailed the methods of testing and the developmental and experimental work in hand in his 1922 report, he remained well aware of the "lessons of Jutland." For example, in describing the tests of major-caliber fuzes, he referenced Admiral John Rusworth Jellicoe's book *The Grand Fleet 1914-1916: Its Creation, Development and Work* (London: Cassell & Co. Ltd., 1919), noting that "from the results of the Battle of Jutland, it was apparent that the Germans possessed a better major caliber fuze than the British did." Greenslade also pointed out that Admiral Jellicoe had complained in his book of the inferiority of British fuzes. Indeed, the delayed-action fuzes the Germans had employed contributed to the sinking of the HMS *Indefatigable*, which caught fire and exploded when its own ammunition detonated seconds after German shells struck one of its magazines. Greenslade had that lesson in mind when he noted that "a fuzed projectile must be capable of penetrating a ship's side armor and of detonating inside the ship before the projectile has time to pass out through the other side of the ship." He gave a detailed, step-by-step description of a fuze test firing, describing all the safety measures, record keeping, and maneuvering of equipment involved. Each single gun firing took an involved schedule of more than 150 man-hours, counting both experts and laborers. He further observed that the U.S. Navy's new fuzes could only be "given the most preliminary tests" by the Experimental Ammunition Unit at the Navy Yard and that the "real tests" had to take place at the Proving Ground by firing projectiles at various thicknesses of armor plate.[71]

Under threat of closure, Greenslade was eager not only to report on the quantity of the work but also to provide details showing how crucial Dahlgren was to the Navy's effort to respond to the changes brought by the growing use of aircraft as weapons platforms. Greenslade's early response had much in common with what a later generation, faced with the Base Realignment and Closure (BRAC) procedures of the 1990s, would define as a "data call"—a request for detailed information that could be used to assess the value of the facility. In a sense, Dahlgren had been only partially finished when it survived its first BRAC, in part because Greenslade showed its value as a location for routine and necessary work. Moreover, he demonstrated that its operations were on the cutting edge of naval technology.

Between the congressional mandate to forestall improvements and the general impact of the Naval Arms Limitation Treaty, both Indian Head and Dahlgren were hit hard. Reporting in July 1923, Dahlgren's new commander, Captain Claude C. Bloch, noted that "the Station force has been upset several times during the year due to necessary reductions and rearrangements brought about by changes in wages and curtailment of funds. These shake ups cause discontent and are a detriment to economical operation and upkeep."[72]

However the congressional debates played out, the distinction between the two facilities was established and quite explicit, with Indian Head referred to as the "Powder Factory" and Dahlgren as the "Proving Ground" in the 1923 Annual Report. Not only had the Inspector of Ordnance moved his headquarters from Indian Head to Dahlgren, but also the new station gradually took precedence in other ways, particularly in personnel. By mid-1923, fourteen officers and sixty-two enlisted men worked at Dahlgren, while the Indian Head Powder Factory counted twelve officers and six enlisted men. Up until 1932, when the two facilities were formally divided, the complement of Navy officers and men at Dahlgren remained much higher than that at Indian Head, suggesting that the Navy's ordnance men preferred firing guns to doing chemistry.[73]

The formal organization of Indian Head and Dahlgren was a complex and overlapping structure, captured in a 1928 publication of regulations governing both installations. The Executive Officer at Dahlgren commanded a Supply Department, a Proof Department, and the Aviation Detail. The Proof Department was by far the largest unit, with separate responsibilities covering postgraduate officers assigned to the proving ground, the Routine Tests Section run by an Assistant Proof Officer, and the Experimentation Office. A Technical Liaison group consisted of the Proof Officer, the Experimental Officer, the Powder Expert, and the Physicist and was responsible for the compilation and analysis of all activities. In 1928 a Disbursing and Time section for civilian employees reported to the Disbursing Officer at Indian Head as well as to the Executive Officer at Dahlgren. A similar dual line of authority existed for the Marine Barracks at Dahlgren, which reported both to the Executive Officer at Dahlgren and to the Marine Officer in Charge at Indian Head. Both of those shared lines of authority were directly addressed and changed in 1932, when Dahlgren became independent. However, between 1922 and 1932, the Inspector of Ordnance of both facilities lived at Dahlgren, and the new station operated independently under its own Executive Officer in other regards.[74]

Through the same period, the administrative structure at Indian Head was more elaborate, as the powder factory required a much larger structure as well as separate administrative units to cover transportation, maintenance, recreation, police, safety, fire protection, supply, housing, and disbursing. In short, the organization chart suggested that by 1928 Indian

Head was a full community with a structured administration and a large civilian complement, while Dahlgren represented a leaner, more military facility. In both locations, however, civilian scientists were already playing key roles.[75]

EXPERIMENTAL PROGRAM

During the 1910s and 1920s, the Navy was just beginning an effort to employ civilian scientists directly as well as to expose a generation of officers to some of the best scientific training in academic circles. Building on a program that trained officers in naval architecture and aeronautics at the Massachusetts Institute of Technology, the Navy began to send ordnance officers for postgraduate ballistics training in a special program at the University of Chicago and chemistry training at the University of Michigan. It is through this concerted effort to tap into academic science that the first civilian scientist with formal academic credentials came to Dahlgren.[76]

In 1923 Dr. Louis T. E. Thompson of Clark University took the post of Chief Physicist at Dahlgren. In addition to serving as administrator of the Physical Laboratory, Thompson served on the Technical Liaison team along with the Proof Officer, the Experimental Officer, and Indian Head Chief Chemist George Patterson.[77]

Thompson had earned his doctorate in 1917 and had taught first at Clark University. During his doctoral research at Clark, Thompson had primarily studied interior ballistics and gun pressure systems under Dr. Arthur Gordon Webster in an innovative program, modeled on German Professor Carl Cranz's Ballistiches Institut, which brought together theoretical training and science education for ordnance officers. The Clark program was funded by the National Academy of Sciences, the American Academy of Sciences, and the Naval Consulting Board and was one of several designed to bring academic science to bear on the problems of the military and on the education of officers. In 1920 Thompson served as a National Research Council fellow at the University of Chicago and subsequently took a temporary teaching position at Kalamazoo College before accepting the appointment at Dahlgren in 1923.[78]

Admiral George Hussey later remembered how Thompson was recruited. Hussey had taken a postgraduate course in ballistics at the University of Chicago in 1921 and then served at BUORD. Commander Theodore S. Wilkinson, Chief of BUORD's Experimental Section, asked Hussey to inquire of professors in the postgraduate course for names of candidates who could fill the newly established billet of physicist at Dahlgren. Hussey wrote to Professor of Astronomy Forrest Ray Moulton at the University of Chicago, who had previously worked in ballistics at the Aberdeen Proving Ground. Moulton immediately replied: "I know exactly the man you need. The man you want is Dr. L. T. E. Thompson of Kalamazoo College, Michigan."[79]

So Hussey wrote to Thompson and invited him to Dahlgren to discuss the new physicist position.[80] Thompson, who was then considering another job offer from AT&T, saw the obvious "opportunities to do extensions of work that had been going on in the past both in interior ballistics and exterior ballistics" and agreed to make the trip down from Michigan. He arrived at Dahlgren on a rainy day in April 1923, "when it was very largely a mud hole," and accepted the offer after Hussey and Greenslade gave him the grand tour of the facility. By Hussey's estimate, Thompson was the first "full-fledged" scientist in the Navy ordnance establishment. Hussey later worked as an Assistant Proof Officer, getting to know and admire Thompson during their work together.[81]

Affectionately known to his colleagues as "Dr. Tommy," Thompson embarked on a vigorous program of experimental work that reflected the Navy's emerging postwar interest in the actual physics and high-level mathematics of naval ordnance. His first project was a program to study the interior ballistics of a 6-inch gun, which involved the development of a specialized pressure gauge, an "extension of one that [he had] been working on at Clark University."[82] This marked the beginning of his personal crusade to bring a more scientific outlook to ordnance research. Influenced by the lessons of Jutland, as well as the development of aircraft, aircraft-dropped weapons, and anti-aircraft weapons, his research agenda subsequently widened beyond large-caliber gun ballistics to include armor penetration mechanics and high-altitude bombing studies.[83]

One item of what Thompson called "foundational work"—what a later generation would call basic research—focused on the erosion of guns. During the mid- and late 1920s, Thompson and his small staff investigated the coppering of the internal bore of guns by shells, an effect that gunners had long believed caused irregularities that affected accuracy. Thompson's team concluded that, to the contrary, copper buildup was actually beneficial by retarding erosion, improving performance, and extending the guns' service lives. Subsequent metallurgical tests revealed that chromium was an even better lining material for gun barrels, leading to a later decision in the mid-1930s to plate gun bores with chromium, especially in the muzzle and breech areas.[84]

Long after leaving Dahlgren, Thompson prepared a retrospective bibliographic record of reports on the erosion problem, hoping to correct an omission in several publications in the period 1946-48 regarding the earlier role of Dahlgren work and to correct the impression that the chromium plating was to protect guns during inactivity rather than from erosion during firing. As Thompson reconstructed the work in later years, he traced the original research back to the period 1927-29 at Dahlgren and showed that it continued there throughout the 1930s.[85]

Thompson's interest in the advanced physics of gunnery resulted in a number of studies and other publications that he produced during the late 1920s and early 1930s. He described a gun as a "heat engine" and analyzed its firings in terms of the number of "cycles" it went through, with tests for uniformity of horsepower generated by the gun.[86] This basic thermodynamic approach reflected the training of a physicist rather than a gunnery officer and led to a series of experiments to try to establish the exact power of a gun. The idea that a gun's performance could be measured in horsepower doubtless struck ordnance officers as unusual. Thompson pointed out that the rapid wear and erosion of guns led to such variation in performance, even through a few test firings, that it was difficult to resolve the sampling and statistical problems sufficiently to come up with accurate predictions of performance for specific powder lots and specific guns. He stated succinctly the long-standing dilemma, one with which officers from Dahlgren through Dashiell were well familiar: "Conditions usually employed are sufficiently extreme, in fact, to render the ordinary machinery of dynamics inadequate as a vehicle for rigorous solution, and, in most cases, difficult of statement. Treatment of ideal or simplified special cases is not often of great practical value because of the extent of departure from actual experience which is necessary in order to accomplish reduction."[87]

Thompson was joined at Dahlgren in 1924 by Nils F. Riffolt, who had worked at Clark University under both Webster and Robert Goddard (later well known for his work on liquid-fueled rockets). Riffolt was a Swedish instrument maker, a degreed physicist from Clark, and a meticulous workman. Thompson remembered him as a perfectionist and sometimes agonizingly slow. But Riffolt was an accomplished technologist, and together the two civilian scientists actively worked through the 1920s and 1930s on a wide variety of practical and theoretical problems in ordnance.[88]

In addition to the thermodynamic and basic research problems, Thompson and the station's ordnance officers carried on a regular program of experimental work with very practical and immediate consequences through the 1920s. Some of the experiments conducted in 1923 and 1924 reflected BUORD's continuing interest in Jutland-inspired issues and focused on such things as fuel oil ignition by projectile bursts, tracer shells, mechanically timed fuzes, and illuminating and marker projectiles. Additionally, in light of Germany's stunning U-boat successes during the war, the Navy became extremely interested in developing countermeasures against submarine threats of the future, such as those of Germany, Japan, and even Great Britain. As a result, anti-submarine ordnance fuzes became an important field of research within Thompson's experimental program during this period. His work would greatly benefit the Navy later in World War II when it had to confront a renewed U-boat menace not only in the North Atlantic but also off America's coasts.[89]

THE AVIATION DETAIL

Contrary to popular belief, as engendered by the events surrounding General Billy Mitchell's 1925-26 court-martial, the Navy and BUORD had been experimenting with air-dropped bombs well before World War I. Two types had been developed, both for anti-troop purposes, and were intended to be dropped by airplanes supporting naval landing forces on enemy coastlines, a concept that foreshadowed the Navy's aviation support operations in the Pacific campaigns of World War II. Indian Head was the first bombing test facility, but as with the gun and powder tests, bomb proving there was both geographically limited and dangerous. The potential danger of airborne bomb testing was amply demonstrated in 1916 when two naval aviators were killed when the bomb their aircraft was carrying prematurely detonated in flight. One can only guess as to what would have happened if it had exploded over the powder factory.[90]

During World War I, the Navy had been impressed by the potential use of bombs to attack submarines (but not battleships just yet). BUORD therefore began experimenting with dual-action bombs that could either explode on contact with a submarine or under the surface as a sort of airborne depth charge. As the war progressed, the bomb-carrying capacity of naval aircraft increased, resulting in greater bomb sizes and destructive power. Early in the spring of 1918, BUORD assigned Naval Aviator Lieutenant Albert J. Ditman to the Anacostia Naval Air Station as ordnance officer to conduct experimental work in bombs and help develop and test the new designs. Since ordnance officers involved in aerial bombing tests had to fly, much of the experimental work was carried out in actual flight. Anacostia soon proved itself unsuitable for the task, so bomb testing was moved to Indian Head and its Lower Station, among other places.[91]

To facilitate the Navy's aircraft ordnance experiments at Dahlgren, BUORD built in 1920 a hangar for land-based aircraft near the station's airfield and a seaplane hangar with a shore ramp on the riverbank. BUORD also established a Naval Air Detail to carry out the experimental work, with a lieutenant serving as the Officer in Charge. Three Curtiss JN-9 seaplane trainers had already been flown to Dahlgren in September 1919 from the Marine Barracks in Quantico, Virginia, and comprised the initial aircraft component of the new Air Detail. Once it was off the ground, the Air Detail's duties soon expanded to include altitude and spotting work, bomb trajectory investigations, transportation and photographic flights, torpedo observation, "test hops" with tow targets, and "pigeon training flights." Appropriately, the pigeons used in naval communications during this period fell under the responsibility of aviation officers. The Air Detail included the Officer in Charge and one enlisted pilot, with another five to nine enlisted men assigned to the maintenance of land planes and nine to fourteen enlisted men working with seaplanes. Including a radioman, a quartermaster for the pigeons, a photographer, and a few others, the total number of enlisted men working in the Aviation Detail ran between twenty-four and thirty-two.[92]

By 1923, the Air Detail, under Officer in Charge Lieutenant (later Admiral) John J. Ballantine, was conducting routine flights in support of Dahlgren's ordnance tests and experiments. It had grown from the initial three planes to include two Hispano-Suiza 2-Ls, one Aeromarine Model 41, one Curtiss JN-4 trainer biplane (the two others had crashed), one Curtiss N-9 seaplane (a naval version of the U.S. Air Service's JN-4, modified with a central pontoon and floats fitted on extended wingtips), one R6L twin-float "torpedo" plane, and two British Airco (de Havilland) DH-4B "Liberty" biplanes. Dahlgren was now firmly in the aviation business.[95]

FLYING BOMBS

Two research projects that were literally decades ahead of their time but technologically premature were the studies of automatically piloted and radio-controlled aircraft conducted at Dahlgren from 1919 to 1925. The original concept was to develop a pilotless, explosive-laden aircraft, or Flying Bomb (with the rather obvious code designation "FB"), also known as an "aerial torpedo," with which to attack ships. The Navy's flying bomb research program had started in early 1915, when noted technologist Dr. Peter Cooper Hewitt (inventor of the mercury-vapor lamp) consulted with Elmer Sperry, a recognized expert in the field of gyro-stabilization and founder of the Sperry Gyroscope Company, about the feasibility of developing such weapons. Sperry was the best choice for possible collaboration. He had helped the Navy develop a reliable motor-driven gyrocompass for its vessels in 1912, and in June 1914 had won first prize in a French technical competition for the most successful aircraft stabilization equipment. Sperry gave Hewitt's idea some thought and decided that his company could do the experiments if Hewitt would pay for them. Hewitt agreed and gave Sperry $3,000 to start the work.[94]

The project soon consumed all of those initial funds, however, and Sperry had to contribute much of his own money to keep the project alive. After the financial risk became too great in early 1916, Hewitt and Sperry, both members of the Aeronautics Committee of the Navy Consulting Board (established on 7 October 1915), appealed to the armed services for support. The Navy Consulting Board was particularly intrigued by the flying bomb's possibilities. Consequently, BUORD Chief Earle, who never shied away from potential new warfighting technologies, sent Lieutenant Theodore S. Wilkinson to Sperry's Amityville field station on Long Island, New York, to observe and report on the tests of a "no-pilot automatic aeroplane," scheduled for 5 September 1916. The tests were delayed twice because of engine troubles but were eventually conducted on 12 September despite bad weather.[95]

The Amityville test results were promising. An aviator aboard actually flew the sea-based "aeroplane" off the water before turning it over to automatic control, since the Navy feared losing the aircraft and its special stabilization and course-keeping gear in a crash. After climbing to and flying at a preset altitude for a pre-determined time, the aviator switched off the automatic control, causing the aircraft to dive straight for the ground before the aviator recovered and landed safely. Despite this partial demonstration of success, Wilkinson concluded in his report to Earle that the system was too inaccurate to hit moving ships. He therefore recommended that the Army take control of the project since the "flying bombs" could be developed to hit large, fixed military targets.[96]

Wilkinson's report notwithstanding, the Naval Consulting Board recommended that the Navy support further flying bomb experiments, this time with the aim of developing weapons that could bombard from the sea large, distant areas such as naval stations, fleet anchorages, and fortified towns. On 22 May 1917, Navy Secretary Daniels formally approved an allocation of $150,000 for the project. The Navy negotiated a contract with Sperry for the new tests, provided five Curtiss N-9 seaplanes fitted with landing wheels, and purchased six automatic pilot systems.[97]

At Amityville in July 1917, BUORD's supervising officer, Commander Benjamin B. McCormick, reported to Earle that the work was being divided into two parts: 1) converting an automatically piloted airplane into an aerial torpedo, and 2) developing a radio control system to allow targeting and guidance from an accompanying aircraft. Early on, McCormick and Sperry agreed that a "catapult" launching system was the best option. To do the necessary calculations and design the new catapult, Sperry turned to an engineer and former employee named Carl Lukas Norden.[98]

Born in 1880 to Dutch parents, Norden had been trained as a mechanical engineer at the prestigious Zurich Federal Polytechnic School before immigrating to America in 1904 and working for a succession of manufacturing firms. He had joined the Sperry Gyroscope Company in 1911 to help design stabilizing gyroscopes for large ships. Though brilliant, Norden was arrogant and volatile and had maintained a tempestuous up-and-down relationship with Sperry. After numerous arguments over patents, he had finally quit in 1915 and established his own engineering business in New York. Despite their personal differences, Norden had continued working for Sperry as an independent consultant on Navy contracts and was thus drawn into the flying bomb project.[99]

Norden soon had a design down on paper, and by mid-August he had his new catapult built and ready for the first flying bomb launch. During that initial attempt, his catapult malfunctioned and the plane never left the ground, but on 23 September he conducted a more successful test. This time the catapult launched the flying bomb into the air cleanly, but once airborne, the plane performed erratically and crashed, possibly because of its flimsy design. Another test three days later ended with the same result. On 17 October, though, Norden finally launched a plane that functioned normally in the air. It climbed steadily and flew in a straight line, deviating only twenty feet from its preset course. However, its distance-controlling device failed to shut down the engine, and the plane was last seen flying eastward at 4,000 feet.[100]

Although Norden had lost the plane over the horizon, Sperry and the BUORD observers considered the flight a big success. Sperry was so thrilled by Norden's "perfect shot" that he wrote Earle that "I believe that the time has practically arrived when we have actually in hand <u>the gun of the future</u>." McCormick was also impressed, particularly by Norden's technical ability. He decided on the spot to cut Sperry out of the project altogether and to turn all of the engineering work over to Norden. Soon after, BUORD contracted with the Witteman-Lewis Company for five flying bombs designed to Norden's specifications and authorized Norden to redesign Sperry's automatic controls.[101]

In late 1918 McCormick requested that the entire flying bomb project be moved from Amityville to a naval station, preferably the new proving ground at Machodoc Creek. Sperry magnanimously agreed to relinquish control, telling Earle that the Navy could better expedite the flying bomb's development by taking over the project and establishing it at the proving ground, "where there is plenty of room and the Naval Officer in Charge of the station could coordinate the efforts of all who might contribute." Earle ordered the move, and by 27 May 1919, Norden, two Naval Reserve Force aviators, the catapult, several automatic pilot systems, three N-9 trainers, two old F-type flying boats, and five new Witteman-Lewis flying bomb aircraft had all arrived at Dahlgren.[102]

At Dahlgren, the project's "check pilots" found the airfield too rough for takeoffs and landings, and also that the Witteman-Lewis planes were tail-heavy and that their ailerons and tail surfaces were too small for safe flight. While Dahlgren's construction crews quickly graded and smoothed the airfield, a concerned Rear Admiral Earle halted all further flights using naval aviators until Witteman-Lewis, with the help of the Bureau of Construction and Repair, fixed the airplanes' design problems and made them safe to fly. Thus, no automatic test flights were conducted at Dahlgren in 1919.[103]

The company modified the planes accordingly over the winter months of 1919-20, and on 30 April 1920 Earle authorized the resumption of pilotless flying bomb experiments. At the time, though, Norden had become preoccupied with another project that would later prove vitally important for American air forces in World War II—the development of a high-altitude, gyro-stabilized bombsight—and could not supervise the next round of flying bomb tests until August.[104]

Taking a break from his early bombsight work, Norden checked out his automatic pilot and catapult equipment and determined that it was all still in good condition, enough for a trial test. On 18 August 1920, he launched his first flying bomb at Dahlgren. After leaving the catapult, the plane flew smoothly for 150 yards, stalled, and then nosed over into the Potomac. On 18 November, Norden conducted his second test, this time after a pilot had pre-flown the plane to adjust its automatic controls. Dahlgren observers reported that "the launching was perfect." The plane climbed slowly, traveled five miles, circled, and, reaching an altitude of 1,500 feet, continued circling until its automatic engine shut off, sending it spiraling into the river. Norden was gratified, writing to BUORD that "a plane, notoriously hard to fly manually and never flown before, has been equipped with automatic control, [was] adjusted according to the information obtained by the Bureau's flight officer, [and] proved capable of perfect sustained flight."[105]

In December, McVay expressed his continuing support for the project and authorized another full-scale test in the spring. On 25 April 1921 Norden launched one more flying bomb, which flew off the catapult perfectly and climbed for a short distance before gently descending onto the Potomac and upsetting end-over. The flight had only lasted one minute and fifty seconds, and Norden attributed its early termination to his own error in pre-setting the plane's horizontal stabilizer. Unfortunately, this was the last flying bomb test conducted before the advent of guided missiles, since BUORD—despite McVay's previous assurance—decided that the flying bombs were tactically dubious and too impractical and halted further development in favor of radio-control technology. As Rear Admiral Delmer S. Fahrney noted in the Navy's official history of pilotless aircraft and guided missiles, BUORD did not necessarily lose interest in flying bombs, but in an era of slashed military budgets in which the Navy needed every cent it could scrape up to maintain a modest fleet, little money existed for experimentation. "If a project was not a complete and howling success on the first trial," he wrote, "it would be dropped." And since the Navy could hardly afford to lose the airplanes that Norden needed to sacrifice in each test, BUORD abandoned the project.[106]

RADIO CONTROL

Radio-controlled aircraft originally had been part of McCormick and Sperry's flying bomb plans in 1917. However, it was not until late December 1920 that a special board appointed by the Chief of Naval Operations to investigate the feasibility of remote-controlled aircraft recommended that the Navy sponsor the research and place it under BUORD's supervision. The project's activation was slow, though, but finally in October 1921, BUORD and Bureau of Engineering (BUENG) representatives visited Dahlgren to devise procedures for conducting the radio-control research project. Once these were approved, BUENG assumed responsibility for designing, installing, and testing the necessary radio equipment, initially done in the Naval Aircraft Radio Laboratory under Dr. A. Hoyt Taylor at the Anacostia Naval Air Station in Washington, D.C. BUORD, for its part, would supervise mechanical equipment installation aboard the project's aircraft and conduct flight tests at Dahlgren.[107]

On 17 January 1922, BUENG advised BUORD that it was ready to proceed. BUORD released funds for the project, and on 28 January BUENG directed the Radio Laboratory to start work. In March the head of BUENG's Radio Division, Commander Stanford C. Hooper, who Admiral Chester Nimitz later called the "father of radio in the United States Navy," persuaded electrical engineer and former reservist Carlo B. Mirick to come aboard as Hoyt's associate radio engineer and to handle the project's actual radio-control engineering. Before starting work, Mirick toured various military and civilian stations and laboratories to identify and select the system's key components. After the Radio Laboratory developed a special relay that could link an airplane's radio receiver and electro-mechanical controls, he spent the fall months intensively testing the equipment and by 28 December had achieved significant progress.[108]

Mirick moved to Dahlgren in early July 1923 and soon mated his radio-control gear with Norden's automatic pilot system in an N-9 seaplane. The Officer in Charge of Dahlgren's Air Detail, Lieutenant Ballantine, became the project's military "safety pilot" and made over thirty manned flights that summer and fall, checking out all of the automatic pilot and radio-control components through simple maneuvers and flying by radio control short distances of up to five miles. During the year's final test, conducted on 14 November, Ballantine's seaplane flew solely by radio control for twenty-five minutes and performed a number of elementary maneuvers before he retook manual control. After the flight and per standard practice, BUORD halted testing for the winter.[109]

During the lull, the new Naval Research Laboratory (NRL), which had absorbed the Radio Laboratory, improved the radio-control system and equipped two Vought VE-7H seaplanes with automatic pilots. NRL also fitted an HS flying boat with an airborne control station to remotely control Ballantine's N-9 from the air and at greater distances. Flight tests using the improved equipment and airborne controls resumed on 24 July 1924, and on the morning of 15 September, two test flights were conducted in which Norden's automatic stabilizer and Mirick's radio-control system worked perfectly together. That afternoon, Ballantine and Mirick decided to attempt the project's first unmanned, radio-controlled flight. In preparation, Ballantine's ground crew lashed a bag of sand into the cockpit to compensate for his weight before cranking the engine, and the flying boat positioned itself over Dahlgren to take control of the unmanned plane if necessary. From the radio-control station near the hangar, Mirick and Ballantine gave the "on throttle" signal, and the crew released the plane into the air. After a forty-minute flight, the controllers landed the plane rather hard and damaged it, but otherwise the flight was a tremendous success. This was the first time in history that a pilotless aircraft had taken off, was controlled remotely through numerous maneuvers, and landed relatively intact.[110]

Following this pioneering flight, which presaged the Unmanned Aerial Vehicles of the modern era, Mirick and Ballantine's technicians transferred the N-9's radio-control gear to one of the newer Vought planes. In December the Vought successfully flew for the first time eleven miles out without an automatic pilot. Flight ranges were increasing so much that NRL had reported on 22 November that radio control was now feasible beyond the range of vision. Test flights again ceased for the winter but resumed on 19 June 1925. By 14 September, Ballantine had conducted twenty-eight more flights before rotating out to another assignment. Ballantine's replacement at Dahlgren, Lieutenant Valentine H. Schaeffer, took over as safety pilot but did not make a completely successful flight until 28 October.[111]

The radio-control project's final flight occurred on 11 December. Assistant Aviation Officer Lieutenant J. E. Ostrander took off in a spare De Havilland (DH) aircraft with his rear cockpit filled with bricks. The idea was that if the unmanned Vought refused to respond to radio control, then Ostrander could fly over it and drop bricks into its propeller, disabling the plane and sending it into the ground. Fortunately for him, perhaps, he never got the chance to test the brick-drop theory since the Vought test plane crashed on takeoff and sank in eight feet of water. Despite improvements made to the controls immediately after the accident and in the years afterward, no further tests were made. Although the project was not canceled, the Navy let it languish until 1936 because of dwindling experimental funding and, perhaps more ironically, because of Norden's parallel bombsight program, which consumed much of the Navy's limited research and development resources during that period.[112]

PIONEER LIFE

Life at the rural proving ground in the early years was later remembered with some nostalgia as rugged, particularly by the small contingent of officers' wives, numbering five or six during the mid-1920s. Except for J. L. Hoge's Store, which had serviced the oystermen before 1918, there was no nearby shopping, so weekly automobile excursions into Fredericksburg, Virginia, some twenty-nine miles away, became necessary for clothing, household items, and other incidentals. Dahlgren's single unpaved road, Thompson later recalled, "shook your eye teeth out" and was passable only in dry weather. By 1923 the wives had organized a group, with two making the drive to Fredericksburg, taking shopping lists for the others. They had to ford two streams en route to Fredericksburg and the same two streams on the way back. After getting stuck a couple of times, the group members worked out an emergency call system, using the base pigeons. They took along four pigeons and released one each time they successfully crossed a creek. Back at the base, the officer in charge of pigeons noted the pigeons' arrival. If one did not return, the truck was dispatched to haul the automobile through the appropriate ford.[117]

The officers at Dahlgren were well aware of the isolation that the base imposed on their spouses and often arranged Friday or Saturday shopping excursions via the steamboats *Grampus* or *Porpoise* to Indian Head, and then a fifty-mile trip by car to Washington, D.C., that could be coupled with an official Saturday visit to BUORD headquarters. Since there was nowhere to take visitors out for a meal in the Dahlgren vicinity, visiting VIPs were invited to the officers' homes for lunch, leading to better liaison with senior officers and some lasting friendships.[114]

The seeds of an outside community were planted in 1922 with the construction of what came to be known as Shelton's Store. However, growth was slow because of Dahlgren's remote location and the lack of private investment, and the government had to step in to develop the outside blue-collar community for the station to survive. This included the provision of employee housing, medical facilities, and a school building, which was constructed in 1922. To educate the station's children, the Navy made special arrangements for the Commonwealth of Virginia to supply one teacher for grades one through twelve. Improving transportation was likewise a priority, and by 1926 the Navy had encouraged Virginia to build dependable gravel roads to accommodate the growing number of automobiles brought to Dahlgren. By 1928, a commercial ferry began operating and connected Dahlgren with Morgantown, Maryland, whose new concrete road slashed the travel time to Washington, D.C., by at least half. The following year, the Navy worked with the Virginia Electric and Power Company to bring electricity to the outside Dahlgren community, which heretofore had not enjoyed the benefit of the station's internal generators.[115]

Although community progress was visible throughout the 1920s, Dahlgren was still primitive by Indian Head standards, but despite the difficult environment, a group of some thirty Indian Head civilian employees transferred there. One was Roger Dement, who had worked at Indian Head since 1907 and had not only signed the petition against the move, but had also testified before the special subcommittee of the House Naval Affairs Committee that Indian Head could still prove all major-caliber guns safely and more cheaply. After reluctantly going to Dahlgren, Dement joined the Armaments Department and later became head of the proving ground's Range Section. As range chief, Dement was not only responsible for maintaining the river range and overseeing the construction of new stations downriver when needed, but also for handling diplomatic relations with the property owners where the stations were built. Maintaining peace with the locals was imperative for the range's successful operation since the Navy did not buy their land for the stations but just obtained their permission to use it. As physicist Donald Stoner remembered, "Sometimes we had complaints that our range vehicles going in and out were tearing up the roads," or that "we had people who went down to man the range stations who would occasionally do things that annoyed the property owner." When grievances arose, Dement always moved quickly to pacify the offended owners, such as bringing them gravel for their rutted roads, or by sternly dealing with range station personnel who failed to maintain a healthy respect for the owners and their property. After a long and successful career at Dahlgren managing both the range and the locals, Dement retired in 1954.[116]

SEPARATION

By 1931, Dahlgren was all but independent, and BUORD finally decided to separate it from Indian Head. In preparation for the division of command, Inspector of Ordnance Garrett L. Schuyler developed a detailed policy statement in October 1931, in conjunction with war planning and in response to a request from the Chief of Naval Operations (CNO). Using the future date "M" to designate mobilization for war, Schuyler pledged that the Dahlgren Naval Proving Ground (NPG) would be prepared to prove newly constructed 16-inch guns at M plus 4 months and new ship projectiles at M plus 12 months. The base would be maintained "in full commission with peace allowance of personnel . . . with plans for rapid expansion of personnel and necessary facilities for the total war complement." In keeping with other war planning documents in 1931, the memorandum was classified "Secret." Schuyler developed a set of changes to the regulations for both the proving ground and the powder factory, dated 3 October 1931, "due to the assignment of a separate Disbursing Officer at Dahlgren." The regulations became effective immediately, and by 1 July 1932, Dahlgren's transition to a separate command was finally completed.[117]

As Dahlgren entered the 1930s, it was well established despite its difficult birth and transition to independence. The site was perfectly chosen for both ordnance and aeronautical work. The open approaches to the aircraft landing field and the surrounding flat country reduced the risks attendant in flying both the landplanes and the seaplanes of the era. The clear, long downriver range, with observation posts strategically situated along the riverbanks, connected by telephone line, minimized the risks that had plagued Lieutenant Dahlgren's Experimental Battery on the Anacostia and the later proving grounds at Annapolis and Indian Head. The seeds of a civilian scientific research and development base were planted by Thompson and his assistant Riffolt, with the technical expertise of Patterson from Indian Head. Increasingly, the Bureau assigned naval officers with postgraduate training in ballistics from some of the best universities in the country to Dahlgren. Despite the early handicap of being established and developing through an isolationist and parsimonious political era, the Naval Proving Ground at Dahlgren, by the year of its independence in 1932, had emerged as a valuable technological center in the service of the nation and the Navy.

3
Chapter

Dahlgren at War, 1932-1945

Cut loose from Indian Head in the summer of 1932, the Dahlgren Naval Proving Ground entered a new era of growth and technological achievement to become a key component in the Navy's rearmament and wartime ordnance program. Dahlgren's focus on mathematics, physics, and ballistic computation not only spawned new weapons systems and research fields but also resulted in the integration of science more fully into the naval establishment, key to the Allied war effort during World War II.

The ten-year "holiday" in capital ship construction, as mandated by the Naval Arms Limitation Treaty of 1922, had essentially crippled the Navy. Worse, President Herbert C. Hoover, supported by anti-navalist majorities in Congress, slashed naval armament expenditures further after the economy crashed in October 1929, and the London Naval Treaty of 1930 not only extended the tonnage limitations to cruisers, destroyers, and submarines, but also extended the battleship holiday until 1937. Consequently, by 1932 the number of ships in the fleet had dwindled to well beneath treaty limits, while the Bureau of Ordnance (BUORD) and the rest of the shore establishment withered to barely functional levels.[1]

Dominated by old-line gunnery officers, BUORD responded to the poor political and economic situation by retrenching into a more conservative bureaucracy and drastically cutting research projects. The Special Board on Naval Ordnance, which oversaw BUORD's technical matters, limited or in many cases simply blocked projects deemed risky or outside the scope of naval gunnery, such as experimentation in aviation ordnance. This conservatism, along with curtailed appropriations, ensured that supply and maintenance of the fleet's existing armament took precedence over experimental work. BUORD's apathy toward new weapons development extended to the Naval Proving Ground, where gunnery officers likewise ran the show. In October 1931 Inspector of Ordnance in Charge Captain Garret L. "Mike" Schuyler (the same Schuyler who had fired on President Wilson's yacht in 1913) described Dahlgren's research role as only supporting the improvement of naval armor and guns. He notably failed to include the development of any new ordnance devices as one of Dahlgren's planned tasks in the event of war. Thus, naval ordnance research and development at Dahlgren remained stunted through his tenure.[2]

With no new capital ships under construction and aging ones being scrapped in increasing numbers, proving activities at Dahlgren ebbed with only 15 officers and approximately 70 enlisted personnel running the station. Civilian employment meanwhile stagnated at roughly 208 workers early in the decade, down from the 694 counted in October 1919. As Captain David Hedrick remembered later, the Main Battery became so understaffed that all hands, military and civilian, had to thoroughly acquaint themselves with all phases of the work just to conduct routine tests. Despite a skeleton crew, the proving ground still struggled with the lagging workload. As a partial solution, proof officers inaugurated a number of time-consuming procedures to keep their batteries busy. These included the practices of ranging most fired rounds and taking velocity measurements using cumbersome Boulange chronographs and screens. According to Captain William Rea Furlong, Dahlgren's Inspector of Ordnance in Charge from 1934 to 1936, much of the latter work was completely gratuitous since the test ammunition produced unreliable velocity data that held "little real value for purposes of record."[3]

A "NEW DEAL" FOR THE NAVY

The outlook for the Navy and Dahlgren improved significantly with the election of President Franklin D. Roosevelt in 1932. The Democrat Roosevelt was an internationalist and navalist, unlike his Republican predecessors. Having served as the Assistant Secretary of the Navy from 1913 to 1920 and a member of the U.S. Naval Institute since 1927, he was firmly grounded in naval matters and distressed by the Navy's nearly moribund condition. Secretary of the Navy Claude Swanson articulated Roosevelt's philosophy in his 1933 Annual Report, noting that "Naval wars are largely fought and decided with fleets existing at the beginning of the conflict" and that a strong Navy was the nation's first line of defense and could not be improvised overnight should a war erupt. Accordingly, Roosevelt launched a deliberate shipbuilding program as part of his New Deal agenda, with the dual purpose of strengthening the Navy and employing the unemployed. He allocated $238 million under the 1933 National Industrial Recovery Act (NIRA) for the construction and arming of thirty-two new ships. The Trammell-Vinson Act, passed in 1934, increased the momentum of Roosevelt's naval rearmament policy by authorizing the construction of enough ships and aircraft to bring the Navy fully up to the allowed treaty size and to replace overage ships. By the end of fiscal year 1934, a total of seventy ships and two gunboats were scheduled for completion over the next thirty months, while seventy-eight additional vessels were slated for construction.[4]

All of the new ships under construction as well as those planned for the future needed guns and armor. Likewise, appropriate powder and ammunition would have to be manufactured in greater volumes. All would have to be proved before entering service. Consequently, BUORD anticipated a more robust testing schedule at the Naval Proving Ground, particularly for the smaller guns needed for the new cruisers and destroyers to be constructed in accordance with the naval treaties. In October 1933 BUORD's Guns and Turrets staff recognized a possible bottleneck at Dahlgren. The officers requested a second 6-inch, 47-caliber pilot gun because, based on the number of tests on cartridge cases, projectiles, and other material produced under the NIRA shipbuilding program, they found one gun simply inadequate for the proving ground's needs.[5]

In December 1935, to come to grips with the sudden cascade of work, Captain Furlong requested changes in the proof regulations to eliminate redundant and nonessential tests. He noted that "In the course of the past few years the firing at the proving ground has practically doubled" and that the "present volume of firing is perhaps three times as heavy as some of the years in the post-war decade." He had already halted the inefficient use of chronograph screens as well as star gauging and bore searching of proof guns, which duplicated tests performed at the Naval Gun Factory. However, even with these laborsaving measures, gun firing continued to dominate battery crews' days while paperwork consumed their off-hours. Furlong recommended that all unnecessary velocity measurements and ranging be eliminated for the building program's duration. Moreover, he warned that Dahlgren must expand and that the testing regimen must be further streamlined to handle additional work.[6]

Furlong's recommendation for change coincided with the Second London Naval Conference of 1935-36, which resulted in the collapse of the arms control system that had governed the navies of the United States, Great Britain, and Japan since 1922. Japan had chafed under its unequal treatment under the system and had announced in 1934 that it would not renew the 1922 Naval Arms Limitation Treaty when that agreement, extended by the 1930 London Treaty, expired on 31 December 1936. At the second London conference, which convened on 9 December, the Japanese delegation rejected all warship ratios and 14-inch gun ceilings and demanded parity with both the U.S. and Royal navies. The American delegation refused to acquiesce, and the Japanese walked out of the conference on 15 January 1936, leaving the United States, Great Britain, and France to sign a weak treaty in March that limited the size of their warships and the maximum calibers of the guns that the vessels could carry. The treaty was soon set aside, however, when it became known that Japan was building "super battleships" armed with 16- and even 18-inch guns. Invoking the treaty's "escalator clause," both the United States and Great Britain began building new battleships once again.[7]

As the battleship holiday all but ended in 1936, BUORD acted on Furlong's advice by overhauling the proof regulations and authorizing new civilian hiring and limited expansion at Dahlgren. Five new range stations were built on the Virginia side of the river in 1936, and by 1938 the Plate Battery had been expanded twice. The Main Battery, which had grown modestly since 1926, was also expanded in 1935, growing from a civilian workforce of 23 in 1937 to a staff of 140 by December 1941.[8]

The increasing numbers of civilian blue-collar employees were vital for Dahlgren's pre-World War II expansion. While the naval officers and professional white-collar staff of physicists and mathematicians supervised the technical aspects of Dahlgren's operations, the blue-collar force, supplemented by enlisted sailors, supplied the necessary muscle to keep the guns firing and the test data flowing. The blue-collar employees were paid cash on a per diem basis and did nearly all of the station's labor—everything from digging ditches to carrying powder bags to assembling shell cartridges and explosive charges. Because of the strenuous physical demands and hazards of proof and testing, Dahlgren's civilian laborers necessarily had to be both hardy and stouthearted. Also, considering the isolated, Depression-era environment in which they worked, they had to possess a certain grittiness formerly found in old frontier towns and wilderness outposts. Accordingly, as physicist Donald W. Stoner later recalled, Dahlgren attracted some "pretty rough characters" for its civilian labor force during this time.[9]

As an example, Stoner remembered one particular weekend trip to Washington, D.C., in which he and another professional colleague traveled with a couple of civilian Plate Battery workers. Crossing the Potomac in a motor launch on a Friday evening, the four men retrieved the station's car for the trip north. Along the way, the two workers stopped "somewhere" to buy some "hooch," oblivious of their two young physicist passengers. After the group's arrival in Washington, the two workers took their hooch and parted company with Stoner for the weekend, but agreed to rendezvous at the Naval Gun Factory at 5:00 a.m. Monday morning for the return trip to Dahlgren. At the appointed time, the two workers came staggering to the car, much to Stoner's astonishment. "Oh boy, they'd had the most wonderful weekend you've ever heard of," he recalled. "They wrecked four beer joints, beat up seven or eight guys, they'd been in a couple of fights that they hadn't won but claimed they were real good ones. They had a great time."[10]

Since "there were quite a few of those characters working around Dahlgren at that time," Stoner noted that it took even tougher foremen to get the work out of them. Like the workers, the foremen were blue-collar, but had risen to their positions after thoroughly learning their trades and demonstrating the reliability and stern self-confidence needed to run the range. Since the foremen often had to deal with "some pretty strange animals" within their respective sections, Dahlgren's military leadership gave them a lot of authority and latitude in enforcing range and shop discipline. Concerning a foreman's purview, Stoner recalled that "he had the power to give and to take away," and "if you didn't leave a certain amount of that with him, some of those characters . . . would be impossible to manage."[11]

While the blue-collar employees could be rowdy off-station on their own time, under the watchful eyes of the range foremen and the station's naval officers, they had to be all business. A glimpse of Dahlgren's blue-collar world can be found in the unpublished memoir of civilian employee and future engineer Charles Roble, who recounted his training and the station's strict working environment when he first started work there in February 1941. On his first day, Roble had to pass a rigorous physical exam, part of which required him to heft a fifty-pound sack of loose shot over his shoulder. After passing the exam, the Civilian Personnel Office assigned him to "Five Weeks School," in which he would gain experience in the five different areas of proof work, be evaluated by the supervisors, and then receive a permanent assignment from Personnel. He worked his first week running bags of 16-inch gun powder from the powder room to the battery at the Main Range, an especially demanding and stressful task. During that week, Roble was exposed to naval discipline for the first time when a coworker pulled a cigarette from his pocket and lit it with a burning twig. As Roble recalled, a naval officer, assigned to direct the work detail, literally smacked it from the offender's mouth and berated him about breaking the very sensible rules against smoking near the powder room. Roble also found the station's military hierarchical relationship with the blue-collar force "very irritating," since the lowest naval rating or marine private could "order around any civilian at whatever level" on the range, resulting in some rather unpleasant encounters early on.[12]

Roble spent his second week in the Shell House, where he assembled live ammunition, helped test a torpedo warhead against a *Liberty* ship hull section, and measured the fragmentation pattern of a 5-inch naval shell. Roble's third week found him at the Terminal Range, where he mostly typed test reports and assembled 6-inch gun charges. Week four took him to the Range Room, which contained a solid brass table-map that served as a scale model of the section of the Potomac River between Dahlgren and Chesapeake Bay. The map was built for the purpose of calculating impact points of shells fired down the Potomac range. It was inscribed with both straight lines, representing distance in yards, and radial lines, so that deviations from a straight course could be determined. During firing tests, a special protractor could be inserted into drilled holes on the map representing the precise positions of range stations. Three appropriate range stations were manned for each test and linked by radio to the Range Room and the firing line lookouts. At each station an observer would level a theodolite, zero in on a common reference, and, once a projectile was fired, record the angle of deviation from that reference for the projectile's observed impact point on the river. After the range stations reported the deviation angles, the Range Room staff marked the designated reference point as '0' on the range table map and duplicated the reported angles on its surface with a very sharp, hard lead pencil. Through triangulation, the Range Room then plotted the projectile's splash position, thereby generating data needed by Dahlgren's physicists and mathematicians.[13]

Roble worked several days as a range station observer before moving to the Range Room, where he learned to set up the map's protractors and to plot splash angles and impact points. During his fifth and final week of "schooling," he worked at the Armor Department's Light Armor Battery and learned the science of projectile velocity measurement. On the last day of that week, he and the other "classified laborers" who had trained at the same time but on different rotations took a placement test at the Civilian Personnel Office based on what they had learned during their five-week training period. Roble was disappointed when Personnel assigned him to the Main Range as a battery attendant, since he did not like working directly with the guns on the firing line. After threatening to quit, he was reassigned to the Velocity Section in the Proof Department, where he worked first as a technician in the chronograph room before rising through the ranks to join the professional staff in the 1950s. Roble retired in 1967 as a GS-13 supervising electronics engineer. However, his training and early work experience typified that of the civilian blue-collar employees who came to Dahlgren to man the range just before America entered World War II.[14]

EXPERIMENTAL RESEARCH ASCENDANT

As naval appropriations increased and the tempo of testing quickened at Dahlgren, so did the scale and sophistication of Chief Physicist Dr. L. T. E. Thompson's work. His research had been somewhat inconsistent since he was dependent upon old proof reports and sporadic test firings for data. In the early 1930s, the slow pace of work did allow him time to write and publish his findings. He tackled such problems as shipboard high-angle gun velocity measurements, gun pressure measurements, the propagation of blast and gas waves, projectile flight characteristics, muzzle flashes, and, perhaps most critically, shot dispersion, which continued to exasperate gunnery officers throughout the fleet.[15]

In 1934, though, Thompson's experimental research gained a powerful patron with the selection of Rear Admiral Harold R. Stark as the new Chief of BUORD. A friend of Thompson's, Stark had served as Inspector of Ordnance in Charge at Dahlgren from 1925 to 1928 and agreed with the physicist that science and mathematics could solve fundamental ordnance problems. At the urging of Furlong and Thompson, he immediately authorized a program to help the physicist confront the nagging problem of shot dispersion in triple gun mounts.[16]

Thompson soon discovered a "second gun effect" and "wing gun interference" in which parallel shock waves threw simultaneously fired projectiles off target and caused abnormally wide shotfall patterns. Other serious problems yet remained to be solved through experimental research, and Furlong backed Thompson's plans to investigate the determination of proper ballistic qualities of all types of guns and shells, development of

Class A armor, development of improved 8-inch armor piercing projectiles, and the development of new fuzes.[17]

While pursuing practical gunnery-related research on these items at Dahlgren, Thompson also turned his attention to a particular ballistics problem that had previously caught his attention but remained a sensitive issue within BUORD. This was the problem of high-altitude bombing.

World War I had revealed the efficacy of aerial bombing. Nevertheless, a much ballyhooed series of experimental bombing tests conducted off the Virginia Capes in July 1921 against stationary target vessels, including the captured German battleship *Ostfriesland*, had stewed resentments not only between the Army and the Navy but within the BUORD itself over the tests' implications. Quite simply, battleship proponents insisted that the tests had been flawed (they were) and that under actual combat conditions maneuvering battleships and fortified shore installations would be largely impervious to aerial bombing (they were not). They further held that naval aviators should only serve the fleet as scouts and observers and not as bombers.[18]

This position contrasted sharply with that held by a few of the more imaginative officers in BUORD, as well as those of the Bureau of Aeronautics (BUAER—established in 1921), who saw aerial bombing as a potentially devastating weapon of war. They realized that a battleship armed with guns ranged at only thirty miles could never match bombers, which could fly more than a hundred miles from their bases or carriers. However, bombing advocates remained a very small minority in BUORD, which did not allocate any funds for bomb research. Dahlgren aviator Sherman E. Burroughs later lamented that "There just wasn't anybody really interested in aviation in BUORD during those years." Although he could get money to build bombs and outfit aircraft, he "never got a nickel" for research and development.[19]

Despite BUORD's antipathy, Thompson believed that aerial bombing would someday be as important as big guns in naval warfare and should not be ignored. In February 1933 he carefully outlined his thinking in a study of the comparative ballistic merits between long-range gunnery and high-altitude bombing. Thompson concluded that aircraft and ordnance technology had not advanced far enough for bombing to supersede gunnery as the primary mode of attack in naval warfare. Nevertheless, he still considered aerial bombing a "very important" method of attack and urged its integration as an equal component of gunnery in American naval strategy and tactics.[20]

During the next two years, Thompson mulled over the problem of aerial bombing and its role within the Navy's tactical and strategic doctrines.

Realizing that his concerns would be ignored within BUORD, he took the unusual step of going outside of regular channels. In a January 1935 memo written directly to BUORD chief Rear Admiral Stark, he expressed his developing ideas on the matter more energetically. He confessed that he had been thinking of expanding his research to include naval aircraft since he had come to "believe that the next decade will see a new type of bombing unit which will be a fair match for battleships." This unit would operate at 15,000- to 18,000-foot altitudes and would carry armor-piercing bombs capable of penetrating battleship armored decks and exploding with greater force than gun projectiles. He suggested the study and possible development of a new type of armored, low-deck aircraft carrier from which the new heavy bombers would operate. This new bomber carrier would not "substitute either for battleships or for present types of carriers," in Thompson's estimation, but it would be comparable to a battleship in combat power.[21]

Thompson had another agenda, however. What he really wanted was to expand his struggling Experimental Department to undertake a broad new research program in naval aerial ordnance. Foreseeing that high-altitude aerial bombing "may be the most important ordnance development of the next ten or twenty years," he asked for the assignment of enough civilian personnel and a number of permanent Engineering Duty Officers (EDOs) who could support the research full time. Thompson believed that this was absolutely necessary since aerial ordnance technology was rapidly blossoming and that the technical work associated with it was becoming too complex for him to manage alone.

Thompson further recommended the reorganization of BUORD's Special Board on Naval Ordnance to include all of his anticipated EDOs. Under his plan, the reconstituted board would act as a "progressive unit" for the systematic study and creation of new trends in naval ordnance, particularly in bombing. In short, he was suggesting nothing less than a revolution within BUORD and at Dahlgren, in which handpicked officers under his guidance would shift the course of ordnance research away from gunnery toward naval bombing and air defense. His Experimental Department at Dahlgren, if so expanded, would harbor the focused research program that he envisioned and integrate science more fully into the Naval Proving Ground's technological culture.

Stark respected Thompson's work, but he was also a professional naval officer and a former battleship captain and shared the world-view of most line officers of the time. So he was not yet ready to embrace most of the physicist's recommendations, particularly shifting BUORD's orientation away from gunnery to bombing. Likewise, he was certainly not about to reorganize the Special Board in such a way that would give the civilian physicist significant influence in naval policy. Stark did authorize the Experimental Department's expansion, though, to include the personnel that Thompson needed as well as a fully equipped Experimental Laboratory, completed in 1936, with which to conduct more coherent and long-term research programs.[22]

THE "PICKLE BARREL" SIGHT

Thompson's growing concern about the possible future dominance of high-altitude bombing in naval warfare sprang from his involvement in the successful development of what ultimately proved to be one of the most important weapon systems of World War II, the Mark XV Norden bombsight. Initial research into bombsight technology had begun during World War I when the Navy became interested in arming its seaplanes with a device that could successfully drop bombs on moving ships. The problem of accurately calculating both a falling bomb's trajectory, particularly at high altitudes, and its impact point presented even greater complexities than those associated with naval gunnery. As a weapons platform, a bomber in motion was anything but steady. Often buffeted by turbulence, it rotated about three axes and flew at relatively high but inconstant velocities in three dimensions.[23]

The bombardier's challenge was to determine the exact point at which to release a bomb in order to achieve the greatest probability of a hit. Theoretically, the problem is comprised of two parts, course and range. The problem, of course, involves maneuvering the bomber so that it and its bombs will follow an imaginary line that will intersect with the target. This was relatively easy in the absence of wind, but with a crosswind, the bombardier had to offset the aerodynamic forces pushing the bombs away from the target by flying a path parallel to the intersecting line at a distance proportional to the crosswind's strength. Range was an even more complex problem. To calculate the correct distance to the target from which he should release the bomb, the bombardier needed to solve a series of complex mathematical calculations. Among the factors to be taken into consideration were the bomber's velocity, altitude and course, wind direction, speed—at both release and impact points, the bomb's ballistic characteristics, and the force of gravity. A moving target only complicated the bombing calculations further, while wind, both track and crosstrack, was one of the biggest sources of error with early bombsights.[24]

Early efforts involving the use of bombardier-pilot teams and visual pilot-director indicator signals to drop bombs on targets at predetermined altitudes and airspeeds were unsuccessful. After some trial and error, aviation officers realized that the complex bombing calculations were beyond the ability of human bombardiers to handle manually, especially in combat. Consequently, both the Allies and the Central Powers strove to develop a mechanical computing device that could do the calculations and derive the correct angles necessary for successful bomb drops. The resulting first bombsights were primitive yet complicated. None of them could be used with any accuracy during drift, and the accuracy of range gained from their complicated computing mechanisms was lost in line error. The Navy's large flying boat bomber, in which the plane's "bomb dropper" was situated forward in the bow while the pilot sat in its waist—where he had no line of vision straight down at the target—proved particularly susceptible to line error.[25]

At BUORD's behest, Naval Reserve Force Lieutenant A. H. Boettcher and U.S. Marine Corps Captain B. L. Smith undertook the task of designing a bombsight that could give a flying boat pilot a physical and visual indication of the relation between his actual course and his target's bearing. They soon produced a pilot-directing sight, called Mark I, that met BUORD's specifications. BUORD tested and approved the Mark I in December 1917. Production started immediately, and BUORD began issuing the sight to naval air units shortly thereafter.[26]

Early in 1918, however, Major H. E. Wimperis of the British Royal Flying Corps developed a new "course setting" bombsight that permitted bombing either with or without drift. A British officer brought a demonstration model to Washington in May 1918, and although it lacked a pilot-directing feature, BUORD immediately realized its superiority over the Mark I. Aerial tests, with Boettcher's pilot directing mechanism attached to Wimperis's bombsight, achieved "astonishingly good results," and as a result, BUORD asked Boettcher to redesign it with his pilot directing device fully integrated into its sighting system.[27]

That summer, Boettcher finished the sight's redesign and dubbed the Pilot Directing Bombsight "Mark III," and by August BUORD started production. Although an improvement over the Wimperis bombsight, the Mark III was still incapable of hitting a moving ship at high altitude. The incorporation of a low-power telescope to the Mark III did not solve the problem. So in January 1920 BUORD asked Carl Norden, who had been working on the flying bomb project at Dahlgren, to determine the feasibility of increasing the Mark III's accuracy by gyroscopically stabilizing it.[28]

Norden studied the Mark III's descriptive papers and drawings and concluded that he could improve the design. As a first step, he mounted the Mark III on a stabilized base and installed it in one of the aviation detail's aircraft. Dahlgren aviators tested it in July and August 1921 with mixed results. The sight was more accurate but it was prone to malfunction and still could not accurately track moving targets. Norden continued to work on the problem—without pay since Navy research funds had grown scarce. He concluded that he would have to design a completely new sight, using a timing mechanism to determine the drop angle, to meet the BUORD's accuracy requirements. In June 1922 BUORD accepted his proposal to design and construct the new sight and issued a $10,700 contract for three experimental models.[29]

In 1923 Norden and partner Theodore Harold Barth went to work, and in the early spring of 1924 they delivered the three prototypes, designated Mark XI, to the Naval Proving Ground for testing. The initial results were disappointing. BUORD maintained its faith in Norden, though, and over the next four years he and Barth tinkered with the design and steadily improved its accuracy.

During this stage of the bombsight's development, Dr. Thompson at Dahlgren began working part time on the project, analyzing test drop data and doing much of the mathematics on Norden's behalf. One vexing problem that Thompson helped solve concerned the bombsight's optical system. While moving the sight about, test officers had noticed an intermittent appearance and disappearance of a parallax, an illusion in which a target's position appears changed because of a change in the sight's perspective. After extensive tests conducted at the Washington Navy Yard and by Norden's optical consultant failed to identify the cause, Thompson suggested that the problem was mechanical and not optical. He thought that rocking the bombsight's case during use physically displaced its crosshairs. To fix the problem, Norden modified the device according to the physicist's recommendations, and the parallax disappeared for good. Thompson's fifteen-year involvement in the project led to a lifelong friendship with Norden and helped strengthen his conviction that high-altitude bombing was the wave of the future.[30]

By 1929 the Mark XI's development had reached a plateau. Though still imperfect, the Navy felt that it was ready to enter fleet service. BUORD accordingly ordered eighty of the devices from Norden and Barth, who incorporated as Carl L. Norden, Inc. to execute the contract.[31]

The Mark XI was a complicated mechanism, and Norden was not happy with its performance, particularly its slow speed of operation, which could be fatal in combat. Likewise, BUORD remained uncertain as to whether the timing system of the Mark XI was better than a synchronous-type system, in which the altitude and airspeed are set and the line of sight takes up automatically a motion which the bombardier can regulate. Thus, even before the first Mark XIs entered service, BUORD, at Norden's urging, authorized Carl L. Norden, Inc. to develop a more streamlined synchronous bombsight as a successor to the Mark XI.[32]

Norden developed the new bombsight, designated Mark XV, in only a year. In February 1931 the first two experimental models arrived at Dahlgren, where aviators conducted flight tests from February to June of that year. The results proved the design's superiority over the Mark XI. Enhanced optics and accuracy, a shorter approach period, simpler operation, and the ability to operate at lower altitudes and at higher speeds were among the Mark XV's advantages. The testing culminated that October when a bomber sporting a Mark XV sight outperformed another bomber equipped with a Mark XI against the target vessel USS *Pittsburgh* (CA-4). Army observers, previously unaware of the Mark XV, were "enormously impressed," and in early 1932 BUORD issued Carl L. Norden, Inc. a contract for an initial production run of fifty-five of the bombsights, thirty-two for the Navy and twenty-three for the Army Air Corps.[33]

Although aviation officers at the proving ground declared the Mark XV perfect, Norden and Thompson continued to tinker with the design over the next few years. Most of their work concerned accessories such as an automatic pilot feature and night and low-altitude bombing equipment, but the sight itself remained unaltered. By 1935 Thompson had developed so much confidence in the system that he informed Stark that Navy bombers, as compared to long-range guns, were three or four times more likely to achieve a hit on a battleship.[34]

Army flyers also learned to love the Mark XV. Dahlgren bombsight mechanic Charles Middlebrook later recollected that they called it the "pickle barrel" sight, after an alleged test in which "Norden-equipped bombers laid their 'eggs' smack on pickle barrels." Whether the story was apocryphal or not, Norden employees adopted a Latin motto: *Cupa fiat melior muriae: per Norden obibit*, meaning "when better pickle barrels are built, Norden will blow 'em up!"[35]

After the Mark XV entered mass production in the 1930s, Middlebrook, who Norden had handpicked and trained to service his bombsights, established a proving regime at Dahlgren in which he thoroughly inspected every device manufactured for the Navy. After inspection, Middlebrook then sent them to the Aviation Detail for flight acceptance testing. Dahlgren aviator Boynton Braun later recalled that "we had to drop eight bombs with every bombsight that came through and make adjustments when the bombsight failed to meet the prescribed standard for accuracy." So many bombsights subsequently came through Dahlgren during that period that additional targets were erected in the Potomac to avoid bottlenecks. The testing procedures remained unchanged until America's entry into World War II, when the sheer volume of manufactured bombsights dictated 10 percent lot testing rather than the peacetime requirement of 100 percent.[36]

The Mark XV proved to be one of the most effective weapon systems of World War II, seeing heavy action over the skies of Nazi Germany and Japan. Ironically, the Navy, which originally had solicited and supported the bombsight's development through the 1920s and 1930s, found little use for it during the war. Quite simply, the Mark XV did not work very well against moving ships despite Norden's best efforts, and combat experience revealed that dive-bombing was far more effective in naval combat than high-altitude bombing. Consequently, of the total 43,292 Mark XV bombsights produced under Navy procurement from July 1939 to September 1945, only 7,920 were allocated to the Navy, while 35,008 were delivered to the Army Air Forces. Carl Norden, Inc. evidently retained the remaining 364. The Mark XV bombsight left a lasting legacy at Dahlgren, where a total of 7,506 of the devices were tested from 1932 through 18 August 1945. Not only did the testing program become the basis of the Aviation Ordnance Department early in the war but it also rooted Dahlgren firmly within the field of mechanical computational technology. In short, the Norden Mark XV bombsight, as a form of analog computer, paved the way for Dahlgren's later evolution from a proving ground into a research and development center, specializing in high-tech computer analysis.[37]

THE EVE OF WAR

In 1939 the Naval Proving Ground at Dahlgren continued its steady expansion, guided by its amiable Inspector of Ordnance in Charge, Captain J. S. "Dad" Dowell. Under Dowell's tenure, the proving ground finally outgrew its original boundaries, largely because of Norden's testing as well as Thompson's naval aviation ordnance studies. Bombing flights over adjacent farmlands presented unacceptable risks to local residents and personnel living and traveling along nearby roads, especially since one or two bombs had already been dropped accidentally. Consequently, Congress appropriated $100,000 to purchase a 6,000-foot "safety zone" around the proving ground's "deck" target, which simulated an aircraft carrier's deck.[38]

A more important development for Dahlgren in 1939 was the arrival of Lieutenant Commander William Sterling "Deak" Parsons as the new experimental officer. A 1922 Naval Academy graduate, Parsons was thirty-seven years old and had earned a reputation as a first-rate ordnance officer. Not only was he a thorough military professional but in many ways he was also an accomplished scientist. His reservist colleague Dr. Charles Bramble believed that Parsons was "the type of scientist that the Navy needed more of. He could stand his ground either aboard ship or with the scientific community in his own right as an equal." Indeed, as scientist Dr. J. E. Henderson observed, Parsons' unique ability to "talk the scientists' language" allowed him to "bridge [the] gap between the scientists and the military" at Dahlgren.[39]

Parsons had first come to Dahlgren in 1930 during his postgraduate "grand tour" and had received Thompson's gospel that science could advance naval weaponry. The physicist recognized Parsons' exceptional qualities and began grooming him, but Parsons had disappointed Thompson by going to sea instead of becoming an EDO within BUORD. After completing his grand tour Parsons had served aboard the battleship USS *Texas* (BB-35) before accepting an assignment from 1933 to 1934 at the Naval Research Laboratory, where he fought unsuccessfully to bring radar detection and fire control to the pre-war Navy. After two additional tours at sea from 1936 to 1939, Parsons returned home to Dahlgren.[40]

As Experimental Officer, Parsons was conscientious and thorough. As Henderson later remembered, "We didn't waste our time when we were down with Deak Parsons." Like most naval officers of the period, he did not have the mathematical background enabling him to do the difficult calculations essential for experimental work. Unlike many of his predecessors and successors, however, he compensated by learning ballistic physics and by working with the civilian scientists to try to understand what they were doing and why they were doing it. Although he was the supervisor of all weapons-related experiments, Parsons also liked to get his hands dirty, no matter how unusual the project was. Dahlgren aviator Horatio Rivero once helped Parsons and rocket pioneer Robert Goddard (a friend of Thompson's then employed at Indian Head) launch one of Goddard's new rocket designs at the proving ground. Parsons also participated in one peculiar test by sitting on a pilot's seat while a burley chief hit its bottom with a sledgehammer. He groaned to his wife, "Wherever it really hurt, we put on more armor."[41]

Although Parsons was technically Thompson's boss, in practice they acted as a team in the Experimental Department and, as the physicist remembered, "spent many, many hours discussing the Navy's program of experimental work and what was needed to make it more effective." As neighbors, they often carried their discussions over outside of normal working hours and were repeatedly seen walking about the station deeply engrossed in whatever problem they happened to be working on at the time.[42]

THE ARMOR & PROJECTILE LABORATORY

Parsons began his tour as Experimental Officer in time to collaborate with Thompson on a project that held lasting importance for Dahlgren and the Navy, the development of a dedicated metallurgical laboratory for improving armor and projectiles. Thompson thought that the Navy should not depend on the private steel manufacturers for advice about the service's armor needs and specifications. Although he had been limited by the available data, Thompson's interest in armor went back as early as 1927 when he had conducted empirical studies in armor penetration mechanics. By 1930 he had derived an all-purpose armor penetration formula based on known armor plate thicknesses, projectile diameters, impact angles striking velocities, and other variables to calculate the required kinetic energy, measured as the coefficient "F," for a particular projectile to penetrate a particular armor plate at a particular angle. Calculated for a wide range of projectiles, armors, and impact angles using real world test data, Thompson's "F" coefficients were compiled into convenient tables for analytical comparison purposes and also to calculate both theoretical and actual Navy "Ballistic Limits." These were the minimum striking velocities of specific armor-piercing projectiles against specific plates under a given set of conditions that would allow projectiles to barely defeat plates using only their non-explosive, kinetic energy. Navy interest in armor mechanics was lacking at the time, though, and BUORD had repeatedly rejected several of Thompson's proposals for additional research in the field. Yet he remained unshaken in his position that the Navy needed an in-house metallurgical research center.[43]

Parsons bolstered Thompson's ideas for a new model laboratory. During their discussions, they developed a scenario in which scientists worked closely with ordnance officers to fulfill fleet armament requirements while enjoying the freedom to conduct fundamental research without military interference. This reflected Thompson's philosophy that scientists should be able to independently introduce a concept and then shepherd it through its research, development, testing, and evaluation stages without external meddling or burdensome contract negotiations until it entered service with the fleet, a process that some later called the "Dahlgren Way."[44]

Thompson proposed conducting the research at a reduced scale by firing 3-inch armor piercing projectiles through very thin armor and scaling the results up. Under this plan, more testing could be done at less cost. In 1940 Thompson and Parsons, with Dowell's endorsement, proposed the new laboratory to BUORD chief Rear Admiral William Furlong. Noting the reduced-scale work at the Naval Research Laboratory and the irregular quality of armor supplied by private industry, they argued that the Navy needed fundamental research to investigate unknown metallurgical properties of armor and projectiles and that Dahlgren was the place to do it.[45]

Thompson's proposal encountered resistance from a number of sources, beginning with BUORD's chief. Ill informed on current armor technology, Furlong hesitated and even consulted with steel manufacturers about the need for such a laboratory, much to Thompson's annoyance. Not surprisingly, the manufacturers unanimously demurred at the physicist's proposal that the Navy should do in-house fundamental armor research. Captain Mike Schuyler, the head of BUORD's Research and Development Section and the Special Board on Naval Ordnance, also loudly questioned why the proving ground wanted to do its own metallurgical research. He feared that Thompson was pushing BUORD into "taking in a tremendous amount of territory without thinking of all the angles involved." News of the proposal likewise offended the Director of the Naval Research Laboratory, Rear Admiral Harold G. Bowen, who was apparently unaware of Thompson's prior collaboration with his staff in the matter. The conflict came to a head when Bowen confronted Furlong in a hallway, exclaiming that "You're not doing any work in this field of penetration mechanics and we're going to take it over. We're going right ahead with that research program." Bowen's prediction notwithstanding, Furlong made a snap decision in Thompson's favor and called him at Dahlgren that same day with the question, "How soon can you get that laboratory built?"[46]

With Furlong's blessing and a congressional appropriation of $300,000 for the project, Thompson proceeded with his plan to construct the new laboratory. Thompson could not give the project his full attention, so in February 1941 he recruited a former colleague named Leonard Loeb, a longtime naval reservist and physics instructor at the University of California at Berkeley, to build, staff, and manage the new Armor &

Projectile (A&P) Laboratory. Loeb went to work and, with Parsons' and Thompson's assistance, constructed the A&P Laboratory in 1941, with the first shot fired inside its enclosed range on 21 November of that year. At that point, according to Loeb, "Dahlgren represented about the only place in the Navy where you had any civilian scientific talent." Unbeknownst to the trio, though, the good times were nearly over, even before they had really started.[47]

"GANG" BUSTERS

Shortly after Loeb arrived at Dahlgren and began building the laboratory, Captain Dowell warned him that "there is trouble ahead." According to Loeb, Dowell, whose tenure at Dahlgren was nearly up, "knew that gang that was coming in." Captain David I. Hedrick, probably the hardest of hardcore gunnery officers, headed up the "gang" and replaced Dowell in April 1941.[48]

Born in North Washington, Ohio, on 31 December 1886, Hedrick had graduated from the Naval Academy in 1909 and rendered long service as a gunnery and engineering officer aboard a succession of warships. An experienced sea captain, he had commanded the light minelayer USS *Burns* (DM-11), the destroyers USS *Marcus* (DD-321) and USS *Talbot* (DD-114), and the heavy cruiser USS *Minneapolis* (CA-36) during his career. He had also served three years at the Naval Academy and later completed the Naval War College's senior course at Newport, Rhode Island. In October 1940, the Navy had assigned him to BUORD so that he could assist in its reorganization and expansion under Roosevelt's emergency National Defense Program. Although Hedrick had enjoyed a model naval career, he had been passed over twice for promotion to admiral, and it was understood that Dahlgren would be his last assignment before he "swallowed the anchor." Naval reservist and physicist Dr. Ralph Sawyer remembered that he was "kind of sour" and "not a very cheerful character." Loeb believed that he was "slightly unbalanced" and a "regular Queeg." Hedrick also engaged in some off-duty activities that Parsons found particularly distasteful, such as flashing piles of paychecks that he had won from younger officers in poker games. Parsons also disapproved of Hedrick's side business of raising and selling chickens from the top floor of the Commandant's House to personnel on the station. Apparently, Parsons' dislike of Hedrick was shared by Dahlgren's aviators, who, as legend has it, enjoyed buzzing the Commandant's House whenever he entertained guests, flustering his chickens and embarrassing him.[49]

"Old Man" Hedrick subscribed to the command philosophy of a strict nineteenth-century sea captain rather than that of a twentieth-century shore establishment manager. According to administrative aide Curtis Youngblood, he believed that "he was the boss of everything that had to do with that Station, everybody that lived on that Station, civilian, military, anything else." His "actions were summary" and he had no use for "advice or counsel of boards or committees," including Dahlgren's school board, which he dissolved.[50]

According to Sawyer, Hedrick believed that Dahlgren "was really a proving ground and that proof and test was our main job," while "the contractors would solve all the problems." Fundamental ordnance research and technical initiative, therefore, were not part of the Dahlgren mission as far as he was concerned.[51]

Hedrick's ascension to Inspector at Dahlgren ignited the long-smoldering rivalry between the military and scientific personnel at Dahlgren. Ever since Thompson had arrived in 1923, the working relationship between the military men and the civilian workers had been complicated, sometimes strained, as both groups struggled to understand one another's means and methods. The friction had been kept to a minimum during the 1920s and 1930s while the numbers of military and civilian personnel remained relatively low. However, America's preparations for war after 1940 had caused an influx of officers and sailors, naval reservists, and civilian scientists and laborers, stirring real trouble between the two factions at Dahlgren.[52]

Hedrick's contempt for Thompson's experimental research program in general, and the new A&P Laboratory in particular, brought the tensions to a head. Backed by Schuyler in BUORD, Hedrick moved to disband the laboratory and drive out Thompson. After unsuccessfully attempting to slash the laboratory's congressional funding, Hedrick reassigned its staff to other duties, effectively gutting it. Loeb was furious. He vowed to put the Inspector "on the rack" if Thompson and Parsons would back him. If not, then he was "getting out."[53]

Thompson and Parsons were hesitant to escalate the conflict, and Loeb returned to his teaching duties at Berkeley. Soon after his departure, though, they finally visited the new BUORD Chief, Rear Admiral William H. P. Blandy, and, as Loeb later recalled, demanded either Hedrick's removal or a BUORD-mandated attitude adjustment. A friend of Hedrick, Blandy refused to intervene and sent Thompson and Parsons back to Dahlgren and an uncertain future for their experimental research program.[54]

In January 1942, with Blandy's quiet approval, Thompson finally left Dahlgren, after nearly nineteen years as chief physicist. He took a much more lucrative position in Indianapolis as the scientific director for the Lukas-Harold Corporation, a subsidiary of Carl L. Norden, Inc., which mass-produced bomb and gun sights for the Navy. Thompson later graciously insisted that Hedrick "was not a factor in the decision to go," but his timing in relation to Hedrick's apparent victory probably was more than coincidental. Thompson's departure chilled the prospects for further scientific research at Dahlgren, and the outlook for the A&P Laboratory appeared bleak.[55]

According to Loeb, Hedrick and Schuyler's own intransigence saved the laboratory. Just before Loeb returned to Berkeley, the director of the British National Physical Laboratory, Ralph H. Fowler, visited Dahlgren to evaluate the Navy's scientific readiness in ordnance matters. Schuyler, an anglophobe, ordered that Fowler "be shown nothing at Dahlgren," especially the A&P Laboratory. Hedrick subsequently kept the British physicist ignorant of the armor penetration research being done at Dahlgren. After returning to Britain, Fowler wrote a blistering report on the sad state of American armor research for Winston Churchill, who then forwarded it to President Roosevelt. As Loeb recounted, Fowler's report puzzled Roosevelt and he called BUORD to learn the truth. After BUORD told him about Dahlgren's armor plate research group, he purportedly replied, "Tell them I want them to go ahead full blast." So directed, Schuyler and Hedrick "put everybody back to work again" and left the laboratory alone for the war's duration.[56]

The A&P Laboratory thus survived Hedrick's vendetta. In February 1942 Sawyer assumed control of the lab by default since both Loeb and Thompson had left. Sawyer was a naval reservist and physics instructor from the University of Michigan who Loeb had brought on board in June 1941 as a spectroscopist. In the wake of the turmoil with Hedrick, Sawyer quietly nurtured the laboratory and later expanded it to twice its original size, building a top staff of metallurgists, physicists, and chemists by 1945. In Thompson's words, the laboratory under Sawyer's direction "did a magnificent job during the war, not only in developing knowledge about armor systems . . . but also in controlling the quality of the armor that was produced." In recognition for his efforts, Sawyer was appointed in December 1944 to the Chief Physicist's position, redesignated as "Officer in Charge of Laboratories at Dahlgren," which had remained vacant after Thompson's departure. Ultimately, Sawyer and the A&P Laboratory preserved for the future a nucleus of pure scientific research at Dahlgren.[57]

THE WAR YEARS

Although Hedrick's idiosyncrasies grated on the scientists, in many ways the proving ground was fortunate that he was in charge after America entered World War II in December 1941. Through his military professionalism and ability to execute BUORD policy to the letter, Hedrick successfully managed Dahlgren's massive wartime expansion program.

Dahlgren's mobilization had originally begun under Dowell's administration with the emergency Naval and National Defense Appropriation Acts, passed in 1940 after the fall of France and the British army's evacuation at Dunkirk. Rather than augmenting Dahlgren to meet the heavy testing requirements under the new National Defense Program, BUORD had first proposed lowering proof standards to expedite the work. Dowell objected, arguing that it was no time to lower standards and that the real question should be, "What can we do to expand the Proving Ground to meet the demands of National Defense expansion?" Accordingly, he submitted a detailed list of suggested improvements and plant extensions including, among others, plans for a new fuze battery, gun emplacements, two 125-ton boom cranes, and range craft, as well as detailed proposals and cost estimates for expanding the facility's military and civilian complement.[58]

Hedrick inherited Dowell's BUORD-approved list and soon demonstrated his skills as a master military builder. Under his watch, the Naval Proving Ground expanded by 3,500 acres to reach its peak size of 5,423 acres (including the annexation in March 1944 of Pumpkin Neck for a new bombing range—the last area so acquired). Additionally, he supervised the construction of a twenty-three-mile rail spur between Dahlgren and Fredericksburg, Virginia, which alleviated transport congestion. Fourteen new magazines, five shops, a new barracks, an additional aircraft hangar, a dispensary, a theater, a technical library, and more than sixty miscellaneous buildings were likewise built. The Main Battery expanded further so that by 1944 it contained ten major caliber gun emplacements, seventeen small caliber gun emplacements, and a vast array of cranes and support structures. In applying some of the new lessons from Pearl Harbor, BUORD authorized Hedrick to build a new and well-equipped Anti-Aircraft Fuze Battery, completed in March 1942, from which the Experimental Officer could help develop and test new anti-aircraft defense weapons. In January 1944 Hedrick also oversaw the establishment of a Gunner's Mates School for training prospective fleet gunners in the operation and maintenance of all calibers of Navy guns and also to safely handle the guns' ammunition.[59]

The influx of new civilian personnel into Dahlgren, from 440 in January 1941 to a peak of 1,856 in June 1945, required additional housing, difficult to come by due to the station's isolation. Hedrick partially solved the problem by building "Boomtown," a temporary off-station community comprised of ninety-four low-cost homes funded by the National Defense Housing Agency. This was still not enough, and in the spring of 1945 Hedrick arranged for the Federal Public Housing Agency to build a trailer park near the proving ground for the civilian employees.[60]

Despite Hedrick's antipathy for in-house research laboratories, the exigencies of war dictated otherwise. Perhaps the most important new lab was the Aviation Experimental Laboratory, which BUORD established in 1943 within the Aviation Ordnance Department. It was outfitted with the latest testing apparatus, allowing its technical staff to develop and test such exotic weapons as 1,250-pound rocket-propelled armor-piercing bombs, experimental target identification bombs, and incendiary bomb clusters. In a high priority project, the staff also developed an experimental armor-piercing 4,000-pound bomb in the latter half of 1944. Ordnance crews tested this monster not by dropping it from an aircraft but by firing it from an 18-inch gun into butt-mounted 10-inch plate armor backed up by a large sand pile.[61]

While Dahlgren grew physically under Hedrick's direction, his ordnance men did what they did best: prove ordnance and armor. During World War II, Dahlgren's testing regime increased at least tenfold, with millions of rounds fired from guns of every caliber and millions of pounds of powder expended. Not only did Dahlgren handle the traditional fleet weapons and mounts but it also hosted tests of new weaponry lines, including anti-aircraft artillery (20-mm, 40-mm, and 5-inch) and rockets, for which a special laboratory was constructed in March 1944.[62]

The proving ground also undertook special studies based on fleet combat experience to solve pressing problems and save American lives in future operations. During the bloody battle of Tarawa in 1943, and despite a heavy preparatory naval bombardment, the Marines had waded ashore into fierce Japanese resistance. Baffled at the intact Japanese coconut log and sand beach defenses discovered after the battle, BUORD asked the proving ground to conduct firing tests against crude structures and emplacements similar to those the Japanese had built on Tarawa. The fleet specifically needed to know what projectile-fuze combination was the most effective against earth-and-log targets. Responding immediately, Hedrick recreated these defenses at Dahlgren and ordered a series of firing tests against them. Within a month, he had the answer. Several different ordnance combinations, including gun projectiles and bombs, were prescribed, forwarded to the fleet, and incorporated into plans for future landings.[63]

Dahlgren ordnance men solved another combat-related mystery that threatened to embarrass BUORD. On 8 November 1942 the French battleship *Jean Bart* threatened Eisenhower's North Africa landing. Though unfinished and confined to her berth at Casablanca, her 15-inch guns were operable and presented a formidable challenge to the American landing force. The battleship USS *Massachusetts* (BB-59), firing salvos of 16-inch armor piercing projectiles, repeatedly failed to put her out of action, seemingly because of defective ammunition. The disappointment heightened later when naval observers reported light damage to the French defenses after the Massachusetts concluded a heavy shore bombardment. After President Roosevelt demanded an investigation, BUORD sent two of Hedrick's "fuze doctors" to Casablanca to investigate the trouble. They discovered, much to BUORD's relief, that the projectiles were fine. The *Jean Bart* had sported only very light armor, and the *Massachusetts'* armor piercing projectiles had sliced clean through her without detonating, leaving her largely undamaged. Hedrick's men also found that the *Massachusetts* had used armor piercing ammunition instead of high-capacity high explosive during the shore bombardment. Consequently, the projectiles just buried themselves deeply into the ground before exploding with a whimper.[64]

During the war, BUORD also implemented at Dahlgren a vigorous program to test captured enemy equipment. Sawyer's staff in the A&P Laboratory subjected many different types of German, Japanese, and Italian projectiles and armor (naval and aircraft) to intensive metallurgical examination and analysis. BUORD then supplied the fleet with the valuable intelligence derived from the experiments, which influenced operational planning and actual combat. In all instances, the proving ground contributed directly to solving combat-related problems, ultimately saved American lives, and thereby continued to demonstrate its value to the American naval establishment.[65]

SECRET WEAPONS

Dahlgren not only hosted the testing of a wide range of conventional naval weaponry and armor during the war but it also played a significant part in the development of some of the country's most secret weapons. The first, of course, was the Norden Mark XV bombsight, which gave the U.S. Army Air Forces the ability, at long ranges and high altitudes, to lay large numbers of bombs within acceptable distances of strategic targets inside Axis-controlled territories.

The second new weapon to cut its teeth at Dahlgren was the radio proximity fuze. Although funded by the Navy, its development was directed by Section "T" of the National Defense Research Committee (NDRC) and later the Office of Scientific Research and Development (OSRD), both of which derived their authority and funding directly from President Roosevelt. The fuze project had been started in 1940 after Roosevelt's scientists recognized the relative ineffectiveness of British anti-aircraft batteries against German bombers during the "Blitz." The British had initiated photoelectric and variable-timed (VT) fuze research in 1939 but had not achieved desired results. They used two types of fuzes during the early stages of the war, impact fuzes and clock fuzes. As its name implies, an impact fuze detonated a projectile upon physical contact with the target, while a clock, or timed, fuze exploded at a preset time after firing. In an age when aircraft technology had reached the point where fighters and bombers could fly at speeds of hundreds of miles per hour, neither fuze was effective. Observers estimated that at least 2,400 shell firings were necessary to achieve a single hit, and some even believed it took 100,000. Hence, British anti-aircraft artillery fire against German bombers had largely been a waste of ammunition. The scientists had determined that the problem could be solved neither by refining existing equipment nor by training gun crews better. A new projectile fuzing technology was required. Specifically, they sought to develop, among others, a new fuze that could trigger a projectile by proximity detection through the emission and reflection of radio waves (radar). The NDRC therefore organized Section "T" to develop and test the new fuzes and placed the program under the direction of Dr. Merle Tuve from Northwestern University ("T" stood for Tuve).[66]

In 1940 the Navy also moved to develop better anti-aircraft technology to defend its ships against the growing threat of aerial bombing (albeit too late to avoid the devastation of Pearl Harbor). BUORD, under the influence of the Research Desk's progressive head and soon-to-be chief Captain William H. P. Blandy, accepted the NDRC's suggestions that applied research could lead to new, more lethal anti-aircraft technology. So the Navy took the lead in promoting the proximity fuze's development, and NDRC scientists going to and from Blandy's office soon became a common sight in BUORD.[67]

Because the fuze could not be safely tested against expensive manned aircraft, BUORD chose Dahlgren as the testing site. The facility was a water range, which was ideal for the project since water reflected the fuze's emitted radio waves and would detonate it in "perfect safety" away from land and personnel. Thus, Tuve found himself on the way to Dahlgren in September 1940 to discuss the new project with the Experimental Officer, Lieutenant Commander Parsons. Tuve's initial idea was a bit bizarre. Because of component space requirements, high "G" forces, and other problems associated with gun-fired projectiles, he first envisioned placing "photoelectric" (PE) fuzes (triggered by a target's shadow rather than by reflected "radio" waves) in 500-pound bombs and then dropping them from a friendly bomber over a large enemy bombing formation and "cleaning it out." Strange as it first sounded, Parsons was intrigued by the concept and read as much literature on the subject as he could find. Throughout the summer of 1941, Parsons worked part time on the project, much to Hedrick's irritation. He left Parsons alone with the NDRC scientists, though, since the fuze work lay outside his realm of authority. Quite simply for Hedrick, as Section "T" scientist Dr. J. E. Henderson recounted, the "word [had] come down from the top 'to cooperate with these boys.'"[68]

Parsons and the scientists successfully developed a working photoelectric fuze, first by building a test model in an old coffee can and having Navy aircraft overfly it to measure the readings on an attached oscilloscope. Then they dropped bombs equipped with the fuzes, which successfully detonated ten feet above the ground. Following this series of tests, Parsons then arranged to drop inert PE fuzed bombs against drones at Cape May, New Jersey, in the first qualifying tests for proximity fuzes in the United States. They worked well.[69]

Parsons was not quite convinced of the tactical prospects for Tuve's "bomber-vs.-bomber" idea with the PE fuzes. Thoroughly familiar with the possibilities presented by radio or "radar" technology from his previous tour at the Naval Research Laboratory, Parsons promoted the radio fuze over the PE fuze, specifically for use in gun projectiles. However, like the scientists, he recognized that the extremely violent forces that acted upon a fired radio fuze would be immensely difficult for the mechanism to withstand. Mass production of the devices in the millions with extremely low tolerances for component error, as well as strict security concerns, would also be problematic.[70]

Despite the seemingly insurmountable challenges and through some fairly crude techniques, Parsons and Section "T" successfully developed a working model of a gun-fired radio fuze. On 8 May 1941, Parsons, Tuve, and three other Section "T" scientists sat in a boat offshore at Dahlgren, listening through earphones, as seven radio-fuzed, 5-inch projectiles were fired. As the projectiles screeched overhead, the men detected electronic signals from at least two of the fuzes. Parsons was so impressed that he subsequently reported to BUORD that delays in getting the fuzes into action "were equivalent to the loss of a battleship every three months, a cruiser a month, and 150 men a day."[71]

In December 1941 BUORD issued the first contract for a pilot production run of five hundred fuzes, labeled VT to keep the secret that they were radar-based and not variable-timed. However, quality differences between the handmade experimental fuzes and the factory-produced service fuzes were significant. Parsons and Section "T" therefore established an intense acceptance regime at Dahlgren before allowing the VT fuze to enter service. By Parsons' way of thinking, the enemy would face stiff odds during an attack if only half of the VT fuzes fired in battle worked, an amazing technological feat in itself since existing timed fuzes only worked one-tenth of 1 percent of the time. Therefore, he set a success goal of 50 percent before declaring it fully operational. Early results showed only 10 percent performed as required, and testing continued, with the success percentage rate gradually climbing to 20 percent.[72]

In January 1942, just as Thompson was leaving Dahlgren, Parsons reported to Blandy that he had finally reached the 50 percent goal and that the fuze was ready for service. BUORD then formally took over the production end of the program but asked the OSRD, the NDRC's newly created project management arm, and its recently appointed director Vannevar Bush to maintain technical control of the fuze's continued development and improvement. Bush agreed but asked for Parsons' full-time services as military liaison between himself and Section "T." Blandy accepted the chairman's condition, and in April 1942 Parsons starting working directly for Bush (and over Tuve) as the "Special Assistant to the Director" and became the project manager of the entire VT fuze program. Parsons' appointment was originally provisional since he was scheduled to go to sea about 15 June, but Bush found him so valuable that he extended his appointment indefinitely.[73]

The VT fuze became one of the most devastating seaborne and battlefield weapons of the war. Its first combat use occurred on 4 January 1943, when a gunner aboard the USS *Helena* (CL-50), with Parsons aboard as an observer, shot down a Japanese bomber with a 5-inch, VT-fuzed shell. Within months, thousands of VT fuzes were being shipped to the fleet, which subsequently decimated the Japanese air forces. Of the various shell sizes for which the fuze was manufactured, the 5-inch, 38-caliber anti-aircraft gun proved to be the perfect match for the fuze. During the war, it accounted for more than half of the Navy's tally of Japanese aircraft shot down by anti-aircraft guns. The fuze also proved vital in defending the fleet against the Japanese kamikaze attacks of late 1944 and 1945. Losses would certainly have been much higher if none of the weapons had been available as American forces moved closer to the Japanese home islands.[74]

The U.S. Army had originally hesitated to use the VT fuze in Europe out of fear that the Germans would recover a dud and reverse engineer it. The service relented in late 1944, though, just in time for the Battle of the Bulge. At that battle's conclusion, Lieutenant General George S. Patton Jr. announced that "the new shell with the funny fuze is devastating . . . I think that when all armies get this shell we will have to devise some new method of warfare."[75]

The OSRD developed VT fuzes for the British 3.7-inch and 90-mm anti-aircraft guns to defend the United Kingdom against the German V-1 "buzz bombs" in 1944. The fuzes were so successfully employed that General Sir Frederick Pile, commander of British Air Defenses, sent his "compliments to the OSRD who made the victory possible." He could have also extended his compliments to Parsons' ordnance people at Dahlgren who had worked hard to prove the VT fuze on behalf of the OSRD.[76]

As the VT fuze helped tip the scales in favor of the Allies in the air and naval wars, Parsons became involved in another vital research effort for the OSRD. In May 1943 Bush recommended him to Brigadier General Leslie Groves to lead the Ordnance Division for the Manhattan Project. Parsons accepted the assignment and subsequently played a vital role in the development, testing, and dropping of the atomic bombs on Hiroshima and Nagasaki.[77]

During the project, two types of devices were developed. The first, a "gun assembly" design, operated just like a normal gun in which a wedge-shaped uranium-235 "slug" was propelled down a converted gun barrel at tremendous velocity to strike a solid core of uranium-235 positioned at the end of the barrel. The impact triggered nuclear fission, and the sudden energy release resulted in an explosion of immense proportions. The second "gadget" utilized a completely different concept, that of "implosion," in which explosive lenses focused tremendous energy inward toward a core of plutonium-239. The resulting compression on the core started a chain reaction and achieved the same effect. The gun-assembly device—code-named "Thin Man" after Roosevelt to confuse enemy agents—measured 10 feet in length, with a varying diameter of 1.5 to 2.5 feet, and weighed an estimated five tons when loaded. "Fat Man," named for Winston Churchill, was nine feet long but much thicker, tapering from 5 to 3 feet long along its axis, and weighed six tons when loaded. Of the two bomb types, Groves' scientists had more confidence in the Thin Man design during the project's first phases, and it accordingly dominated the early research efforts through 1943. Thin Man was later rechristened "Little Boy."[78]

Dahlgren's physical role in the development of the atomic bomb was relatively minor due to security concerns and space limitations associated with VT fuze testing. However, an early part of Parsons' job as the Manhattan Project's ordnance officer involved testing the ballistic qualities of the Thin Man gun-assembly bomb. Beginning in July 1943, Parsons and a former colleague of Tuve's, Dr. Norman Ramsey, first tested the Thin Man design at Dahlgren using scale models. They were chiefly interested in learning whether Thin Man would drop straight down or fall head over heels in the air. Using Section "T's" shops at the Applied Physics Laboratory, Parsons and Ramsey cut ten 500-pound bombs in half and welded long pieces of sewer pipe between the ends. Interested colleagues, not knowing the ultimate purpose of the work, dubbed the models "Sewer Pipe Bombs."[79]

To test the bombs, Parsons and Ramsey filled them with sand and debris and had the Applied Physics Laboratory cast lead billets so that they could precisely adjust the bombs' centers of gravity. Arthur Breslow at the laboratory then drove the billets down to Dahlgren in his car, which was so burdened by the lead that, as he later recalled, his tires were "squashed down halfway" and it could not top twenty miles an hour. They then had an aviator in a twin-engine Navy torpedo plane drop the bombs at Dahlgren's bombing range from an altitude of 20,000 feet.[80]

Between tests, Parsons and Ramsey tinkered with the fin design in one of the airfield's hangars to increase a model's stability while Breslow measured its center of gravity using cranes to swing it back and forth. The bombs eventually exhibited satisfactory ballistic performance, but Parsons also wanted to test the design's ruggedness, which required scale model drops on water. Due to Dahlgren's crowded gun and air test schedule, shallow river bottom, and primitive and inadequate recovery facilities, Parsons recommended that testing be moved away from the proving ground to Muroc Field (now Edwards Air Force Base) in California. Groves agreed, and Parsons had the remaining sewer pipe bombs transported west, putting Dahlgren out of the Manhattan Project.[81]

Although Dahlgren's role as a test facility in the Manhattan Project was very limited, it ultimately contributed much to the project in the way of technical knowledge in ordnance-related research and ballistics. Many of Groves' key Los Alamos personnel had ties to the Naval Proving Ground. Parsons was foremost among these and rose to be the number three man in the project's overall management. He subsequently served as the "weaponeer" aboard the *Enola Gay* during the Hiroshima mission, arming

Little Boy just after takeoff. During the project, and Hedrick's angry protests notwithstanding, Parsons had recruited at Dahlgren under Class "A" Priority certain individuals capable of solving the sundry problems encountered in the bomb's design and delivery. Among them was Dr. Norris E. Bradbury, who served at Dahlgren as a reservist in the first half of the war before going west to help design the plutonium bomb's explosive shell. He also led the project's implosion field testing program and supervised the assembly of the TRINITY device, before later succeeding J. Robert Oppenheimer as the director of Los Alamos National Laboratory. Other former Dahlgren personnel who found themselves involved in the project included the aviator Commander Frederick Ashworth, who armed Fat Man aboard the *Bock's Car* during the Nagasaki mission, and even Thompson himself, who worked as a consultant analyzing and improving the ballistic characteristics of both Thin Man and Fat Man. After the war, Dr. Ralph Sawyer left his position as the Officer in Charge of Laboratories at Dahlgren to become the technical director for Operation CROSSROADS, working under proving ground veterans Parsons and Blandy.[82]

The trouble between Hedrick and the scientists aside, Dahlgren emerged from World War II much larger and better equipped to conduct new research into naval weaponry and technology. From 1941 to 1945, the exigencies of war had forced the old guard to accept the presence of new research labs at the proving ground, fulfilling the agenda set by Thompson in the1920s and 1930s. Despite the effort by Hedrick and others to maintain the tradition of strict military shore establishment, Dahlgren was well on the way to being transformed by its researchers and its research assignments into a civilian-dominated research and development center as it stood on the threshold of the Atomic Age.

4

Chapter

Numbers Over Guns, 1945-1959

By late 1945, despite a divisive conflict between military conservatives and progressive scientists over the role of fundamental, in-house research, the Naval Proving Ground at Dahlgren emerged as a budding Research, Development, Test, and Evaluation (RDT&E) center. Wartime exigencies had ultimately vindicated the scientists. The proving ground established a number of important new experimental laboratories that contributed significantly to the Allies' victory. However, conventional wisdom within the Navy held that air power had largely eclipsed gunnery during World War II and would continue to do so in the new Atomic Age, leaving the future of shipboard guns very much in doubt as the nation reorganized its defense apparatus and shifted its strategic focus from conventional warfare to nuclear warfare. Dahlgren physicist Donald W. Stoner recalled the sentiment of the time: "Guns and ammunition were obsolescent. You could more or less just look forward a certain number of years and say there just wouldn't be any more guns. Ordnance would be all bombs, missiles, and rockets."[1]

From 1945 to 1959 Dahlgren was in a transitional period in which a new generation of civilian scientific leaders assumed control of its technical direction, modernized its organization, and redirected its mission. These scientists drew upon Dahlgren's experience in ballistic computation and steered its laboratories into the dynamic new fields of computer science and *geoballistics*. Their efforts not only saved Dahlgren from closure but also made it the Navy's foremost weapons laboratory, ultimately responsible for trajectory computation for the Fleet or Submarine-Launched Ballistic Missile program.

NORMAL PEACETIME LIVING

On 19 August 1945 Captain Hedrick officially announced V-J Day to Dahlgren's military and civilian personnel. He congratulated the station on its contribution to the defeat of the Axis powers and noted with pride that Dahlgren had become a "major activity" in the United States' naval establishment. However, Hedrick warned that Dahlgren, along with the rest of the country, would soon be entering a transition period from wartime conditions to "normal peacetime living." Although few in 1945 could envision the changes that the Cold War later wrought, Hedrick prophetically added that "the 'normal' will, no doubt, be different in many respects to pre-war times" and would "be occasioned by a new worldwide economy and international alliances and relationships which have heretofore had relatively little effect on our American way of life."[2]

Dahlgren's demobilization began even before the ink dried on the Japanese surrender documents. In mid-September, while the station's big guns stood silent, Hedrick announced the names of forty-four civilian employees who would be laid off because of the sudden work curtailment. More layoffs followed those, and the civilian complement dropped to 1,513 by 20 October, down 16 percent from 15 August. In accordance with the Navy's demobilization mandate, the station's complement of 540 enlisted personnel began dropping in the fall of 1945. Likewise, reservists, who had comprised nearly 80 percent of Dahlgren's officer corps at the height of the war, began returning to their former civilian careers or moving into the government's atomic testing program. Many regular officers at the station, meanwhile, faced either formal separation or reassignment to sea duty by the Navy, which had to scavenge its shore establishment for every sea-qualified officer who would volunteer in order to get most of America's servicemen home and the first decommissioned ships into mothballs. With his wartime mission complete, Hedrick himself finally retired from the Navy in June 1946 and lived another nineteen years before succumbing to a heart attack in March 1965. When his successor, the decorated combat veteran Rear Admiral Charles Turner Joy, arrived to take command, Dahlgren was well on the way to reverting to its pre-war size.[3]

Surprisingly, demobilization opened the door for the advancement of Dahlgren's research and development capability. While military and civilian staff left, Dahlgren's small scientific cadre remained, protected by the decision to maintain and expand the United States' research and development establishment in the event of a third world war. This decision owed much to studies by former NDRC and OSRD head Vannevar Bush, who maintained that Allied science had won the war and argued that only a government-funded, civilian-controlled scientific establishment could keep the nation prepared for future adversaries. The Navy welcomed this message. It had already established an interim Office of Research and Inventions in May 1945 under the former head of the Naval Research Laboratory (NRL), Vice Admiral Harold G. Bowen. That was replaced in 1946 with a permanent Office of Naval Research (ONR), also under Vice Admiral Bowen. That same year, the Navy also created the Naval Research Advisory Council (NRAC) to engage civilian scientists as ONR advisors and to screen research proposals submitted to the new office.[4]

The Bureau of Ordnance (BUORD) aligned itself with the Navy's new course toward fundamental science in weapons technology by shifting its major emphasis from production and maintenance to research and development. In the Secretary of the Navy's 1946 annual report, BUORD announced that its postwar *modus operandi* would be anchored on the enlightened premises that 1) naval ordnance must keep abreast of world scientific developments, 2) civilian scientific talent should be used to the greatest possible extent to augment and supplement the naval research facilities, 3) long-range programs for the future Navy are far more productive and efficient than hurried development or improvement of interim equipment during a crisis, and 4) research coordination among interservice and intraservice programs is more economical and produces an integrated approach to future weapons development. Placing science firmly at the center of its new mission, BUORD thus began investing in a number of long-range programs that built on the "Buck Rogers fantasy" weapons that were developed during the war, including jet-powered aircraft, guided missiles, rockets, and the atomic bomb. As a measure of its commitment, BUORD began projecting future technical development in terms of years rather than months. It also responded to the government's postwar perception that technical competence lay outside of its Washington hallways by decentralizing R&D responsibility to defense contractors or Navy field laboratories.[5]

As BUORD put the finishing touches on its new policy, Dahlgren's scientists continued working on a number of experimental projects that were carried over from the war, particularly in the preeminent Armor & Projectile (A&P) Laboratory. The A&P Laboratory was now led by Dr. Russell H. Lyddane. Considered a "super scientist" by his colleagues, Lyddane received his Ph.D. in physics from the Johns Hopkins University in 1938 and taught at the University of North Carolina-Chapel Hill before coming to Dahlgren in July 1941 to work as a civilian physicist under Dr. L. T. E. Thompson. After Dr. Ralph Sawyer left Dahlgren to serve as the technical director for Operation CROSSROADS, Lyddane assumed his duties as the head of the A&P Laboratory. As a civilian, he could not take Sawyer's former title, "Officer in Charge of Laboratories," and since BUORD made no effort to establish an equivalent civilian title, that position officially remained vacant. Lyddane nevertheless became the station's senior physicist.[6]

Under Lyddane, the A&P Lab, in coordination with the Experimental Department, conducted a number of comparative ballistic and metallurgical studies of captured German and Japanese materiel as part of the Navy's ongoing general investigative program on foreign naval technology. The Japanese hardware generated the most discussion. BUORD was particularly interested in the heavy armors and large caliber projectiles recovered by the U.S. Naval Technical Mission from the Kure Naval Arsenal and Dahlgren's Japanese counterpart, the Kamegakubi Naval Proving Ground at Kurahashishima. Post-impact metallurgic analysis indicated that Japanese armor was "definitely inferior" to American armor, while firing tests of Japanese armor-piercing projectiles revealed that the Japanese had generally sacrificed plate penetration power for underwater ballistic stability, leading Lyddane to conclude that Japanese naval gunnery doctrine focused, dubiously, on hitting American ships below their waterlines rather than topside, as conventional sea sense dictated. It appeared after all that the once-vaunted Japanese capital ships had been no match technologically for their American counterparts.[7]

As they tied up these loose ends, Lyddane and his colleagues realized that Dahlgren's longer-term survival would hinge on its ability to adapt to BUORD's new research and development orientation. Under the Navy's decentralization plan, research installations were to become semi-autonomous. They would have to compete fiercely for both funding and work and continuously justify their existence to both BUORD and congressional budget cutters.   Therefore, the scientists began seeking ways in which the proving ground could carve out and keep a slice of the BUORD's R&D pie. One obvious field suitable for Dahlgren was advanced gun and projectile development. In one early postwar project, the station's Experimental Department prepared a historical survey and feasibility study on hypervelocity guns, which BUORD considered as a possible defense against future high-altitude supersonic bombers and guided missiles. The survey presented a variety of exotic concepts that had been investigated by German scientists during World War II, including advanced "sabot" and rocket-assisted projectiles (RAPs) and magnetic guns.   Contrary to the experimental staff's positive conclusions, however, hypervelocity guns proved impractical, and BUORD subsequently adopted a more conventional 3-inch, 70-caliber rapid-fire anti-aircraft gun to meet high-speed aerial threats.[8]

In any event, staking a claim on advanced gun and projectile research was a risky proposition upon which to gamble Dahlgren's survival. After V-J Day, it seemed to many Truman administration officials and defense strategists that air power and the atomic bomb had rendered all guns, and even the Navy itself, obsolete.   The feeling was so pervasive that congressional budget cutters forced the service to decommission twenty-one battleships by 1949. Meanwhile, more than 2,000 additional ships of all types left active duty.[9]

Not surprisingly, then, Dahlgren's scientists quickly dropped further advanced gun and projectile research. For a time they found a niche in the improvement of existing conventional guns and ordnance for the Navy's remaining warships and aircraft. More significantly, Dahlgren hosted the postwar unguided rocket work of Dr. Charles J. Cohen. Cohen had been a geological engineer for the Bureau of Mines before coming to Dahlgren as a reservist in 1944 to work in the exterior ballistics group. An uncommonly gifted mathematician with keen foresight and initiative, Cohen tackled the complex problem of catastrophic instability in a 12.75-inch-diameter antisubmarine rocket called Weapon "A," or "Able."   Collecting data from test firings, wind tunnel experiments, and spark ranges at Dahlgren and the White Oak Naval Ordnance Laboratory, he developed in 1950 what is thought to be the world's first operational six-degrees-of-freedom trajectory simulation, based on the simultaneous linear (vertical, lateral, and longitudinal) and rotational (pitch, roll, and yaw) motions of an unguided rocket. Despite Cohen's important breakthrough, which made the successful development and deployment of guided ballistic missiles possible, Lyddane knew that weapons and ordnance improvement was only a short-term solution to Dahlgren's long-term problem of finding its permanent place within the BUORD's growing RDT&E structure.[10]

With this in mind, Lyddane took pains to expand Dahlgren's technical capability through the gradual replacement of blue-collar workers with professionals by attrition. As he later recalled, "We *had* to increase our technical staff, and we took every step we possibly could." To economize his precious billets, Lyddane also ruthlessly "starved" his people of engineering assistants, compelling them to do their own mundane chores around the office. Complaints were greeted with a stern speech that he had burned into memory from repetition, "I can't give you an engineering aid, because that's one billet I could put a professional in, and what we're going to live or die by is not how many engineering aids we have, but how many engineers we have." Lyddane's farsightedness proved fortunate when the true crisis engulfed Dahlgren in the mid-1950s.[11]

ELSIE

The Dahlgren Naval Proving Ground reentered the atomic weapons business in 1948 after the Atomic Energy Commission (AEC) tasked BUORD with developing a lightweight, ground-penetrating atomic bomb for carrier-based tactical strikes against hardened underground targets. Called Project ELSIE (for LC, or "Light Case" bomb), the program was an outgrowth of the Navy's struggle to reserve a nuclear mission for itself within the Truman administration's grand defense strategy, which a newly independent, aggressive U.S. Air Force threatened to monopolize.[12]

Ever since 1946 when Operation CROSSROADS demonstrated that open formation fleets could survive an atomic attack and conceivably launch a counterstrike, the Navy had been fighting for a role in U.S. nuclear strategy. The practical problem that the Navy faced, however, was the fact that atomic bombs were extraordinarily heavy, and it had no aircraft available capable of delivering them. The Air Force, conversely, had a large B-29 fleet and was developing an enormous new intercontinental bomber, the B-36 *Peacemaker*. To build its own airborne nuclear weapons capability, and also to improve its sea control and tactical air support capacities, the Navy proposed building a new flush-deck "super" carrier, the USS *United States* (CVA-58), for deploying both modified nuclear-armed and more advanced conventional naval bombers and attack fighters. While the Navy had envisioned the ship as a multipurpose weapons platform capable of a broad range of missions and had intended it to supplement land-based strategic forces rather than supersede them, the Air Force saw the *United States* as a clear challenge to its preeminence in nuclear warfare. Its ensuing opposition to the vessel's construction fueled an ongoing and increasingly bitter interservice feud that pitted the Air Force and the Army against the Navy over the questions of defense reorganization and roles and missions in the post-World War II world.[13]

At a pivotal conference held in Key West in March 1948, Secretary of Defense James Forrestal met in seclusion with the Joint Chiefs of Staff to clearly define the services' roles in the postwar world. During the conference, the Navy agreed to forego its own strategic air arm, while the Air Force conceded the Navy's right to continue operating carriers and to attack with tactical atomic weapons inland targets that threatened its ships during a war. Despite the agreement, the Navy ultimately lost the *United States* in April 1949 when new Secretary of Defense Louis Johnson summarily canceled it, but the service remained free to develop tactical nuclear weapons for deployment aboard its existing carriers.[14]

The Key West compromise was timely since intelligence suggested that the Soviets were undertaking the construction of a huge submarine force, based on the advanced German Type XXI U-Boats that the Red Army had captured in 1945. Naval strategists believed that striking the new submarines inside their concrete-hardened underground pens would be the most effective means of preemption in case of war. Conventional bombs did not have the necessary penetrating power, however, and nuclear weaponeers suspected that airdropped, contact-fuzed implosion bombs could not do the job. What was needed for this specialized naval mission was a rugged, ground-penetrating, relatively lightweight device that could be fitted to naval aircraft and deployed at sea—hence, ELSIE was born.[15]

The AEC and DOD authorized the project in October 1948; the Navy promptly chose a gun-assembly-type weapon over the bulkier but more fragile implosion design and distributed ELSIE's R&D elements among several different BUORD and AEC installations. BUORD managed the device's overall design and development, while the Naval Gun Factory built the device's gun barrel at the Washington Naval Yard and White Oak took responsibility for its safing, arming, and fuzing. The Sandia National Laboratory, just spun-off from the Los Alamos National Laboratory as a separate lab, designed the handling equipment and an aircraft saddle to carry the bomb, and the AEC supplied uranium from its Oak Ridge, Tennessee, plant.

BUORD chose the Dahlgren Naval Proving Ground as the primary test and evaluation facility for the ballistics end of the program based on its expertise in conventional gun and bomb testing. Deak Parsons' World War II experiments with "sewer pipe" bomb casings, very similar in size and shape to the planned penetrator bombs, also played a part in the decision to bring ELSIE to Dahlgren. To manage the secret work, a Special Projects Division was organized, and one of Lyddane's top metallurgists and ordnance engineers, Wesley W. Meyers, took charge of the ballistic testing. A former reservist, Meyers had remained at Dahlgren as a civilian scientist after the war and was serving as the head of the Plate Battery Division when the ELSIE project landed in his lap. During Parsons' "sewer pipe" bomb experiments in 1943, Meyers had tested "peculiar little pieces of elliptically shaped armor plate" that were incorporated into the Thin Man's casing. As one of the very few men left at Dahlgren who had directly worked with the "sewer pipe" bomb, he was uniquely qualified to oversee ELSIE's ballistic tests.[16]

After ascertaining the project's need for special facilities at the proving ground in which to conduct the testing and analysis, BUORD acquired from the War Reserve a special 40-foot by 100-foot windowless, prefabricated "Butler" hut and authorized its assembly at Dahlgren several hundred yards away from the Main Battery. A concrete vault for test bomb storage was constructed within the new laboratory (now Building 492—Dahlgren's current Mail Room). BUORD also authorized an extension of the Plate Battery's bomb proof, which would be needed during the firing tests, as well as additional fencing and security.[17]

To simulate the launching of ELSIE penetrator bombs from an altitude of 50,000 feet, the tests called for gun-firing full-scale devices into concrete targets rather than dropping them from the air as Parsons had done. As the test bombs contained components made from expensive and rare uranium, Meyers had to fire all of the projectiles inland. "We couldn't afford to lose them out in the water," he recalled. He therefore supervised construction of enormous thirty- or forty-foot-thick concrete targets and a test butt well away from the river. Meyers also helped modify several 16-inch guns to accelerate the bombs to their free-fall terminal velocities without inflicting the high G-forces that the bombs were incapable of sustaining.[18]

Meyers recalled that security for the project was tight, bordering on paranoia. The laboratory was surrounded by a tall chain-link fence and floodlights, and it was equipped with a variety of security features and elaborate alarms. All personnel not involved in the project, even commanding officers, were barred. When the station's fire chief once appeared demanding to inspect the building, he was told that "if it caught on fire to watch it burn and not let the fire spread." The battery workers who loaded the guns and recovered the test bombs likewise "knew that they were not supposed to discuss even the size, shape, or length and diameter of whatever it was they were working with." The Marine detail that guarded the hut twenty-four hours a day was particularly menacing. At night, the Marines hid along the dark road. The usual procedure for those wanting to access the laboratory was to "go down and honk the horn and then get out and stand in the headlights." "Pretty soon," Meyers shuddered, "some guy would come out of the darkness with a .45 in his hand." The Marine guards also played a dangerous cat-and-mouse game with the "Invasion Team," a security force operating out of Indian Head that would occasionally infiltrate Dahlgren from the river and attempt to hang a flag on the laboratory's fence undetected. Meyers remembered that the Invasion Team was successful several times, but in his opinion its operatives were fortunate to have incurred no casualties during the project since he had no doubt that the Marines "would have shot at them had they seen them."[19]

Since stable uranium-238 (not the weapons-grade U-235) was used in the test bombs, AEC inspection teams periodically and without warning arrived to audit the project's books and ledgers and also to weigh all of the metal on hand. The Special Projects Division had to account for all of it, down to the fraction of a gram. To protect Dahlgren's personnel from radiation exposure, the AEC assigned special teams from Los Alamos to monitor radiation levels and to clean up accidental spills. Cognizant of the dangers associated with radiation leaks, the scientists and technicians kept buckets of axle grease around the laboratory. When an accident did occur, they would "jam" the leaking material into the bucket. The grease then checked the radiation long enough to implement more thorough remediation procedures.[20]

Tests ran smoothly from 1949 through 1951, although with renewed urgency after the Soviet Union unexpectedly detonated its first atomic bomb, JOE-1, on 29 August 1949. The finalized ELSIE design, designated Mark 8, was a much lighter and more efficient version of the Mark 1 "Little Boy." Measuring 9.7 feet long and fourteen inches in diameter, the Mark 8 weighed only 3,230 pounds due to its "Light Case" external shell, as compared with the heavily armored, twenty-eight-inch-thick, 9,000-pound Little Boy. Its relatively light weight meant that it could be carried externally by the next generation of naval attack aircraft, including the AD-4B *Skyraider*, the AJ-1 *Savage*, and the FJ-4B *Fury*, all of which could operate from existing 45,000-ton *Midway* class aircraft carriers. The Mark 8's suitability as a subsurface weapon was confirmed not only by the Dahlgren ballistics tests but also by the atomic test shots of Operation BUSTER-JANGLE conducted at the Nevada Proving Grounds in the fall of 1951. The Mark 8's yield probably ranged from 20 to 50 kilotons. It went into production in February 1952, with forty of the bombs ultimately entering the stockpile before its replacement in 1956 by a more streamlined, Dahlgren-tested improvement, the Mark 11, later renumbered "91."[21]

Though never used in war, the Mark 8's design was so successful from a Navy point of view that BUORD asked the Special Projects Division to consider new designs for gun-assembly devices. However, the advent of guided missiles armed with nuclear warheads sounded the death-knell for gun-assembly bombs in the late 1950s, and the Mark 91 was the last of its kind, remaining in service until its retirement in 1960. After the Mark 91 bombs entered service, the station's role in the Navy's first and only atomic weapons development program ended. By that time Dahlgren had capitalized on its core competency of ballistics calculation to enter a more promising long-term research field, computer science.[22]

MAKING NUMBERS

Carl Norden's bombsight work had exposed Dahlgren to computational technology as early as 1923. In the 1930s the station's familiarity with automatic computing equipment broadened to include a variety of mechanical tabulators, sorters, collators, and punched-card calculating machines manufactured by the International Business Machines Corporation (IBM). These primitive calculators could do basic arithmetic operations but not the longhand calculus required for accurate ballistics computation. Consequently, after America entered World War II, advanced computing technology found an advocate at Dahlgren in the person of Captain Hedrick. In the ensuing years, he doggedly fought to bring improved calculating machines and trained support personnel to Dahlgren to better fulfill its military mission of ballistic computation. In so doing, he placed the station squarely on the road to becoming the Navy's premier computing facility— and ironically hastened the overturn of his beloved military shore establishment tradition.[23]

Hedrick's campaign began in the months after Pearl Harbor, when BUORD assigned Dahlgren the task of computing new range tables for high-capacity projectiles and powder charges for anti-aircraft ordnance. BUORD's action coincided with Hedrick's high-priority request in April 1942 for a differential analyzer of the type that Vannevar Bush had developed at MIT in 1930 and an "appropriate staff" for operating it. Bush's analyzer was a mechanical, or analog, integrating device capable of solving ordinary differential equations using variable-speed gears that could be interconnected in a variety of configurations. Frustrated by the limited availability of MIT's analyzers for proving ground work, Hedrick wanted a machine of his own to improve existing ballistic data, which was imperative for newly deployed fleet radar-ranging systems. He also wanted to launch a program of ballistic "refinement" of existing resistance functions and erosion and powder temperature data, which required what he called "better computing methods."[24]

To lead this staff, and to take up some of the computational work that the recently departed Thompson had left behind, Hedrick tapped Dr. Charles C. Bramble, a brilliant Naval Academy mathematician and reservist. Bramble had earned his Ph.D. in 1917 at the Johns Hopkins University and had started teaching at the Academy that same year. He had first visited Dahlgren in 1924 out of general interest in ordnance problems and casually consulted with Thompson. As a Naval Post Graduate School ordnance instructor, he taught ballistics and gun design to a number of young students who became key flag officers in BUORD, including Deak Parsons and future Dahlgren commanding officer Turner Joy.[25]

BUORD approved Hedrick's recommendations on 1 May 1942—with the exception of a full-time appointment for Bramble. BUORD deemed him "virtually irreplaceable" at the Post Graduate School. A compromise was shortly arranged in which Bramble would work four days a week at Dahlgren and teach two days in Annapolis. Although BUORD recognized that it might take considerable time to obtain a differential analyzer from MIT, it was eager to get the program up and running. After contracting with MIT to do the basic computations on Dahlgren's behalf and to start the initial design work for a new machine, BUORD permitted Hedrick to form a new "exterior ballistics group" under Bramble. Its job was to "polish" MIT's data and to generate the final range tables until such time as MIT could deliver an analyzer to the proving ground.[26]

When Bramble officially started work in September 1942, he was mortified to find that "there were only two desk-type calculators in the place and two mathematicians to operate them." The other IBM punched-card machines at the station were unsuitable for the types of calculations that his group needed to do, so he quickly requested five more desk calculators as well as sufficient mathematically trained reservists and WAVES to run them. When the new personnel arrived shortly thereafter, Bramble lectured them on the science of ballistics and set up computational procedures to get his laboratory going. Once under way, the program expanded quickly due to the sheer volume of work. Within a few months, Bramble had about fifty desk-type calculators that Hedrick had acquired from the recently defunct

Mathematical Tables Project of the Works Progress Administration. Still, the anticipated analyzer design from MIT had not materialized.[27]

By autumn 1943, the exterior ballistics group was straining under BUORD's insatiable demands for range tables and other calculations involving rocket ballistics, antisubmarine ordnance trajectories, and airborne fragment retardation and distribution. The "routine" preparation of bombing tables, assigned in December, only burdened the group further. While still waiting for the differential analyzer, Hedrick realized that war-driven advances in weapons technology were rendering the machine obsolete. Even if one became available for the proving ground, it would be wholly inadequate for future computing needs. Therefore, in October Hedrick asked BUORD to explore the possibility of developing a more advanced analyzer, one capable of a wide variety of calculations beyond the scope of Vannevar Bush's single-method integration machine. Stressing the desirability of Dahlgren doing all of the computation rather than MIT, he insisted that if a survey demonstrated the technological feasibility of an advanced analyzer type, then one should be immediately designed, constructed, and installed at the proving ground.[28]

BUORD responded by commissioning the NDRC's Applied Mathematics Panel to investigate the current state of computer technology and determine its availability for Dahlgren. The panel's Committee on Computing Aids for Naval Ballistics Laboratory reported in late April 1944 that technology existed for improving existing computer designs but that it had not progressed far enough to fully satisfy the proving ground's need for a new generation of calculating machines. The committee identified three private companies that were interested in participating in a BUORD-sponsored computer project and could supply the necessary technology for interim machines. All three, General Electric, Bell Telephone Laboratories, and IBM, submitted proposals to Bramble for consideration.[29]

From the start IBM appeared to have an advantage. At a conference held at Dahlgren on 11 September 1944, Hedrick and Bramble met with military, academic, and private sector experts to discuss the preliminary designs of new computing equipment. Among those consulted was Commander Howard H. Aiken. Previously associated with IBM, Aiken was an electrical engineering professor from Harvard and a reservist assigned to the Bureau of Ships (BUSHIPS). Born in 1900, he had earned his bachelor's degree in electrical engineering from the University of Wisconsin in 1923 and had successfully worked as a power engineer in the private sector before entering graduate school at Harvard in 1934. He took a master's degree in 1937 and earned his Ph.D. in communication engineering two years later.[30]

At the time of the conference, Aiken was already a renowned force in computing technology. While working on his master's thesis at Harvard, he had wrestled incessantly with time-consuming mathematical problems. Deciding that he could mechanize the tedious calculations through electromechanical technology, he had submitted a proposal to the Monroe Calculating Machine Company in 1937 for an "automatic calculator." After Monroe declined to back the expensive venture, Aiken went to IBM, where Chief Engineer James Wares Bryce and Chief Executive Officer Thomas J. Watson Sr. grasped the design's potential and agreed to build it. During the project's first years, Aiken spent his summers at the company's Endicott, New York, plant developing his automatic calculator in collaboration with IBM's engineers. After he was called to active service in 1941, though, the company had to complete and program the machine without his assistance. In early December 1943 IBM successfully demonstrated the calculator to a Harvard delegation and the following February, shipped it to the university for installation. The Automatic Calculator became operational on 15 March 1944 and underwent a short period of intensive testing before it was transferred to naval control in May under Commander Aiken, who had just returned to command the Harvard Computation Laboratory for BUSHIPS.[31]

Aiken's Automatic Calculator was a behemoth. At fifty feet long, eight feet high, and almost three feet wide, it weighed nearly five tons and contained 530 miles of wire within its stainless steel body. It also contained 765,299 parts, including 3,300 relays and 2,200 counter wheels. Up to that time, it was the largest, most complex electro-mechanical device ever built. The Calculator was very slow in comparison with later electronic machines, but it could solve ten or fifteen equations simultaneously and, according to Aiken, could produce as much work in a single day as a human could in six months. What it sacrificed in speed, it gained in reliability, and Aiken soon had it working around the clock, seven days a week.[32]

Unfortunately, during the Calculator's formal dedication ceremony in August 1944, the caustic and headstrong Aiken clashed publicly with his IBM patron Thomas Watson Sr. The controversy concerned a Navy-approved press release issued by the university's public relations office that gave Aiken virtually all of the credit for the machine while ignoring IBM's financial and technical contributions. Watson was outraged, and bitter words were exchanged when Aiken stubbornly refused to correct the mistake. Thomas Watson Jr., who was present at the confrontation, later remembered, "If Aiken and my father had had revolvers, they would both have been dead." Both Aiken and Harvard's relationship with IBM suffered irreparable damage, and from that point forth IBM referred to the machine as the IBM Automatic Sequence Controlled Calculator while Harvard and the Navy referred to it as the Harvard Mark I.[33]

Whether or not Hedrick and Bramble knew about Aiken's rift with IBM, they were undoubtedly impressed with the Mark I. During the conference, Aiken had expressed his willingness to design another calculator specifically for Dahlgren and estimated that he could complete the preliminary design within four months of authorization. Although Hedrick and Bramble still had their hearts set on a next-generation differential analyzer, they became determined to acquire an Aiken Calculator also. On 22 September Hedrick proposed that BUORD either extend the old 1942 contract with MIT for the preliminary design of a differential analyzer or authorize the proving ground to execute and supervise a new design contract with the institute for that purpose. Furthermore, he requested the Bureau to ask the BUSHIPS to rewrite Aiken's duties to include the preliminary design of a Controlled Sequence Calculator under the proving ground's direct supervision. To support his request, he cited not only the NDRC report and the results of the conference but also referenced the entire chain of correspondence in the matter, beginning with his original "high priority" request in April 1942. In putting BUORD on the spot, Hedrick promised that the machines were "complementary rather than competitive" and that the exterior ballistics group urgently needed them both to provide the required speed and versatility for future calculations.[34]

Stirred by Hedrick's relentless drumbeat, in early October BUORD granted him the authority that he had requested concerning MIT and also moved to secure Aiken's services for Dahlgren from BUSHIPS. BUORD also authorized Hedrick to contract with Harvard University—and not IBM—for the design and construction of a controlled sequence calculator, with a scheduled completion date of 30 June 1947. Aiken's colleague Robert Campbell later suggested that this was part of the deal with the proud professor, who "could not imagine going to IBM, hat in hand, to ask for a renewed form of collaboration." Moreover, Campbell believed that Aiken had something to prove since he "had been so dependent upon IBM for the technology of Mark I" that he "needed to demonstrate that he could design and build a machine on his own." Aiken's new calculator would henceforth be a purely Harvard-Navy effort and be built without any IBM parts or assistance.[35]

The contract with Harvard University was signed on 1 February 1945, and Aiken began augmenting his Dahlgren staff to support the new project and to ensure technical continuity from the start of development through reassembly and operation. He hired ten civilian technicians from Boston who agreed to transfer to Dahlgren as Navy civil service employees after the new machine was ready for shipment. This crew of young guns was led by Ralph A. Niemann, a mathematician who had earned his master's degree from the University of Illinois in 1942 and would later become a giant in Dahlgren's computer laboratories.[36]

Reinforced by Niemann and the other nine specialists, Aiken wasted no time building the new calculator, designated Mark II. Aiken deliberately took the electromechanical approach over electronics even though digital technology was then becoming available. The Moore School of Electrical Engineering at the University of Pennsylvania was just completing the new Electronic Numerical Integrator and Computer (ENIAC) for the Army's Aberdeen Proving Ground. Constructed with vacuum tubes, ENIAC was a thousand times faster than the Harvard Mark I, but Aiken believed that the technology had not yet reached his exacting standards. Aiken acquaintance Henry Tropp later noted that Aiken's philosophy dictated that each model should be designed around a specific and reliable level of technology. Shortly before his death in 1973, Aiken defended his decision not to leap the electronic divide: "By God, we had to have complete machines and they had to compute. And within that framework, we didn't give a damn whether we did it with carpet tacks or electronics or what. It didn't make any difference." The Mark II, then, was largely an improvement over the Mark I rather than a radical new design.[37]

Some of the differences were significant, however. Among other things, the Mark II used large relays rather than the Mark I's electromechanical rotary decimal counters to store and add numbers. For this reason, the Mark II would later be christened the Aiken Relay Calculator (ARC) at Dahlgren. Further, the Mark II was a floating-point machine unlike its fixed-point predecessor. It was also comprised of two separate computers capable of operating independently on separate problems or in tandem on a single problem. Technicians could also run the subassemblies against each other to check their data or to locate malfunctions. Looking from the front, they were respectively called the "Left (L) Machine" and the "Right (R) Machine," depending on which side of the room they were installed. The Mark II was twice as large as the Mark I, consisting of 13,000 relays, a cam unit, a large front panel and operator's panel, six large walk-in relay racks, four teletype printers, and other auxiliary devices. It also required more than 4,000 square feet of floor space and was built in sections to facilitate disassembly and transport. Further, it was about six times faster and could do basic addition or subtraction at .2 seconds, reciprocal square root extraction at 6 seconds, and arctangent calculations at 9.5 seconds. It still could not match the lightning speed of ENIAC, but it was an advancement in electromechanical computing technology.[38]

By the summer of 1945, Aiken, soon to be released from active service, had the Mark II up and running at Harvard. The basement of the Physics Research Laboratory, already housing the Mark I, was cramped, so Aiken had assembled his new calculator in an austere World War I-era outbuilding without air conditioning. Not surprisingly, the infant Mark II was full of "bugs," but in more ways than one. On 9 September, a particularly hot day, several of Aiken's programmers had the windows open in their sweltering building when a moth flew inside and straight into one of the Mark II's relays. After the machine malfunctioned, they investigated the cause and found the squashed moth. Removing it with tweezers, technician William Burke carefully placed it in the Mark II logbook with the notation, "Relay #70, Panel F, (moth) in relay." Much later, someone added "First actual case of bug being found" to the entry, highlighting for posterity the incident's significance. When Aiken stormed in demanding to know why they were not "making any numbers"—his peculiar phrase for "computing," they told him that they were literally "debugging" the machine. Although the incident entered Dahlgren lore, the term "bug" had been coined much earlier to describe mechanical breakdowns—even Thomas Edison was known to have documented "bugs" in his many inventions. At any rate, the job of debugging the Mark II was an enormous undertaking, and it would be some time before it was ready for service at Dahlgren.[39]

While Aiken and his people were debugging the Mark II in Cambridge, Hedrick's efforts to shake a differential analyzer out of MIT stalled for good. With the apparent success of ENIAC at Aberdeen, an exasperated Hedrick finally gave up on MIT and recommended in September 1945 that a fully electronic calculator be considered in lieu of a differential analyzer for Dahlgren's postwar exterior ballistics computation program. Informal discussions among Bramble, Aiken, and BUORD led to the final cancellation of the MIT contract in January 1946 and an amendment to the Harvard contract in April to include the development of an electronic computer under Aiken's supervision, even before the Mark II was ready. ENIAC had at last convinced Aiken that digital technology was the "way to go," and he jumped into the parallel project with as much fervor as he had with the Mark II.[40]

At Dahlgren, Bramble left active duty in 1946 but remained employed at the station as a civilian scientist in charge of the exterior ballistics group. Aside from frequent consultations with Aiken at Harvard, he prepared for the Mark II's arrival by supervising the design and construction of a new wing of the Proof Building (Building 218), which required a large removable skylight so that the machine sections could be lowered through the roof by crane. The computer space within the new wing was air conditioned to minimize humidity and to cool the hot-running Mark II. (The offices were not, however, since Navy policy provided only for the air conditioning of equipment and not civilian employees.) In 1947 Bramble organized the exterior ballistics section into the permanent Computation and Ballistics Department to manage the computer and its scheduling, and he became its first head. To gain some practical experience with large computers, Bramble and his staff experimented with one of IBM's smaller and inferior 799 punched-card, plugboard relay calculators, which were then relegated to secondary computing chores after the Mark II arrived.[41]

One thing that Bramble did not expect was competition for the Mark II from White Oak. Because of the great expense of computers in the postwar period, BUORD had decided to centralize all Navy computing into a single complex where all scientific calculations for the Navy would be done. Seeing an opportunity to acquire an insurance policy for itself in the increasingly cutthroat RDT&E world, White Oak, led by Technical Director Dr. Ralph D. Bennett, physicist Dr. Raymond J. Sieger, and mathematician Dr. Harry Polachak, approached BUORD with a strong pitch for the machine. They argued that White Oak had a more legitimate claim to large computing facilities than Dahlgren due to its own computational problems in aeroballistics and hydrodynamics. Backed by Rear Admiral Joy, Bramble countered White Oak's attempted hijacking of the Mark II by lobbying his high-ranking friends in BUORD to affirm Dahlgren's right to the machine. As Niemann recalled, the competition "became bitter at times and surfaced in meetings," particularly on the BUORD's Aeroballistics Advisory Committee, where both White Oak and Dahlgren personnel held key positions. After some difficult closed-door negotiations, Bramble finally cut a deal in which Dahlgren agreed to halt its work in specialized spark photographic technology that had previously been White Oak's exclusive domain if White Oak agreed not to pursue the Mark II. The conflict touched off over the Mark II signaled the start of a lasting, often intense rivalry between the two facilities that lasted for another three decades.[42]

The Mark II was finally shipped to Dahlgren in March 1948. Niemann's support team accompanied the machine and installed it in the newly renovated Proof Building and, as agreed, remained at the proving ground as Navy civil service employees to operate and maintain it. The difficult reassembly and recalibration process took about nine months before the Mark II began operating on a regular production schedule. As Bramble recalled later, "The programming was rather difficult. There were lots of failures, inaccuracies, and a great deal of difficulty in troubleshooting." Despite the early bugs, the Mark II grew more reliable over time. In August 1949 Aiken proudly reported that the machine had proven very satisfactory in handling Dahlgren's required engineering calculations and had delivered useful data 92 percent of the time during that month. Over the next two years it settled into an average efficiency rating of 78.4 percent, with a total number of 24,747 "good-time" machine operating half-hours, as compared to 6,817 "down-time" half-hours.[43]

To manage the Mark II, now called the Aiken Relay Calculator (ARC), the Computation and Ballistics Department established several long-standing policies that had lasting significance for Dahlgren. The first involved placing an overhead charge on each computer hour used to support numerical analysis and programming research. This measure effectively subsidized further mathematical and computer research that was applicable across a broad range of projects. Next, the Department recognized that Dahlgren had to stay current in computer technology and hired a small group of engineers to ensure this by making appropriate hardware modifications as programming methods and technology improved. Finally, the department expanded early requirements for computer program documentation to include mandatory technical memoranda from every program author to familiarize colleagues with the addressed problem and to prevent duplication. Collectively, these policies nurtured an in-house technical expertise that few other Navy RDT&E labs enjoyed and also encouraged the development of a dedicated, research-oriented engineering cadre that could knowledgeably advise the Navy on computer questions and quickly respond to fleet problems. As a result, Dahlgren's technical wizards quickly earned reputations within the fleet as innovators and crack troubleshooters.[44]

ARC's sister machine, designated Mark III but called the Aiken Dahlgren Electronic Computer (ADEC), was completed at Harvard in the fall of 1949 and shipped to the station the following March. Like ARC, ADEC's reassembly required nine months. BUORD had wanted an all-electronic computer, but Aiken, apparently still leery of all-electronic machines, had designed ADEC as a hybrid. While containing some 4,500 vacuum tubes, it also housed some 2,000 mechanical relays. ADEC used a magnetic tape input system, stored data and instructions on eight internal electromagnetic drums, and could multiply in thirteen-thousandths of a second. Unlike ARC, the new machine was a disappointment. Bramble had had Aiken ship it to Dahlgren untested and with inordinate haste after DOD's Research and Development Board proposed in December 1949 that ADEC be left at Harvard indefinitely for general purpose computation. Consequently, while completing its first trial problem on 31 January 1951, Niemann's technicians found it not only unreliable but nearly inoperable.[45]

Working through the spring and early summer, the support team improved ADEC, but only slightly: in July, its downtime amounted to 65 percent of its scheduled operating time. Although originally advertised as ten times faster than ARC, it in fact operated at less than half its design speed. Aiken called it "the slowest all-electronic machine in the world," while Niemann morosely announced the "unanimous verdict" that it had not lived up to its original expectations. The $1 million ADEC only operated until 1956, when Niemann's staff scrapped it for $60. Electromechanical technology was at a dead end, and Aiken developed no more computers for the Navy. He returned to teaching at Harvard and built a virtual copy of ADEC for the Air Force in 1952, designated Mark IV. Establishing Howard Aiken Industries Inc. in 1963, he worked as a consultant the last ten years of his life before passing away in 1973.[46]

Because of ARC and ADEC, Dahlgren was well on its way to becoming the Navy's centralized computer complex. Moreover, through its computers, the Computation and Ballistics Department quickly superseded the A&P Laboratory as the preeminent research activity at Dahlgren, reflected by Bramble's appointment over Lyddane as the new "Director of Research." Bramble's appointment was in place just in time to grapple with a host of grave new challenges in the 1950s.[47]

A NEW LOOK

Research and development work at Dahlgren suddenly diminished when the Soviet-backed North Korean Red Army crossed the 38th Parallel and invaded South Korea in June 1950. Much to its chagrin, the Truman administration realized that nuclear weapons were undeployable in the "limited" conflict, for both political and practical reasons. The Navy had to quickly remobilize its moth-balled battlewagons and lesser warships to conduct shore bombardments and support U.N. amphibious operations on the Korean peninsula. The reactivated ships needed guns and ammunition for the mission, disproving predictions of their obsolescence after Hiroshima and Nagasaki. Rearmament meant proof and test at Dahlgren. For the Korean War's three-year duration, Dahlgren returned to a wartime footing. This entailed a tremendous increase in workload since Navy ships and planes ultimately fired more ammunition tonnage in the Korean War than in all of World War II. Blue-collar workers returned to man the station's reawakened gun line, while proof and test preempted experimental work.

Dahlgren's scientists adjusted easily, with Lyddane succinctly summing up their prevailing attitude: "You've got to fight this war, so we'll be here tomorrow to do our R&D on a reasonable kind of time scale."[48]

In the post-World War II years, jet propulsion technology had profoundly changed both airborne weapons development and flight testing. As a result, some of the most important projects that Dahlgren undertook during the Korean War involved operational problems associated with jet aircraft. The sheer speeds that they were now capable of achieving had already prompted the development of exterior-mounted Low Drag Bombs. Bombing tables and aircraft fire control data likewise had to be adjusted to compensate for the higher velocities. The theory of "toss" bombing, in which a jet pitches its nose at high angle and lobs a bomb like a football, was pioneered at Dahlgren, while studies of high-velocity aerodynamic flowfields around external bombs and rockets vastly improved the safety of aircraft store separation (the actual release of ordnance onto a target), a potentially hazardous occasion under high-velocity flight conditions.[49]

Dahlgren's ballistic engineers also solved an extremely dangerous problem facing fighter pilots. The Naval Ordnance Test Station at Inyokern, California, had reported instances in which accelerating fighter jets had run into their own decelerating 20-mm projectiles a few seconds after firing. Using ARC and trajectory theory, Dahlgren's engineers confirmed that this was possible under certain firing conditions and aircraft maneuvers. They solved the problem by calculating firing constraints (points where a pilot could not safely fire his guns), which were immediately incorporated into combat flight procedures. Despite its important wartime contributions to jet-based ordnance, Dahlgren could not host jet aircraft since its airfield was too short. By 1957 flight operations had been transferred to the Naval Aviation Test Center (NATC) at Patuxent River, Maryland. Dahlgren retained responsibility for bomb ballistic test planning and analysis, bombing table computation, and ordnance tactical manual preparation until these activities were finally transferred to NATC in 1983.[50]

Following the cease-fire on 27 July 1953, the proof and testing bubble burst again as quickly as it had after World War II. Although the gun line kept firing for some time to replenish the Navy's ammunition supply, layoffs and slashed funding soon rocked the station. Closure rumors were rampant. Even before hostilities had ended, the Navy had to dispel a news report that it planned to shut down Dahlgren and move its facilities out west into the desert. Despite repeated denials, closure seemed imminent as Pentagon technocrats began questioning whether or not Dahlgren was needed anymore. Much of the uncertainty resulted from the incoming Eisenhower administration's promised "New Look" for the American military machine, based heavily on strategic nuclear weapons rather than conventional sea power. After DOD revised the 1948 Key West agreement in 1954 to permit the Navy a strategic nuclear capability, the pendulum of support quickly swung back to guided missile programs. Sensing that recent history was repeating itself and that the proving ground's gun heritage had become an albatross, Lyddane grimly concluded that "Dahlgren was going to have to find something else to do because its old sources of funds were simply drying up."[51]

As Dahlgren's outlook remained bleak in the mid-1950s, Lyddane inventoried Dahlgren's sources of strength, identifying Bramble's Computation and Ballistics Department as the station's most prominent asset. Not only did ARC and ADEC represent the pinnacle in Navy computing technology, but Bramble's staff was the only group in the country that could get useful data out of them. Lyddane recognized that the loss or breakup of the entire department, particularly its highly trained staff, would be catastrophic. He was determined to make this clear to the headquarters admirals, many of whom believed that a computer was simply a black box with a button on it that any sailor could push to have his question promptly answered.[52]

Lyddane also counted his own A&P Laboratory as a significant asset since the expertise applied to designing better armor-piercing projectiles could be equally applied to designing better warheads. "If the Navy was going along the route of guided missiles," he asked, "why couldn't we do the warhead work?" Additionally, he had built by this time a solid technical corps in the laboratory. This had been difficult since low government pay coupled with job insecurity made the booming defense contractors much more attractive to career-conscious scientists and engineers. Furthermore, Dahlgren's rural isolation made it unappealing to career men with families. Not surprisingly, Lyddane had a "terrible time" recruiting technical personnel. Worse, both he and Bramble had to defend their turf against both private contractors and rival government laboratories, especially Los Alamos, that trolled at Dahlgren for trained personnel. Despite the challenges, Lyddane had been successful in increasing Dahlgren's professional scientific staff, and he thus counted it as the third tier of the proving ground's technical triad.[53]

With Bramble's early support, Lyddane continued the reconfiguration of Dahlgren, begun after World War II, from a test station into a science-based weapons laboratory. This required, in his words, gaining a "greater, broader responsibility from the Bureau of Ordnance" in the weapons development arena. A good start had been made with the installation and operation of

Aiken's computers, but Lyddane, Bramble, Niemann, Cohen, and others realized that it was essential to stay abreast of the rapidly advancing state of computer technology and to expand the capabilities of the Computation and Ballistics Department. They therefore sought to bring to Dahlgren the Navy's latest and most advanced computer yet developed, the Naval Ordnance Research Calculator (NORC).[54]

BATTLE ROYAL

Stung by Dahlgren's victory in the 1947 skirmish over the Mark II, Bennett and Sieger at White Oak had engaged IBM in a series of low-key discussions concerning the "state of the art in computers." After learning that IBM had made significant progress with an electrostatic memory tube and had developed an exceptionally fast arithmetic unit capable of 10,000 multiplications per second, White Oak expressed an interest in procuring a high-speed electronic computer based on this technology for the laboratory's aeroballistics and hydrodynamics programs. In October 1950 IBM therefore proposed to build the new computer for the Navy, including the special arithmetic unit and high-speed magnetic tape devices for input, output, and all auxiliary storage. Following up in November, IBM estimated that the total package would cost $1,300,000, with an additional $30,000 to $50,000 per year for maintenance, but offered to build the computer at cost plus $1. White Oak recommended that BUORD accept the contract, but its new chief, Rear Admiral Malcolm F. Schoeffel (commander of Dahlgren's Aviation Detail and acting fire chief from 1931 to 1932), was taken aback by White Oak's effort at a *fait accompli* and would not be rushed into something that he knew little about. Instead, he appointed an ad hoc committee to spend two weeks assessing the need for an advanced computer and considering alternatives. The committee, which convened in January 1951, included not only White Oak's Sieger but also Dahlgren's Bramble, among others. According to Niemann, Schoeffel appointed Bramble to avoid partisanship, but in effect his participation made true cooperation unlikely and may have even tipped the scales in Dahlgren's favor from the start.[55]

The ad hoc committee visited IBM and Remington Rand, where the machine and its peripheral equipment would be built, and solicited advice from government and private sector experts. Concluding that the new computer was indeed needed, the committee recommended that BUORD contract with IBM for the machine. To supervise the computer's design and development, the committee suggested the creation of another, more permanent technical committee, comprising representatives from BUORD,

White Oak, and Dahlgren. BUORD accepted the recommendations but made no decision concerning the computer's final destination. Niemann, who sat on the technical committee from 1952 to 1954, later wrote that BUORD believed that an early announcement would diminish the losing party's interest in the project and negatively affect the end product's quality. He believed that Schoeffel had essentially played White Oak off against Dahlgren to get the best computer possible for the Navy.[56]

During NORC's development phase, the technical committee meetings escalated from a "tug-of-war" between White Oak and Dahlgren to what mathematician Dr. Armido DiDonato later called a "battle royal." As voices were once being again raised behind closed doors, Bramble summoned his BUORD contacts for help one last time. Fortunately, his old student Rear Admiral Deak Parsons, who had risen to prominence in the late 1940s as the "atomic admiral," was then serving as Schoeffel's Deputy Bureau Chief. Schoeffel had delegated to Parsons, perhaps with a wink and a nod, the decision of where to install NORC. Bramble enthusiastically argued that if the Navy wanted a new central computing facility, Dahlgren's prior experience with the Aiken machines as well its experienced and available staff made it the most obvious choice. Parsons did not need much prompting, and he decided that BUORD should go with the outfit that "appeared most likely to succeed"—Dahlgren. As it happened, the decision to award NORC to Dahlgren was Parsons' final and perhaps most invaluable service to Dahlgren before his sudden death from a heart attack in early December 1953 at the age of fifty-two.[57]

BUORD announced Parsons' decision about eight months before NORC was completed, giving the Computation Department sufficient time to prepare space for the machine and to work with the IBM development team at the Watson Scientific Computing Laboratory in New York to gain some operational experience before shipment. A month before completion, though, both BUORD and Dahlgren were shaken by the unexpected appearance of a new heavyweight contender for NORC, Dr. Edward Teller. As the "father of the hydrogen bomb," Teller was a physics prima donna and the head of the Lawrence Livermore National Laboratory in California. He had heard about NORC's fantastic capabilities and believed that his thermonuclear calculations at Livermore were more important to national defense than Dahlgren's ballistics research. He thus wanted NORC shipped to California for his own use. BUORD summoned Niemann and Dr. Eugene Ritter from Dahlgren to Washington to help draft a reply to Teller's request. They maintained that NORC would not be transferred to Teller's Livermore lab but softened the news with the compromise that his staff could use NORC at Dahlgren for one shift on a three-shift operating basis whenever needed. Teller agreed to this arrangement, and Livermore subsequently used NORC to perform nuclear calculations during the third shift for an entire year.[58]

The Navy accepted NORC at a dedication ceremony in New York in December 1954. It was shipped to Dahlgren the following March and was operating by July. As the world's first supercomputer, NORC was awesome. Incorporating 264 Williams Tubes (special cathode ray tubes used for random access storage), 9,000 vacuum tubes, and 25,000 diodes, it could run 15,000 operations per second with high precision. It also incorporated a labor-saving automatic error-checking feature that Bramble's engineers, particularly Niemann, had insisted on adding during the development phase. IBM had thought automatic checking impossible, but for an additional $400,000 it was proven possible and would later be vital for ballistic targeting during the POLARIS program. The feature subsequently became a key selling point in all of IBM's successor machines, beginning with the 7090 STRETCH computer.[59]

NORC immediately made its mark in the RDT&E establishment as BUORD encouraged other labs, both from the Navy and civilian agencies, to bring their calculations to Dahlgren. The new machine was so powerful that both the Mark II and Mark III Aiken computers were retired shortly after its installation in 1955. Bramble later mused that because NORC was "so far beyond anything else at that time," its arrival was clearly "one of the turning points in the history of the Laboratory."[60]

On 9 September 1955 Schoeffel's successor, Rear Admiral Fredric S. Withington, another former Bramble student, officially designated the Naval Proving Ground as BUORD's prime agency in the Naval Ordnance Establishment for the respective scientific fields of computation, exterior/rigid body/terminal ballistics, and warhead characteristics. He also authorized the creation of a new Computation and Exterior Ballistics Laboratory ("K" Laboratory) and a Warhead and Terminal Ballistics Laboratory ("T" Laboratory) to help the proving ground better execute its enhanced responsibilities. Finally, and most importantly, he authorized Dahlgren's laboratories to initiate and plan their own research and development programs and later supported the creation of the Weapons Development and Evaluation Laboratory ("W" Laboratory). To Lyddane's delight, Dahlgren had taken a giant step closer to becoming a dedicated science installation.[61]

Bramble did not remain at Dahlgren to savor the fruits of his labors. He retired from government service in January 1954 before IBM transferred NORC to the Navy and subsequently went to work for Carl Norden.

Thompson's old instrument maker, Nils Riffolt, replaced him as the new Director of Research, but his was more of an honorary appointment since Riffolt was more interested in lab work than doing administrative chores. Bramble remembered that "he was the type that either worked in a laboratory or sat thinking at his desk." As a result, Lyddane slid effortlessly back into his former role as chief scientist, confirmed in September 1956 when he succeeded Riffolt as the station's new part-time "technical director."[62]

As technical director, Lyddane was well positioned to spark the final phase of Dahlgren's transformation. He realized that the impetus had to come from the bottom up, so he mobilized his senior personnel and launched a campaign to convince BUORD to give Dahlgren a chance to modernize and diversify its product line for the Navy's benefit. In late 1956 Lyddane pushed his advantage in a lengthy presentation to Rear Admiral Withington, carefully outlining Dahlgren's history, its three sources of strength (the Computation and Ballistics Laboratory, the A&P Laboratory, and his premier technical and scientific cadre), and its computing capabilities, highlighted by NORC. Concluding, he presented BUORD's chief with four options: 1) leave Dahlgren as it was and let it slowly wither on the vine as financial support dwindled each year; 2) convert Dahlgren into a field station of White Oak; 3) rejuvenate Dahlgren by funneling new weapons projects into its laboratories, or 4) close Dahlgren altogether.[63]

Withington was suitably impressed by both Lyddane's candor and Dahlgren's promise as a weapons development laboratory. Choosing Lyddane's third option, he transferred the troubled Hazards of Electromagnetic Radiation to Ordnance (HERO) program to Dahlgren on the spot. This was a safety and reliability program that the Navy had earlier established at White Oak to study and deal with premature detonations of electrically triggered ordnance, particularly on aircraft carriers. Tests had revealed that the inadvertent explosions were caused by the electromagnetic radiation (EMR) generated by shipboard radio communications and radar systems. Some of the symptoms of this phenomenon were weird. For instance, some aircraft parked on carriers were found so energized by the electromagnetic field that flashlight bulbs either lit up or burned out when their contact terminals touched the airplanes' exposed metal surfaces. Likewise, the structures of an energized aircraft were occasionally so "hot" that crewmen could suffer electromagnetic burns at a touch. The problem was a big one and had caused several fatal accidents in the fleet, and the Navy did not yet have it under control.[64]

Following Lyddane's meeting with Withington, "W" Laboratory chief Donald Stoner and division head James N. Payne assumed control of HERO.

They founded a new department in which to manage the work and, together with senior project engineers Dick Potter, Charlie Hinkle, and Gil Gilbertson, set up a simulated flight deck at Dahlgren where they tested every ordnance design used by the Navy in a realistic EMR environment. The engineers soon found that the solution lay in making the ordnance invulnerable to EMR through proper shielding and the avoidance of radio frequency coupling. "W" Laboratory tackled the EMR problem so successfully that the Navy expanded it from a small $50,000 a year project in 1956 to a $1 million project by 1959. HERO ultimately became one of Dahlgren's hallmark missions and proved that the facility could in fact achieve excellence outside of its traditional gun testing role.[65]

FROM OUT OF THE DEEP TO TARGET

After Withington approved Dahlgren's diversification, Lyddane and his senior staff began actively soliciting new sponsors for work at Dahlgren. The Navy's Special Projects Office (SPO) was the most important of these. Established in November 1955, its mission was the high-priority development of Fleet Ballistic Missiles (FBM), later called Submarine-Launched Ballistic Missiles (SLBMs). The Navy had been developing this technology since the end of World War II, conducting a number of shipboard tests using captured German V-1 and V-2 rockets. The volatility of the weapons' liquid fuel, however, demonstrated with shocking effect during Operation PUSHOVER in 1949, had appalled the Navy's steely-eyed missile men. Subsequent Navy efforts to develop cruise missiles during the LOON, REGULUS, and TRITON programs enjoyed some success. However, in light of the 1954 revisions to the Key West agreement, the Navy was now interested in developing a long-range ballistic missile of its own to counterbalance the Air Force's vigorous ATLAS, TITAN, and THOR programs— hence, the FBM program.[66]

Lyddane and Niemann approached SPO in 1956 with an unsolicited proposal to undertake FBM's trajectory calculations, fire control, and guidance based on Dahlgren's heritage of ballistics analysis, its exceptional computer capability, and Cohen's leadership in six-degrees-of-freedom missile flight simulation. Despite sharp competition from rival laboratories, such as White Oak and the Naval Ordnance Test Station at Inyokern, California, Dahlgren seemed to be the natural choice for the mission. It was a problem reminiscent of the old high-altitude bomb-dropping problem during Norden's era. The calculations required to successfully hit a target more than a thousand miles away with a warhead launched from a specific point either on or in the sea reached a new order of complexity that only the latest computers could handle. Not only did the three classic launching platform motions of pitch, roll, and yaw have to be taken into account but also surge (wave motion), sway, and heave, as well as the global forces of rotation, gravity, and atmosphere that would affect the reentry vehicle (RV), as the warhead was euphemistically called. Moreover, the calculations had to be done well beforehand. As Dahlgren guidance engineer Rob Gates explained later, "When you wanted to launch a missile, you wanted to be able to do it in a relatively few minutes. You didn't want to wait days to do a computational solution, which wouldn't have been right anyway because you would have moved some more." Although this preliminary computation, or presetting, was enormously difficult, SPO apparently failed to fully appreciate Dahlgren's ability to handle this type of work. Additionally, SPO was more interested in adopting the Air Force system of using private contractors rather than involving Navy laboratories in the project, and so Lyddane and Niemann's proposal met with initial rejection.[67]

Dahlgren's Cohen and his colleague David R. Brown Jr. better understood the FBM's unique technical and mathematical complexities well before anyone else in SPO. They continued to lobby for an opportunity to demonstrate Dahlgren's capabilities. SPO finally agreed in 1957 to assign Cohen and Brown a reentry study for the Missile Branch. Cohen subsequently conducted the first presetting studies for real-world operational conditions using a Q-Matrix guidance system developed at MIT. When confronted with the results at a technical coordination meeting at Lockheed headquarters in Burbank, California, David Gold, SPO chief engineer for the Guidance and Fire Control Branch, recognized the value of Cohen's achievement and brought "K" Laboratory into the project under a consulting contract. By 1958 Dahlgren had earned the responsibility for preparing the presettings for all Navy guided ballistic missiles.[68]

As Dahlgren assumed the central computational role for FBM, NORC began producing a veritable cascade of punched cards with targeting solutions. These were arranged in order according to a launch area's latitude and longitude, with prospective targets in numerical sequence. Launch point areas were divided into twenty nautical-mile squares, and target areas into thirty nautical-mile squares, as set by DOD's Joint Strategic Planning Staff (established August 1960) at Strategic Air Command headquarters in Omaha, Nebraska. Aboard ship, the cards were used to manually set knobs on the fire-control panel—several boxes of cards were needed for the ship to cover its assigned operational area. Later, storage problems led to the generation of microfilm, which was produced and proofed at Dahlgren before it was shipped to the fleet. Each film frame contained three launch point-target point combinations, and a crew could choose a desired launch point using lines on the side of the film. Targeting cards could be produced as needed using small microfilm readers and keypunches that were installed aboard nuclear-powered ballistic-missile-carrying submarines (SSBNs).[69]

The SPO had originally planned to deploy the Army's large, liquid-fueled JUPITER missiles aboard surface ships as FBMs. Lyddane called this the "the most cockeyed scheme I'd ever heard in my life," and the plan was scrapped when sudden breakthroughs in warhead miniaturization, solid propellants, and high-accuracy gyroscopes made smaller missiles possible that could be safely deployed on stealthier nuclear-powered submarines. The Navy therefore abandoned the JUPITER for the underwater-based POLARIS system. SPO accordingly tasked Dahlgren with the production of all trajectory calculations for the new program, which contemplated an interim missile with a 1,200-nautical-mile capability (POLARIS A1) operational by late 1960, a missile with a full range of 1,500 nautical miles (POLARIS A2) by mid-1962, and one with an advanced 2,500-nautical-mile capability (POLARIS A3) by late 1964.[70]

POLARIS was enormously successful. The first underwater test launch of the A1 missile was conducted on 22 July 1960 from the USS *George Washington* (SSBN-598) off Cape Canaveral. After the missile lurched out of the water and roared more than 70,000 feet into the air, sending its dummy warhead a thousand miles into the South Atlantic, the *Washington's* captain, Commander James Osborn, signaled Eisenhower, "Polaris—from out of the deep to target. Perfect." Nearly four months later, on 15 November, the *Washington* departed Charleston, South Carolina, for its first operational patrol carrying sixteen POLARIS A1 missiles and some 300,000 targeting cards prepared at Dahlgren. The viability of the punched-card technology was subsequently proven on 6 May 1962 during Operation FRIGATE BIRD, when the USS *Ethan Allen* (SSBN-608) launched a POLARIS A1 missile with a live warhead toward an open ocean nuclear testing area near Christmas Island. The missile flew over one thousand miles and reportedly detonated "right in the pickle barrel." This was the only live "end-to-end" test of any U.S. land- or sea-based strategic missile system ever conducted, since the Partial Test Ban Treaty of 1963 prohibited further nuclear testing in the atmosphere, in outer space, and underwater. Dahlgren produced the target cards for test, and as Rob Gates later said, "You can believe that they were very well checked out!" The *Washington's* triumphal deployment, followed by the successful FRIGATE BIRD test, marked an auspicious start to "K" Laboratory's long-term mission of providing the fleet with precision SSBN fire control and targeting products.[71]

Additionally, POLARIS gave Cohen and Brown in "K" Laboratory the opportunity to break ground in physics research. The applied science of geoballistics represented an extension of classical exterior ballistics. Now, however, the earth's curvature and rotation, atmospheric density, and gravity fluctuations were incorporated into trajectory calculations for ballistic missiles and their warheads. By taking ballistics to the global level, Cohen, Brown, and NORC put Dahlgren on the map within the naval RDT&E establishment, just as the Space Race was starting.[72]

FENCING THE HEAVENS

In October 1957 the launch of Sputnik I came as a discomforting surprise to the American people, who had blithely assumed that the United States was well ahead of the U.S.S.R. in rocket and space technology. The launching, a month later, of the much larger Sputnik II carrying a cosmo-dog named Laika reinforced the perception that the Soviets had surpassed the United States technologically. Frightened citizens imagined that they would soon be sleeping under the light of a Communist moon, or worse, that "Sputniks" could drop atomic bombs from space onto American cities with impunity. The Eisenhower administration was compelled to take immediate action to dampen the outcry and began by funding and implementing a number of space initiatives through the armed services and civilian agencies such as the National Advisory Committee for Aeronautics (NACA), later reorganized and renamed the National Aeronautics and Space Administration (NASA) on 1 October 1958.[73]

The Navy had its own worries about Sputnik, specifically that succeeding Soviet spy satellites could locate and observe U.S. naval forces at sea. Consequently, in early 1958 DOD asked the Naval Research Laboratory (NRL), which had already been involved in developing a tracking system for the VANGUARD satellite program, to assess the problem of a defensive detection system to identify and track nonradiating, "dark" satellites. Soon afterward, NRL submitted a proposal to DOD's Advanced Research Projects Agency (ARPA) calling for a radar "fence," comprised of transmitter and receiver stations arranged alternately and running along an east to west line at 33.5 degrees latitude, from Fort Stewart, Georgia, to San Diego. On 20 June 1958 ARPA ordered NRL to develop the fence and soon after amended the order to include construction of a complete Space Surveillance (SPASUR) System, including the station complexes, transmission lines, and an analysis center supported by high-speed computers. SPASUR came together quickly, and in August it detected the first confirmed satellite signal. Since each station recorded information on a twenty-four-hour basis, the data processing load increased substantially as new stations came on line and new satellites entered orbit. By the end of 1958, NRL realized that it needed Navy computing power to continue the SPASUR program.[74]

As part of their diversification program, Lyddane and his senior scientists had been seeking a suitable military tenant to share the proving ground's excess facilities and to absorb part of its overhead cost and help support community activities and services. When Dahlgren's new commanding officer, Captain M. H. Simons Jr., learned about the SPASUR program in late 1958, he proposed that NRL establish its analysis facility at the proving ground and utilize NORC for orbital computation and data distillation from the transmitter and receiver stations. His case was bolstered by Cohen and Brown's FBM experience as well as their early work in celestial mechanics and satellite geodesy, another new science pioneered at Dahlgren, that of calculating orbital trajectories. NRL accepted the proposal, and on 20 February 1959 "K" Laboratory personnel began working with NRL's Data Processing Group to learn SPASUR analysis methods.[75]

On 24 May the new Space Surveillance Operations Center opened at Dahlgren, and surveillance operations started a week later when four channels of data came in by telephone lines, from each of the four receiver stations. The system was improved and further automated through the remainder of the year. On 19 April 1960 Navy Secretary William B. Franke issued Instruction 5450, which formally established the U.S. Naval Space Surveillance Facility (NAVSPASURFAC) under Officer-in-Charge Commander D. Gordon Woosley, a veteran aviator and missile range operations officer fresh from a planning assignment with Project MERCURY. In October Secretary of Defense Thomas S. Gates transferred management responsibility for SPASUR from ARPA to the Department of the Navy, which then delegated control to the Bureau of Naval Weapons (formerly the Bureau of Ordnance until 18 August 1959). NRL retained technical direction of the systems improvements and additions. On 1 February 1961 the U.S. Naval Space Surveillance System (NAVSPASUR) was finally commissioned as an independent command under Woosley, now promoted to captain, and assigned to the Commander-in-Chief of the North American Air Defense Command (CINCNORAD) for operational control. The Navy was now fully in the space business and Dahlgren had its first major outside military tenant activity.[76]

A PEACEFUL REVOLUTION

By 1959 the Dahlgren Naval Proving Ground, through the strenuous efforts of Lyddane and his scientist colleagues, had completely changed its character, its mission, and its outlook for the future. Lyddane and his staff had not only worked to modernize Dahlgren's mission and diversify its product line but also to change its management culture from a military shore establishment to a civilian-style corporate model. One important step had been taken previously in April 1952 when BUORD approved a proving ground request to establish an Advisory Council, similar to a board of directors, composed of the Navy's leading civilian scientists and private industrialists to assist Dahlgren with its administrative and technical problems and long-term strategic issues. Dahlgren's Advisory Council met for the first time on 7 May 1953 and included Dahlgren's former chief physicist, Dr. L. T. E. Thompson, who was then technical director for the NOTS at Inyokern, California, and Dr. Ralph Sawyer, who had returned to teaching at the University of Michigan after Operation CROSSROADS. The change in management culture became more apparent in the administrative language used by Dahlgren's administrators and scientific community. "Capital investment," "salesmanship," and "tenant" entered the Dahlgren lexicon, while "sponsors" became "customers" and rival laboratories were styled "competitors," all very much in a business sense. Lyddane later summarized this new mentality: "We always took the attitude that the Fleet was, after all, our customer, and if you are going to stay in business, you'd better worry about and respect your customer."[77]

To modernize Dahlgren's management structure, the Advisory Council pushed hard in the late 1950s for a reorganization plan that called for higher "supergrade" civil service positions for the technical director and the directors of both the Computation and Warhead Laboratories. After the council's ninth meeting in May 1958, Chairman Ralph Sawyer highlighted for BUORD the pressing concern that civilian management was not adequately represented on Dahlgren's organization chart: "While recognizing that command in such an establishment is an obvious military function, its professional work depends fundamentally on civilian staff. The council believes that appropriate recognition of the official staff and responsibilities of the principal civilian officers is of great importance to the best morale and productivity of the civilian staff." In the spring of 1959 BUORD finally agreed with the council's assessment and approved the desired reorganization plan. Lyddane was therefore elevated to full-time technical director and assumed parallel authority with the deputy commander in Dahlgren's new organization chart.[78]

Reorganization brought a name change that both reflected Dahlgren's new mission and confirmed the primacy of civilian science over military establishment at the facility. Lyddane and his colleagues wanted most to bury Dahlgren's stereotype as a gun testing station and to convince its "customers" that Dahlgren was "a factor in the world of today and not yesterday." A great deal of thought went into the new name. "Naval Ordnance Laboratory" was already taken. Besides, no one wanted Dahlgren to be mistaken as a mere field station of White Oak. "Naval Ordnance Test Station" was likewise unacceptable since Lyddane and his people wanted to get away from the "test station" concept. The Advisory Council preferred "Dahlgren Naval Laboratories" to maintain the historically significant "Dahlgren" name and to convey the breadth of its "naval" research activities. In the end, the slightly more specific "Dahlgren Naval Weapons Laboratory" was chosen. The new designation went into effect on 15 August 1959.[79]

Just before the name change, the facility's Advisory Council commented on the fulfillment of the final phase of Dahlgren's transformation:

> Three years ago, Dahlgren was basically a fine facility and team serving a partially obsolescent mission of warfare whose continuation could not but be adjudged unwise by any sound top navy management. Today an entire reorientation and new pattern of activity has again placed Dahlgren in the main stream of defense effort with a clear future based upon serving needs. We sense this peaceful revolution as the outgrowth of wise, imaginative, and vigorous leadership in both the military and the civilian roles, including mutual recognition of the areas in which each must have freedom and authority for responsible action. The result is indeed impressive.[80]

Thus, Dahlgren had finally become a key science facility in both name and mission within the Navy's RDT&E establishment. The new weapons laboratory had overcome potential obsolescence and strong competition for its facilities, personnel, programs, and funding. Although Dahlgren was firmly embedded within the RDT&E structure, the laboratory would continue to face inter-laboratory competition and Cold War exigencies— particularly a simmering "brush-fire" conflict in Southeast Asia—as it moved into the turbulent 1960s.

Captain John Adolphus Bernard Dahlgren (1809-1870), known as the "Father of U.S. Naval Ordnance," shown here during the Civil War aboard the USS *Pawnee* with a 50-pounder Dahlgren gun. Note the characteristic "bottle" shape of the barrel. (Library of Congress)

The explosion of the "Peacemaker" aboard USS *Princeton* on 28 February 1844 killed seven and injured 20 people. Among those killed that day were U.S. Secretary of State Abel P. Upshur, U.S. Secretary of the Navy Thomas Gilmer, and Chief of the Bureau of Construction Captain Beverly Kennon. The event spurred the Navy to adopt a more scientific approach to naval ordnance testing and development. (Naval Historical Center)

This inert 16-inch projectile blasted through a reinforced abutment at the Indian Head proving ground, tumbled a mile downrange, and smashed through the Swann farmhouse. There were no injuries, but such incidents underscored Indian Head's shortcomings. (National Archives and Records Administration)

This 14-inch naval railroad gun was one of eight built for service on the Western Front in WWI. Five saw combat, but the war ended before this one ever fired a shot at the Germans. After returning to the United States, it became a terminal ballistics test bed at Dahlgren. BUORD's need to safely test such long-range guns led to Dahlgren's establishment. (Naval Historical Center)

Rear Admiral Ralph Earle, Chief of BUORD, persuaded Navy Secretary Josephus Daniels and Congress to approve the creation of an auxiliary Lower Station to Indian Head. He named it "Dahlgren" after Rear Admiral John A. Dahlgren. (Naval Historical Center)

Commander Henry E. Lackey, Inspector of Ordnance in Charge at Indian Head, chose the site for the new Lower Station and became its first commanding officer. (Dahlgren Historic Photograph Collection)

This 7-inch, 45-caliber, tractor-mounted gun is prepared to fire the first shot at Dahlgren on 16 October 1918. (Dahlgren Historic Photograph Collection)

The cost and size of the commandant's home sparked a congressional investigation that nearly led to Dahlgren's closure in the early 1920s. (Dahlgren Historic Photograph Collection)

The Main Battery, looking northeast, about 1925. The Potomac River flows by the station in the background. (Dahlgren Historic Photograph Collection)

Naval Proving Grounds
Dahlgren Virginia

Dahlgren's rural isolation is captured in this aerial shot of the station, taken circa 1930. Looking southwest, the commandant's house, administration building, and officer's quarters lie along the roads near the mouth of Machodoc Creek, away from the Main Battery. The dock and seaplane hangar are to the left on the riverbank, while the proof building sits below the crossroads, with the gun line to its immediate front, facing the river. The airfield is to the right of the crossroads. (Dahlgren Historic Photograph Collection)

Dahlgren hosted early studies of aircraft vulnerability to projectiles. After this test, conducted in March 1922, ordnance officers circled with white chalk the fragment holes made by 3-inch high explosive projectiles on this target aircraft. (Dahlgren Historic Photograph Collection)

Early experiments with radio-controlled aircraft culminated in the 15 September 1924 flight of this N-9 seaplane, the first aircraft ever launched, maneuvered, and then landed intact under full remote control. (Dahlgren Historic Photograph Collection)

Engineer Carl L. Norden, an inverterate inventor and promoter, poses next to an automatic pilot installed in the midsection of an aircraft. (Naval Historical Center)

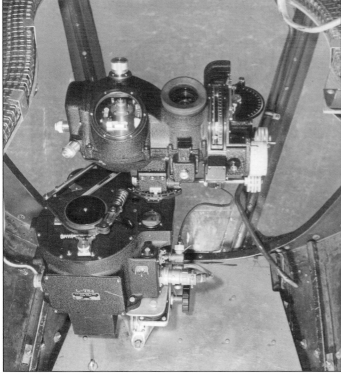

Carl Norden's Mk 15 bombsight, developed and tested at Dahlgren, was an early analog computer and one of the most important Allied weapons of World War II. This bombsight is mounted in the plexiglass nose of a B-24 and flanked by two ammunition belts for the bomber's nose guns. (National Archives and Records Administration)

As the Roosevelt Administration stepped up naval rearmament in the late 1930s, this team of ordnance officers and civilian workers installed an advanced base mount on the main range, 30 June 1939. (Dahlgren Historic Photograph Collection)

Ordnance workers fire a 14-inch gun during a test on Dahlgren's Main Range. (Naval Historical Center)

In 1945, inert projectiles are being prepared in the Shell House for testing.

Former BUORD Chief Rear Admiral William H. P. "Spike" Blandy (left) and Dahlgren's commanding officer Captain David I. Hedrick (right) examine an armor plate pierced by a major-caliber projectile, 25 May 1944. (Dahlgren Historic Photograph Collection)

During World War II, women ordnance workers (WOWs) bolstered the labor force at Dahlgren, here assisting in the test of a 40-mm twin-mounted anti-aircraft gun. (Dahlgren Historic Photograph Collection)

William S. "Deak" Parsons served as Dahlgren's Experimental Officer from 1939 to 1943. He bridged the professional and cultural gap between the station's scientific and military personnel and helped develop the "VT" proximity fuze at Dahlgren before becoming the Manhattan Project's ordnance officer. He is shown here as a Rear Admiral after World War II. (Los Alamos National Laboratory)

The advent of more advanced, faster, and heavier aircraft during World War II led BUORD to develop 3-inch, 50-caliber anti-aircraft guns, which did not enter service until the late 1940s. This 3-inch gun was installed at Dahlgren in 1952 for testing. (Dahlgren Historic Photograph Collection)

Dr. Howard H. Aiken demonstrates his electromechanical Mk II Relay Calculator to Dahlgren's commanding officer Rear Admiral C. Turner Joy in 1947. (Grace Murray Hopper Collection, Archives Center, National Museum of American History, Behring Center, Smithsonian Institution)

Dahlgren technicians "make numbers" on the Mk II Aiken Relay Calculator's control console. (Naval Historical Center)

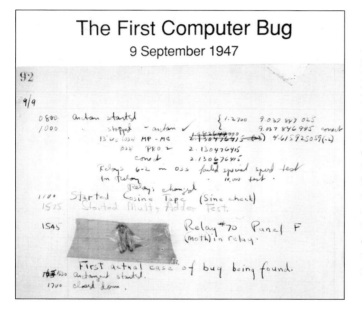

A page from the log book recording testing of the MK II computer, the Aiken Relay Calculator, at Harvard University just prior to delivery to Dahlgren. Technician Bill Burke was searching for the cause of a computation error in the machine on the afternoon of 9 September 1947. He traced the error to a moth caught in a relay. Mr. Burke removed the moth, checked to see if the computer worked properly, and taped the moth into the daily maintenance log book.

The Mk 8 Light Case (LC, or "ELSIE") ground penetrator atomic bomb. Dahlgren scientists fired ELSIE bomb casings into thick concrete targets and analyzed the resulting ballistic data using the Mk II Aiken Relay Calculator. (Los Alamos National Laboratory)

An 800-pound, 12.75-inch rocket leaves the muzzle of Weapon "Able" at Dahlgren's rocket battery. While working with Weapon Able in 1950, Dr. Charles J. Cohen developed the world's first operational six-degrees-of-freedom trajectory simulation, making possible the development of guided ballistic missiles. (Dahlgren Historic Photograph Collection)

The Naval Ordnance Research Calculator (NORC) arrived at Dahlgren in 1955. At the time, the most powerful computer in the world, it used over 9,000 vacuum tubes and could run 15,000 operations per second. (Dahlgren Historic Photograph Collection)

Dahlgren's first chief physicist, Dr. L. T. E. Thompson, during a visit in April 1967. His pride is evident as he stands by the road named for him near the Main Battery. (Dahlgren Historic Photograph Collection)

5

Chapter

Rebels and Revolution, 1959-1973

The newly christened United States Naval Weapons Laboratory (NWL) at Dahlgren had weathered the early Cold War years, and more importantly lived down its former reputation as a "vestigial remnant of an earlier type of warfare." Under Research Director Dr. Charles Bramble and then Technical Director Dr. Russ Lyddane, Dahlgren had made computer technology the centerpiece of its operations. The pioneering mathematical and computer research of Bramble and Lyddane's staff, including most notably Ralph Niemann and Dr. Charles Cohen, launched Dahlgren into the space age as "K" Laboratory assumed primary responsibility for target analysis, guidance, and fire control for the POLARIS program, and the Department of Defense (DOD) established the Naval Space Surveillance Command as an outside tenant activity. As Dahlgren moved into the 1960s, however, there would be new Cold War challenges, both internal and external, as the Navy grappled with DOD reforms, the management of its research laboratories, and the combat realities of Vietnam.[1]

THE MAIN STREAM

On 5-6 November 1959 Dahlgren's Advisory Council met for the twelfth time. Chaired by Dr. Ralph Sawyer, with Dr. L. T. E. Thompson in attendance, the distinguished group collectively noted that "Strong leadership, both civilian and military, together with forward looking and imaginative planning have placed Dahlgren in the main stream of defense work." The council members were particularly "gratified" with NWL's progress in developing opportunities for leadership in three major areas of importance in the Navy's operations for the foreseeable future. These included analysis and computation in "K" Laboratory; exploratory and development work in "T" and "W" Laboratories; and analysis, development, and test of special aircraft and ground systems components, as well as safety and reliability research at the facility as a whole. Looking ahead, the council saw potential for further long-range work in the field of analysis and "massive computation" and set an ambitious goal for NWL to go beyond the limits of the Navy as the "lead" *national* analysis and computation center within ten years. The council accordingly encouraged Lyddane and his staff to develop a clear, orderly plan with costs, size, schedule, and space requirements to pave the way within the Navy and DOD bureaucracy for realizing this goal.[2]

Although the council's desire that NWL become the nation's leading computer center was an admirable goal, it was somewhat unrealistic. Dahlgren had already undertaken more computation work than it could immediately handle with its existing facilities. During the Advisory Council meeting, Lyddane noted that the Naval Ordnance Research Calculator (NORC) was "fully saturated" and was running twenty-four hours a day, seven days a week. Not only were the POLARIS targeting analysis and the Navy's satellite surveillance program consuming copious computing hours, but wargaming and operations research for the Chief of Naval Operations (CNO) as well as orbital analysis for the Navy's new navigation satellite program (TRANSIT) also left little computation time available for other projects. And the load was increasing rapidly. Lyddane, in fact, estimated that Dahlgren would probably need computing power equivalent to two NORCs by the middle of 1960, and two more in 1961.[3]

But Lyddane and Niemann were ready to go beyond NORC, supplementing it with IBM's latest electronic mainframe computer, the 7030 STRETCH. STRETCH was the world's first mainframe computer designed with high-speed transistors. Its central processing unit (CPU), measuring approximately thirty feet by six feet by five feet, contained approximately

169,000 transistors. Incorporating IBM's advanced high-speed magnetic core memory and newly invented random-access drives with multiple read-write arms, the machine could do five-digit addition in 3.5 microseconds, multiplication in 40 microseconds, and division in 65 microseconds. STRETCH had been conceived in 1955 (just as NORC was completed) at the behest of the Los Alamos National Laboratory and the AEC. Originally expected to be at least a hundred times as powerful as NORC, its design and development had been troubled with performance problems, cost overruns, and delays. These early problems notwithstanding, Lyddane and Niemann believed that STRETCH would supersede NORC and accordingly requested permission to order one of the machines as soon as it became available. In the meantime, they rented IBM's technologically similar but more limited 7090 computer, developed hastily in 1958 for the Air Force's Ballistic Missile Early Warning System and just introduced into the commercial market.[4]

By October 1960, Lyddane and Niemann had installed the 7090 and assigned it to the Naval Space Surveillance Command on a full-time basis. Unfortunately, Lyddane's estimates for the required additional computing power proved to be inadequate: additional incoming satellite and astronautics work quickly strained NORC and the 7090 beyond their capabilities, causing a backlog. To take up the slack, "K" Laboratory acquired yet another 7090 while awaiting the delivery of STRETCH.[5]

But a hitch appeared when Dahlgren and IBM came into dispute over a contract clause guaranteeing the computer's performance to specifications and the efficiency of its compiler. Lyddane and Niemann insisted upon the clause, which levied financial penalties in the event that specifications were not met, but IBM balked at the idea. Tensions were further aggravated by Niemann's swaggering engineers. They had worked closely with IBM to develop NORC and felt that they could significantly contribute to the STRETCH design. The chill in the relationship between Dahlgren and IBM was palpable during the first contract negotiation session, which ended when Lyddane, nettled by IBM's refusal to accept the clause, abruptly rose from his chair and stalked out of the meeting. A few weeks later, a chastened IBM agreed to compromise language—it would provide an efficient compiler to the best of its ability.[6]

When STRETCH finally arrived at Dahlgren in 1962, Lyddane's obstinacy proved justified. Its compiler was not even as efficient as that of the 7090. The technical director, backed by Niemann, made his case to IBM, and despite some resistance the company agreed to design and install a satisfactory replacement compiler. To compensate for lost time, IBM agreed to give Dahlgren a single shift of rent-free computer time for a full year.[7]

Operating at full power, STRETCH soon struggled like its immediate predecessors with the soaring demands of POLARIS. The demands for targeting calculations reached new heights when new Secretary of Defense Robert S. McNamara, perusing strategic warfare plans, noticed target "gaps" in both the Air Force and Navy's ballistic missile forces. He ordered a five-fold increase in the POLARIS target list, creating much more work for Dahlgren's computers. The subsequent advent of POSEIDON dictated the acquisition of a third-generation computer with even greater speed and storage capacity than STRETCH. NORC was due for retirement in 1968, but the Johnson administration's cumbersome regulations for procurement of large computers delayed the acquisition process by three years. In 1969 the Navy finally approved Dahlgren's request for a Control Data Corporation (CDC) 6700 mainframe computer, designed by CDC chief engineer Seymour Cray. Delivered in 1972, the 6700 contained two CPUs and 1.3 megabytes of memory, and integrated some twenty remote consoles located in different buildings at Dahlgren. A dial-in networking feature allowed users at the consoles to "call" the computer, do their work, and receive answers back over the phone lines. The 6700 remained in operation until 1985, but computer technology advanced so quickly that it was soon eclipsed by CDC's "Cyber" series, introduced in 1976.[8]

While Dahlgren's computer capability expanded further in the 1960s, so did the breadth and scope of the research programs conducted within its three sub-laboratories. "W" Laboratory, led by Donald Stoner and employing approximately four hundred civilians and more than twenty military personnel, continued work on a number of key programs, including research and development of Cartridge Actuated Devices (CADs). These small, self-contained energy sources were used in emergency cockpit ejection systems, missile stage separation systems, self-destruct mechanisms, and ordnance release systems. Stoner's engineers also continued developing the Navy's Guided Missile Safety Program for both surface- and air-launched weapons systems, including the TARTAR, TERRIER, and TALOS missiles. Their accomplishments included a fast-actuating sprinkler system for ship missile magazines, rocket blast tests aboard missile ships, and an environmental evaluation test program for Submarine-launched Anti-submarine Rocket (SUBROC) motors. In addition to determining safe firing and aiming zones for missile launchers and furnishing the control cams, "W" Laboratory also took responsibility for planning and scheduling the entire Guided Missile Launcher Control Program, assisting in installation and evaluating shipboard systems.[9]

"W" Laboratory's Electromagnetic Compatibility (EMC) Branch also supervised the largest "at sea" laboratory test ever conducted in the history of the U.S. Navy. Led by branch head David B. Colby, the Fleet Research (FR)-69 project involved nine major warships and numerous support vessels and aircraft. Its purpose was to not only determine the high frequency interference generated among the fleet's various search, acquisition, and fire control radar systems, but also to identify and develop ways of alleviating that interference. For three weeks in the spring of 1965, the ships and aircraft illuminated one another with their respective radar systems while performing a wide range of tactical maneuvers off the southeastern U.S. coast. CNO Admiral David L. McDonald was particularly interested in the problem and was present throughout the testing. While McDonald watched, Colby and his major project engineers James H. Mills, Jr., Charles Yarborough, and Lee A. Clayberg measured radio frequency noise and recorded data aboard the ships. They then developed a number of new guidelines to help the fleet keep its radar systems from jamming each other out, which the office of the Chief of Naval Operations (OPNAV) immediately adopted and are still in force today. FR-69 thus helped solve a vexing technical problem that could have conceivably cost lives in a war situation.[10]

As an outgrowth of its exterior ballistics heritage, "T" Laboratory, led by Richard I. Rossbacher and staffed by more than two hundred civilians and roughly ten military personnel, focused on nonnuclear guided missile warhead development and ballistic and explosive effects. The laboratory's scientists began an early incarnation of the 5-inch rocket-assisted projectile (RAP) program during this period, undertook applied studies of target vulnerability, and computed the probabilities of lethal damage to aircraft when flying through fragments from warhead detonations. "T" Laboratory also retained responsibility for the test and evaluation functions of the old Experimental Department and A&P Laboratory, including ordnance improvement and metallurgical research. Because its functions descended directly from Dahlgren's old gun-testing role and were perhaps not as glamorous as missile development, the laboratory did not enjoy the same RDT&E status as its sister labs at Dahlgren. Consequently, it struggled to develop long-term projects that were relevant to a Navy then enthralled with missiles.[11]

In contrast, the third point on the NWL triad, "K" Laboratory, was well established with more than 235 civilians and approximately five military personnel tending and operating the Navy's supercomputers. Under Niemann's gentle direction, an all-star team of scientists, engineers, and mathematicians, including Dr. Charles Cohen, David R. Brown Jr., Richard

J. Anderle, Robert T. Ryland Jr., Gene H. Gleissner, Dr. Allen V. Hershey, and Dr. Armido R. DiDonato, generated all fire control and guidance data for the Navy's POLARIS strategic strike system. Cohen's work in the fields of geoballistics and satellite geodesy was particularly outstanding. In 1957 he had developed the first rigorous mathematical descriptions of the Earth's gravitational field, which DOD had adopted for all original long-range missile trajectories. Further research using Doppler observations of the TRANSIT 1B satellite verified that the gravitational field was "pear shaped," a fact Cohen and Anderle announced in a 1960 issue of *Science* magazine. The subsequent development of a "General Geodetic Solution" for calculating orbital trajectories led to the standard DOD gravity model—the World Geodetic System 1966 (WGS-66), which was later revised in the WGS-72 and WGS-84 versions.[12]

KENTUCKY WINDAGE

"K" Laboratory moved into the 1960s intent on expanding its fire control product line as computer technology rapidly advanced. In 1961 the Navy charged "K" Laboratory with developing and supporting shipboard digital fire control computers for the POLARIS system. This was a critical task since true targeting computers had been too large to install aboard the first ten POLARIS A1 submarines. Instead, they had gone to sea carrying hundreds of thousands of punched cards containing Dahlgren-generated preset targeting data. Microfilm was introduced in 1960 and used aboard both A1 and A2 class SSBNs, but it was only a temporary and ultimately unsatisfactory fix.[13]

The greater striking area of the A3 missile, carrying Multiple Reentry Vehicles (MRVs)—three 200-kiloton warheads that were aimed around a single target and released shotgun style in flight, demanded a new approach. Accordingly, the laboratory developed the stand-alone Mark 148 POLARIS Target Card Computer System (PTCCS). Installed aboard the A3 boats to supplement the Mark 80, PTCCS averaged 66,000 operations per second and generated its own punched cards. This eliminated the need for microfilm and the laborious card preparation and organization process but still was less than ideal for real-time fire control. In 1963, "K" Laboratory scientists finally developed the Mark 84 digital fire control system, which incorporated a central Digital Geoballistic Computer along with two Digital Control Computers, each of which could handle some 87,000 operations per second. Using real-time navigation inputs, the Mark 84 electronically generated targeting data at a moment's notice, eliminating the need for punched cards altogether.[14]

POLARIS's successor system, the POSEIDON C-3, presented "K" Laboratory with new fire control and guidance challenges. Conceived by Lockheed and SPO in 1962 and approved for development in January 1965, POSEIDON had a 2,500-nautical-mile range and carried up to fourteen low-yield, Multiple Independently Targetable Reentry Vehicle (MIRV) warheads. Unlike MRVs, which could only strike around a single target, MIRVs from a single missile could be individually programmed to strike separate targets simultaneously. Each MIRV warhead had to be assigned a calculated target, or "aimpoint," within a missile's collective impact/detonation pattern, or "footprint," as developed by the Joint Strategic Targeting Planning Staff. Multiple footprints could then be grouped into a "target package" to inflict an extensive yet precise pattern of devastation from a single SSBN. Precision deployment of MIRVs therefore required fire control presetting calculations that were an order of magnitude more complex than those for POLARIS. As POLARIS itself had proven inherently inaccurate at longer ranges since the primitive guidance computers of the 1950s could only assume a simple earth (and gravity) model, a vacuum (no atmosphere), and a point mass reentry simulation, "K" Laboratory developed a much more sophisticated guidance system for POSEIDON. The system used a realistic "round earth" gravity model for in-flight course corrections and calculated target *offsets*, which guidance engineer Rob Gates wryly called "Kentucky Windage—from the old Kentucky long rifle shooting days." During reentry, target *offsets* guide the missile away from the target in order to *hit* the target, specifically by compensating for the earth's atmospheric density, gravitational inconsistencies, and aerodynamic forces. The concept, according to Gates, was much like "an old sharpshooter aiming away from his opponent to account for wind and gravity."[15]

This advanced guidance scheme was bundled into a new Mk 88 Fire Control System, essentially an upgraded Mk 84 but with doubled storage capacity and a keyboard interface. "K" Laboratory also developed a special computer operating system called POSEIDON SUPERVISOR that controlled managed memory and allowed the simultaneous preparation of all sixteen missiles for launch. Flight tests started in August 1968 and the system reportedly attained a Circular Error Probable (CEP) accuracy of 600 yards (which meant that 50 percent of the warheads aimed at a specific target would strike within a 600-yard radius). POSEIDON became operational on 31 March 1971 when the converted POLARIS submarine USS *James Madison* (SSBN-627) left port for an operational patrol carrying sixteen of the new missiles, each with a payload of six 40-kiloton MIRVs. Later, designers increased the payload to a standard of ten warheads.[16]

Dahlgren's navigation through the mainstream of defense work during the 1960s was not without its turbulence. In early 1960, as part of its investigation of Bureau of Naval Weapons (BUWEPS) research and development facility requirements, a study team from the Naval Research Advisory Committee (NRAC) submitted to the Office of Naval Research an unflattering confidential report on NWL. The team was unimpressed by Lyddane's diversification efforts and shameless "job shopping," and was concerned by the divergent goals of Dahlgren's three laboratories. It also saw no single cohesive objective for the laboratory as a whole other than simple survival. While recognizing that "K" Laboratory was "the outstanding computer facility in the Navy, and perhaps in the country," the team disparaged both "W" and "T" Laboratories as "weak in depth" and insignificant.[17]

If this were not bad enough, the NRAC team judged Dahlgren's scientists and engineers as largely second rate. To account for this supposed lack of technical talent, the team cited NWL's isolation, its undeveloped community, aging buildings, and poor housing, all of which prohibited the recruitment and retention of high-quality technical personnel. Suggesting the relocation of Dahlgren's computing center to an academic institution such as Johns Hopkins University and the transfer of its weapons work to other Navy installations, the team recommended closure as perhaps "the best solution for the long term."[18]

Fortunately for NWL, the Navy did not accept the NRAC team's recommendation, but the chill cast by the report still made things uncomfortable at Dahlgren's laboratories over the next few years. An indignant Niemann later summed up how NWL's community accurately caricatured NRAC's general attitude: "Dahlgren is out in the sticks. You can't get any professional people to work there. After all, the educational system is no good. They're not close to universities. Who would want to work at Dahlgren?" Contrary to NRAC's assessment of Dahlgren's isolated environment, many did find it an inviting place to work. Dr. Armido DiDonato later reminisced: "It was very small then. Your kids could play in the street with no problem and it was a very nice place to live. I knew that I would be here the rest of my life because it had clean water, it had athletic fields, and a gym, and tennis courts, and the work was great. It was just beautiful here."[19]

Lyddane and Dahlgren's commanding officers sensed that part of Dahlgren's image problem stemmed from its ancient physical plant, comprised mostly of World War II-era special-purpose buildings situated on the gun line and river front. Dahlgren still looked like a backwater gun testing station rather than a state-of-the-art weapons laboratory and computation center. Lyddane later noted the prevalent attitude toward Dahlgren among Navy RDT&E policymakers and advisors, many of whom had come through Dahlgren as postgraduate ordnance officers before World War II: "When I was at Dahlgren as a JG [Junior Grade Lieutenant] in the early 1930s when they didn't have anybody, it was a very remote, primitive, picturesque isolated spot. My goodness, you wouldn't think about putting anything modern and new there. Hasn't that been closed yet?"[20]

To counter these perceptions, Lyddane, Niemann, Cohen, and the NWL commanders resolved to modernize the facilities along with the mission. The most conspicuous example of the early 1960s effort to make Dahlgren look more like a modern science installation rather than a gun range was the construction of the Computation and Analysis Building (Building 1200). "K" Laboratory had been in need of office space for some time. Despite an expanding workload, Niemann and his staff had remained in the old Ordnance Proof Building 218, which housed the computers. As Niemann later related, "They weren't good office buildings. For air conditioning, we had to go through several inches of concrete. Lighting wasn't good, and we needed new floors." Moreover, Building 218's location directly behind the gun line was potentially hazardous for the scientists working inside. Dr. DiDonato recalled one incident in which a shell accidentally exploded outside of Cohen's office "and a piece of the scrap metal hit his window and tore out part of the woodwork." Consequently, Niemann and Cohen, with the early support of NWL commander Captain Manley H. Simons Jr., began lobbying for a new office building at Dahlgren, using POLARIS, Naval Space Surveillance Command, and TRANSIT as justification for the additional work space. The process was slow since congressional approval and funding were necessary, but construction finally started on 1 April 1963 and was completed the following year.[21]

Designed by Dahlgren engineer Robert Ryland, the Computation and Analysis Building was (and remains) situated near the station's front gate, well away from the Potomac and the gun range. There was no mistaking it for a testing shed. It really looked like a science building with its graceful lines and large windows, standing in sharp contrast to the rest of NWL's research plant. It was no mistake that the building was at the front gate, as it was intended to instill visitors coming to Dahlgren with a sense of scientific enterprise. The stratagem worked. "Once the building was constructed," said Niemann, "then the issue about closing Dahlgren sort of went away because when people would come down, they'd see a new building. They'd figure things were going good, and maybe Dahlgren shouldn't be closed."²²

Construction of the Computation and Analysis Building was by no means the only measure that Dahlgren senior management and command staff took to revamp the station's image in the 1960s. Efforts at building a self-sustaining community outside the base were redoubled, with private housing developers being approached for proposals to construct new employee homes. Progress was rapid. In 1966 NWL's Advisory Council applauded the senior staff for finally solving the troublesome housing and electrical power problems. The closure of the ramshackle "Boomtown" outside the base was recognized as a triumph of progress for Dahlgren. In conjunction with these practical improvements, NWL also launched a public relations campaign that included a slick new brochure and a movie to tell Dahlgren's "story" to the technical and lay public. NWL also sponsored a "Classified Open House" in which security-cleared personnel from other defense and contractor laboratories visited to learn about Top Secret research projects. This campaign to improve Dahlgren's scientific and technical image extended through the 1960s and ultimately did much to polish NWL's RDT&E reputation just as the Navy's laboratory system entered a turbulent period of reorganization and consolidation.²³

A REBEL

In December 1963, after seven years as technical director and twenty-two years as a government scientist, Lyddane retired to accept a position with General Electric's Engineering Service Group. The Bureau of Naval Weapons was left looking for a replacement, since Niemann, Cohen, and Hershey all declined to take the job. The Bureau compiled a list of forty other possibilities, but Dahlgren commanding officer Captain Robert F. "Mike" Sellars had his own candidate, a rocket engineer who had impressed him at Inyokern, Bernard "Barney" Smith.²⁴ Born in the Lower East Side of New York City in 1910 to a poor Russian immigrant family, Smith (Americanized from the Russian Smeed) was a blunt, dour, and utterly fearless engineering manager. As a youth, he held down odd jobs to help support his family but spent the rest of the time engaged in his true passion, designing and building rockets. In early 1933 he enjoyed some fame by conducting the first-ever public launch of a liquid fuel rocket, but his rocketry waited while the Great Depression sent him west in 1935 to seek a living in California. For the next nine years, he worked alternately as a mechanic, welder, blacksmith, and locksmith before finally deciding that college was the best route to success.

Smith earned a B.A. in physics from Reed College in 1948 but could not get into graduate school because of his age. Instead, he went to work as a rocket engineer at the Naval Ordnance Test Station (NOTS) at Inyokern, California. While there, Smith came under the tutelage of Dr. L. T. E. Thompson, who had been the station's technical director from 1945 to 1951. Thompson had successfully transplanted to NOTS the unique approach to research that he and Deak Parsons had formulated years earlier at Dahlgren. After chalking up some impressive engineering and managerial achievements in a number of priority missile projects, Smith left NOTS in 1961 to become the Bureau of Naval Weapons' (BUWEPS) chief engineer. At Smith's own request, BUWEPS detached him for a one-year stint in London with the Office of Naval Research (ONR), where he was working when Captain Sellars added his name to the candidate list without his knowledge. Sellars was so eager to bring Smith to Dahlgren that he personally flew to London to formally invite the engineer to be Dahlgren's next technical director. Smith, who had grown contemptuous of the continued jostling for influence at the Bureau, itched to get back into a laboratory environment and gladly accepted the captain's offer.[25]

Smith arrived at Dahlgren in August 1964 with little fanfare. An outsider unfamiliar with NWL's past experiences and traditions, he saw the research work currently under way in Dahlgren's three laboratories from a completely different perspective than Dahlgren's old guard. "I thought I could help the efforts here with what I knew, with the ideas that I had, and with the experience that I had in management, to look at all aspects of it," Smith later remembered, "but I wasn't exactly prepared for what greeted me after I arrived." Indeed, he was disappointed by what he saw. Contrary to the expectations and hopes that Lyddane, Niemann, and others had earlier expressed for Dahlgren's RDT&E future in the 1960s, Smith sensed that NWL might be falling into stagnation. "K" Laboratory notwithstanding, research was confined largely to analysis without any practical hardware developments for the fleet and was in danger of becoming irrelevant. Smith also frowned on NWL's retention of its traditional responsibilities for naval ordnance safety, which he believed had been over-researched and should have been kicked out to the fleet. He later mused that "the organization had done such a good job in this area that I often wondered if the ammunition presented any hazard to the enemy." Most importantly, he believed that Lyddane's diversification program, however necessary in the 1950s, had left Dahlgren too dependent upon marginal "job shop" work and had stripped it of its self-sufficiency, a mortal sin according to the gospel of Thompson. Hence, Smith concluded that Dahlgren had lost its "way."[26]

But there was yet another unpleasant surprise. Shortly after becoming Dahlgren's technical director, Smith learned that the station and some other Navy laboratories were on the "chopping block" as part of the reform package of Secretary of Defense Robert S. McNamara. A businessman and former president of the Ford Motor Company, McNamara had joined the new Kennedy administration in 1961 and had remained with Lyndon Johnson after Kennedy's assassination. Bringing a more sophisticated, business-oriented approach to DOD, McNamara sought comprehensive reforms of defense program management, especially in RDT&E, through centralization, computer analysis, and cold number crunching. He advocated keeping in-house laboratories as the incubators of defense research programs, although with clear lines of management and responsibility for each in-house lab. But he also insisted on cost efficiency and consolidation if necessary and made his decisions based on numbers alone. Those programs that failed to achieve their expected efficiency and productivity levels were either liquidated or consolidated. Moreover, as Smith's colleague and future technical director Jim Colvard described it, McNamara's overly linear "numerology world" fostered an overarching concern for neatness, orderliness, predictability, and lack of surprise in the R&D process. This flew in the face of the whole innovative process, insisted Colvard, since "scientific breakthroughs can't be scheduled." Not surprisingly, Navy RDT&E was in for a rough ride.[27]

By early 1965 McNamara had compelled the Navy to reevaluate its RDT&E establishment and to eliminate laboratory costs, inefficiencies, and redundancies. This once again placed Dahlgren in mortal danger, much to Smith's mortification. "Apparently I had been allowed to enter the scene in time to officiate at a decent burial," he later wrote. He resolved on the spot that an important Navy resource like Dahlgren would not be "liquidated" on his watch. Calling upon influential Navy contacts (including Admirals Frederic Ashworth and Edwin B. Hooper) and silently assisted by the sleek new Computation and Analysis Building, he quashed the immediate talk of closure. He fully understood, however, that the status quo was working against NWL and that the laboratory would have to change. He later summed up the situation as he saw it: "Clearly it was essential to become more essential. The good, honest work of the laboratory was insufficient to ensure its survival. New blood and new ideas were needed in order to give Dahlgren a unique standing in the Navy." His solution was simple: revitalize the laboratory and bring the "Dahlgren Way" back to Dahlgren.[28]

A REVOLUTION

As Lyddane had done under similar circumstances, Smith embarked upon a deliberate campaign of reorganization, or, as he characterized it, outright "revolution." Under his plan, though, the revolution would take place in an "evolutionary way"—it would require time and patience to accomplish. Accordingly, he spent the next four years studying Dahlgren's operations and determining who would accept his planned changes without complaint. His first major step was to shake up the lower echelons of Dahlgren's laboratory management and instill his management principles into his project managers. Smith's philosophy was derived from Thompson but with a tougher edge: Give people complete authority and turn them loose. If they cannot hack it, then relieve them. Slavish obedience to higher authority in RDT&E matters was anathema to Smith. He believed that a good leader and project manager should be bold and fearless. "He has to have something of the rebel in him," Smith once explained, and be willing to "fight for the permission to change to meet new objectives, or else have the guts to run the risk of doing what he thinks is right," even if it ran counter to the wishes of superiors in the laboratory or an admiral in the Pentagon. As far as Smith could see, Dahlgren's lower management ranks contained no rebels but mostly sheep, complacent with safe, comfortable, and nonessential government projects, always on time and within budget, and guaranteed a line item in the next appropriation, whether it was needed or not. This could not endure if Dahlgren expected to survive McNamara's regime.[29]

A stark realist as well as a rebel, Smith understood the daunting task that lay ahead of him. He would have to recruit several hundred new young professionals, inaugurate a good indoctrination program, and redirect older staff into new research and management approaches. Those that could not adapt or just did the bare minimum to draw a salary would be phased out. He knew that almost everyone involved would experience some pain, but in his opinion, the alternative— Dahlgren's closure—was worse.[30]

With a certain ruthlessness, then, Smith informed the laboratory's leadership that the good old days were over. From that point forth, he announced, all promotions and rewards would not be based on how well an edict from Washington was carried out, but on how well individual project managers and engineers judged the value and suitability of assignments for the Navy or the laboratory. Research for the sake of research was out, since Thompson's dictum required Navy labs to conduct research in connection with bona fide Navy problems. If a manager deemed a project inappropriate, then Smith expected proposals for better use of the Navy's resources at

Dahlgren. He also wanted to force his laboratories to be more selective in the work they accepted. As for personnel, he demanded over-achievers who were "unafraid to pioneer into areas of pressing needs, untouched because of high risk and lack of vision." Smith's dictates predictably brought about retirements, transfers, demotions, and dismissals, a few of which he noted were "accompanied by an undying hatred in some individuals, voiced occasionally to this very day."[31]

This was only the beginning. In the following years, Smith conducted a thorough housecleaning, terminating the short-term nonessential projects that he deemed either marginal or simply "busy work." When he was finished, some $11 million worth of projects had been axed, often to the outrage of former sponsors, who were powerless against Smith because of his friends in Washington (he was, after all, BUWEPS' former chief engineer) and simple force of personality. To reinvigorate Dahlgren, Smith recruited seasoned senior personnel who understood his ideas about the process of military technical development. For candidates, he looked no farther than his old home facility, the NOTS at Inyokern, California, near China Lake, where a large crop of engineering "rebels" sharing his Thompson-inspired philosophy were ripe for the harvest. Putting the lie to the old myth that no one wanted to work at Dahlgren, Smith noted with pride that "they came in sufficient numbers to make of Dahlgren a second China Lake, which, by a trick of fate, returned to its home base the philosophy cast on the waters twenty years earlier." Two China Lake engineers were particularly influential among the growing management cadre, James E. "Jim" Colvard and Charles "Chuck" Bernard. In the years ahead, Smith would rely on them heavily to help oversee his revitalization program, and later accorded them his highest praise: "Without them, the revolution at Dahlgren simply would not have been possible."[32]

Smith's most important reform was mandatory department head rotation, officially implemented in 1968. During his lengthy evaluation of Dahlgren's management structure, Smith had observed a disturbing insularity among the laboratory departments. He specifically found that his department chiefs did not have the slightest clue about what was going on outside of their respective departments. Even worse, some actively tried "to knock down all other departments so they would shine." Complacency was also rife within the management structure. Many longtime Dahlgren managers believed that they were "indispensable." As a result, innovation and ambition had both stagnated, and serious lapses in leadership had occurred when key individuals had departed. The trouble that BUWEPS had encountered in finding a replacement for Lyddane was, in fact, symptomatic of the management problem that Smith now confronted.[33]

At China Lake, Smith had learned that the first thing a good manager should do is train his replacement. At Dahlgren, he allowed younger managers to fill in for him while he was away on official trips. After considering the current problem in this context, he found a simple yet elegant solution: "Why just have them rotate to my job? Why don't they rotate to each other's job—all of the department heads? Then nobody is indispensable."[34]

Smith therefore ordered in 1968 that all department heads rotate on a regular basis into different departments for extensive periods. Not surprisingly, he was confronted with all manner of excuses and foot dragging, but he refused to yield: "I said we are going to jump into this thing and see if cold water really kills us." As it turned out, cold water did not kill him or his department heads. He later recalled:

> I had a good bunch. And they took the risk. And they found out that they could manage in other departments, and could learn quickly who was doing what, and they could see what was wrong in the other department with fresh eyes, which the old department heads couldn't see anymore, and how certain guys had done a pretty good con job. . . . The new guy wasn't so easily fooled because he didn't have these ten-, fifteen-year associations.[35]

Smith's radical rotation plan became so successful and widely accepted that not only were department heads rotated but division and branch heads also. As a result, the NWL organization became very flexible, with no component member becoming too closely identified with his home laboratory or department. Smith was also gratified that his mandatory rotation policy buried destructive intrigues since department heads plotting the downfalls of their counterparts would invariably draw the stricken department as their next assignment.[36]

Smith's evolutionary "revolution" not only generated personal friction with his department managers but also with his new skipper, Captain William A. Hasler Jr. Captain Sellars had actively solicited Smith for the technical director's job in 1964, but his tour had ended just before Smith's arrival. Hasler, who commanded the station from August 1964 to July 1968, was more conservative in his approach to running the military side of NWL and not as receptive to the changes that Smith wanted to make. Furthermore, Hasler had been an Engineering Duty Officer (EDO) in the electronics and ordnance fields and had served in a number of technical positions in BUORD and BUWEPS. He came to Dahlgren with his own ideas about how research should be conducted at NWL and what technical personnel were needed. Additionally, Hasler was loathe to make any waves within BUWEPS and intended to follow his orders to the letter. The captain particularly opposed hiring new civilian professionals to beef up NWL's three component laboratories. It was not long before Smith and Hasler clashed in private.[37]

Smith, who had witnessed a disastrous collision between a former NOTS technical director (post-Thompson) and that facility's commanding officer, did not want a similar cataclysm to befall Dahlgren and moved to resolve his differences with Hasler. Satisfied with Smith's conciliatory gestures, the captain relented and agreed in 1966 to a major civilian personnel increase within Dahlgren's allowance. This helped initiate what Smith called "the great change at Dahlgren" and left the technical director wiser in his dealings with future commanders.[38]

Unlike the research director and chief physicist roles played by his predecessors, Smith saw his primary job as preserving the laboratory's technical continuity, which a commander could not influence because of his routinely short tour of duty. Since Dahlgren usually represented a brief stopover in the careers of officers destined either for flag rank or retirement, Smith developed a surprisingly simple system of dealing with his nominal superiors: "Never bring up who the hell the boss is. Never bring that up." Additionally, whenever a new commander reported to Dahlgren for the first time, Smith always met with him and laid out all of NWL's projects, going through each and every one of them in overwhelming detail. Counting upon the new skipper's technical ignorance, he would cynically then ask him, "Now, what is your wish on this?" He invariably received the same reply, "Carry on! Carry on!"[39]

EMPIRE BUILDING

In 1966, as Smith's revolution was aborning, McNamara's reforms finally rattled the Navy from top to bottom. That year witnessed a complete overhaul of the Navy's material management system, including its RDT&E establishment. Under pressure to centralize and systemize its R&D structures, the Navy abolished the technical bureaus and replaced them with lesser systems commands (SYSCOMs) under the Chief of Naval Material (CNM). The Bureau of Ships became Ships Systems; BUWEPS, which had been created in 1959 as an amalgamation of the Bureaus of Ordnance and Aeronautics, was split into Air Systems and Ordnance Systems; and the Bureau of Yards and Docks became Facilities Engineering. Additionally, the

Naval Material Support Establishment became the Naval Material Command (NAVMAT), also under the CNM, who in turn would report to the Chief of Naval Operations. The Navy's fifteen primary RDT&E centers, including Dahlgren, presumably rose to the same level of independence as the SYSCOMs and closer to DOD management. However, the civilian Assistant Secretary of the Navy for Research and Development (ASN(R&D)), who coordinated Navy research programs, and the military SYSCOMs, which funded the programs, argued over who would actually control the labs. As a compromise, Secretary of the Navy Paul H. Nitze placed the laboratories in NAVMAT, under the control of the CNM.[40]

McNamara approved this major reorganization on 7 March 1966, subject to congressional approval, and on 15 March Secretary Nitze formally transferred Dahlgren and the other laboratories to NAVMAT. To preempt complaints that the SYSCOMs, which still controlled project funding, might favor contractors over Navy laboratories, Assistant Secretary of the Navy for R&D Robert Morse and CNM Admiral Ignatius J. Galantin agreed, after some debate, to establish a new Director of Naval Laboratories (DNL) position. Under the plan, the DNL would report directly to the ASN(R&D), protect laboratory interests against possible neglect or misuse by NAVMAT and the SYSCOMs, and also shape, influence, and direct the course of the Navy's R&D laboratories in the future. Nitze approved the arrangement and formally established the DNL on 20 December 1965.[41]

Afterward, McNamara's Director of Defense Research and Development (DDR&D) Dr. John Foster asked Dr. Leonard Sheingold, a vice president of Sylvania Electronic Systems, to chair a Task Force on Department of Defense (DOD) In-House Laboratories. The resulting report, submitted on 31 October 1966, proposed that individual laboratories be grouped into weapons centers, with each laboratory within a center focusing on a specific military system problem area. Expertise from separate labs could be concentrated on individual system projects as necessary and marginal labs could be eliminated. Each center would then possess a "critical mass" of at least one thousand specialists involved in research and development and prototype testing. The plan called for 70 percent of the work to be devoted to in-house R&D rather than contract management. Ultimately, the idea was to bring management closer to technology and promote a new emphasis on systems engineering. McNamara concurred with Sheingold's report, and within days of the report's appearance, DDR&E Foster directed the Navy to establish weapons systems development centers.[42]

The resulting Navy plan, approved by CNM Galantin on 24 January 1967, called for laboratories with related missions to be merged into *warfare centers.* Within each center, one lead laboratory would dominate the others by controlling project funding and setting broad research agendas. The DNL, Dr. Gerald Johnson, agreed wholeheartedly with the center concept. His office argued that "a grouping of related scientific talents is a more powerful and versatile tool than any of its parts alone. A Center which provides such a grouping should therefore be capable of doing a broader and a better job for the Navy than any of its components alone could do." ASN(R&D) Robert A. Frosch had some reservations about this, but he nevertheless approved the plan on 21 March 1967.[43]

One ambitious advocate of merging laboratories into centers was White Oak's technical director, Dr. Gregory K. Hartmann. For ten years, Hartmann had been frustrated as White Oak repeatedly lost projects and pieces of projects to other laboratories, especially Dahlgren. When he first learned that a major laboratory reorganization was afoot in late 1966, he recognized an opportunity to redeem White Oak's fortunes, perhaps by capturing a lead lab role in Anti- Submarine and Underwater Warfare R&D. Hartmann was soon stunned to learn, though, that White Oak was out of the running. Recovering quickly from this setback, Hartmann submitted a counterproposal for the establishment of an all-encompassing East Coast Weapons Systems Center, comprised of White Oak, Dahlgren, Panama City, the Underwater Systems Laboratory at New London, Connecticut, and the Naval Underwater Weapons Research and Engineering Station at Newport, Rhode Island, with White Oak as the lead lab. Frosch did not fully support Hartmann's grand scheme, but he was convinced that merging NWL and the Mine Defense Laboratory (MDL) at Panama City, Florida, under White Oak's management would be "an appropriate and necessary grouping" that would result in a "first-class ordnance center." Accordingly, in his 21 March memorandum that approved the creation of centers and lead labs, Frosch ordered a study of a possible "Naval Ordnance Center, White Oak." He asked for a report by 1 July 1967. Hartmann appeared to be very close to bringing White Oak's old rival under his thumb, but the proposed Naval Ordnance Center, White Oak, soon encountered fierce resistance from its would-be subordinate, Dahlgren.[44]

Word of the merger plan reached Dahlgren in late 1966. Captain Hasler was perhaps the first to receive the news. "Deeply concerned" about the implications of the proposed laboratory "unions," he enlisted the aid of the venerable Dr. Thompson. In a letter to Thompson dated 22 December 1966, Hasler confided that he was "at a loss to see where anything can be achieved other than 'layering'" if the merger scheme was carried out. He was "confident that laboratories subordinated by this move will experience a tremendous morale problem." Hasler further believed that if NWL fell under White Oak's domain, then "much of our past efforts to improve the image and to attract and keep quality civilian and military personnel will go out the window." Consequently, Hasler urged Thompson, who still sat on a number of key Navy R&D committees and was about to attend an important advisory group meeting in New York, to throw his full weight against the plan. To arm the retired physicist for bureaucratic combat, the skipper supplied materials supporting the establishment of NWL as a "Warfare or Systems Analysis Center."[45]

In early May 1967, NWL's Advisory Council, chaired by Dr. Norris Bradbury with Dr. Bramble in attendance, closely examined the plans to create Warfare Centers and merge White Oak, Dahlgren, and Panama City. Not surprisingly, the council arrived at the same conclusion as Hasler. Those in attendance understood that if one laboratory were selected as a control center, then the others would be reduced to mere satellites, seriously undercutting the morale of the satellite staffs since the lead lab would almost certainly take care of itself first in both funds and in the choice of projects. Deeper analysis also revealed potential violations of the principles of command and management by the anticipated "Naval Ordnance Center." Under the proposed structure, technical directors of satellite labs would become second tier "associate directors" while satellite commanders could conceivably become deputies to the commanding officer of another organization. Under this organization, the council noted, "an impedance gate of intolerable proportion" would deny direct access to a satellite lab's most important customers, the commanders of Naval Ordnance Systems Command and the Naval Air Systems Command. The council therefore considered it "possible (and very likely desirable) to improve coordination between these three important activities [NWL, White Oak, Panama City] without taking the drastic step of merger and consequent subordination."[46]

NWL Technical Director Barney Smith's reaction to the proposed merger was less reflective. Although he acknowledged that consolidation made a "little sense" to save money, manpower, and resources, he considered the plan to be nothing more than "empire building" by Hartmann. When he asked the rival technical director where the headquarters would be located and where all the present directors would fit, Hartmann blithely "expressed his willingness to have them all work for him at his home base." Unmoved by Hartmann's "most admirable sacrifice," Smith assured him that he "would be happy to comply whenever he would crawl on hands and knees from his laboratory to mine to make the request."[47]

Telling off Hartmann was one thing, but convincing higher authority of the fallacy of the merger was quite another. To present NWL's case, Smith wrote ASN(R&D) Frosch an extraordinary letter on 20 April 1967 protesting the proposed plan with customary passion. Smith acknowledged that there was no doubt that Frosch had a strong inclination to consolidate White Oak, NWL, and Panama City, but insisted that Frosch get all the pertinent facts in the matter as well as Smith's interpretation of them before a final decision was made. The technical director first noted that three times over the past ten years, new layers of administration had been imposed upon defense R&D establishments, each having the character of "a wave which originated in the DOD and which is now breaking on the banks of the laboratories, with tremendous potential for disrupting the ongoing work." In each case, the new layers were introduced with the hope and the promise of greater autonomy, greater support, and better understanding for the in-house naval laboratories. Whatever else may have been accomplished by creating the DDR&E, NAVMAT, and DNL offices, Smith argued, the benefits to the laboratories had been trivial at best. As he saw it, yet another layer of management through the creation of centers only complicated matters further and would render the DNL—charged, after all, with defending the interests of all laboratories—completely ineffective in dealing with the smaller labs.[48]

Additionally, Smith complained that he could not quite understand what the master plan was, or even if one existed. He claimed to be perplexed by the talk of "problems with the laboratories, magic numbers tossed about on the minimum complement of professionals required for a first rate laboratory, and some persuasive but unfounded conjectures that bigger development programs will come to the laboratories if they are combined under Warfare Centers." Smith was particularly vexed by a veiled reference to "the problem at NWL" (possibly whispered by Hartmann), which he knew nothing about. Whatever the supposed problem was, he argued that the history of the three impositions of management showed that "trying to reorganize around them, or reducing the problems to a quest for the magic organization chart from which all the good things will automatically flow, is not the best answer and no answer at all if the game of organization chess is played too often."[49]

After attacking the reorganization, Smith shifted to NWL's defense, pointing out that it was the fastest-growing Navy in-house laboratory involved in weapons research and development, and that in the ten years since the decision was made to become a laboratory, its productive effort had increased _five_-fold, from $6.6 million to $33 million. Smith estimated conservatively that over the next five years the effort would increase to $45 million, and that under current plans NWL would reach its "critical mass" of one thousand professionals by the early 1970s. He justified this projection by citing NWL's aggressive leadership in breaking new ground in new technology areas needed by the Navy's newer weapon system concepts, which had been in such demand that there had not been enough uncommitted resources to permit the luxury of undertaking a particular weapon system development in a particular warfare area.[50]

Confronted by a wall of resistance from Smith, Hasler, Thompson, and the NWL Advisory Council, Frosch hesitated, allowing the debate over the proposed White Oak/Dahlgren merger to continue through 1967. In January, DNL Johnson met with the laboratory commanders and technical directors. NWL and MDL preferred an arrangement in which they, along with White Oak, would form equal divisions of a center headquarters at White Oak. However, they reached an agreement with DNL to support a "Naval Defense Center, White Oak," with subordinate Divisions at Dahlgren and Panama City, if White Oak would agree. Stung by Smith's earlier rebuke, Hartmann refused to consent to both the "equal entities" arrangement and the "subordinate Divisions" compromise. He stubbornly insisted that the best consolidation scheme would be "to take the White Oak Laboratory as a nucleus and absorb the other two laboratories into it in a cooperative manner." An impasse ensued. In response to Frosch's request for a report that would break the stalemate, DNL Johnson established a working group that visited the three laboratories in question and interviewed key individuals from NAVORD, NAVAIR, NAVSHIPS, and NAVMAT. Johnson's study found that "the viewpoint at these levels is, in general, that the value and desirability of forming a Center of these three laboratories is not clear, and that until it is clear, there is no great enthusiasm for making a change."[51]

Consequently, by January 1968, DNL and ASN(R&D) had deferred merging White Oak and Dahlgren indefinitely. The third participant of the merger plan, the Panama City MDL, was consolidated into the Naval Ship Research and Development Center, Carderock, Maryland (Panama City was subsequently detached and renamed the Naval Coastal Systems Laboratory in February 1972), while most other Navy laboratories, including NOTS Inyokern (merging with the Naval Weapons Center, Corona, to become the China Lake Naval Weapons Center), were likewise consolidated into new warfare or weapons centers. NRL, which had operated outside the Bureau system since its inception in 1923, remained under the control of ONR. For the foreseeable future, though, White Oak and Dahlgren would remain completely separate and independent laboratories.[52]

HITTING THE BULL

Fortunately, the case for maintaining Dahlgren's integrity as an independent weapons laboratory was strengthened by the Vietnam War. The Navy had been the first to "officially" engage the armed forces of North Vietnam during the Gulf of Tonkin Incident in August 1964, when North Vietnamese patrol boats attacked the American destroyer USS *Maddox* (DD-731). After hostilities escalated into open warfare, the Navy discovered that missiles were neither as effective nor as efficient as expected. Indeed, the emphasis on missile technology over gunnery had completely changed the surface fleet from an offensive to a defensive force. As Dahlgren weapons engineer Carl Wingo Jr. later lamented, "We lost almost any capability to conduct offensive strike warfare." This became immediately apparent when Navy gunners embarrassed themselves during early shore bombardments of the North Vietnamese coast. Thoroughly disgusted, CNM Admiral Isaac "Ike" Kidd Jr. growled that "the Navy couldn't hit a bull in the ass with a shovel!" Gun breakdowns and equipment failures only exasperated the admiral further. After Kidd dressed down Barney Smith during a presentation on gun accuracy improvement, Smith organized a task force at NWL to investigate. Given the task force's findings, Smith's conclusion was more precise than Kidd's but just as cogent:

> It turned out that for too many years the exercises at sea were organized for attaining high scores in anti-air missile firings. So much effort was put into these self-defense monsters that shipboard training with guns for attack purposes was given short shrift. As a result, the crews really were unprepared. Moreover, because the guns were not exercised as they should have been, their defects and idiosyncrasies were unknown and showed up only in a real fight.[53]

Not only were training and gun testing sorely lacking, but institutional knowledge about major caliber guns had also diminished. When the battleship USS *New Jersey* (BB-62) was recommissioned and refitted for combat duty, Wingo recalled that "most of the people that knew anything about that kind of naval gunnery work were gone. We had to go back and get some of them out of retirement." Fortunately for the Navy, there were still old-timers who had not retired at Dahlgren. As a result, NWL was instrumental in getting the *New Jersey* back into service.[54]

After the Navy's discovery that the fleet needed more guns and fewer missiles, Dahlgren's gun line roared back to life. The NWL Advisory Council soon boasted that "Guns have returned and NWL is playing the predominant role, not only in testing and proofing, but also in actual battlefield performance evaluation." To further meet the fleet emergency, Dahlgren resumed its old projectile trajectory computational function by producing new ballistic data and tables, which had lapsed over time.[55]

Dahlgren's role in the Vietnam War was not confined solely to renewed proof and testing and ballistic computation. On 18 November 1966, DNL Johnson requested that Navy laboratories apply a greater portion of their skills and technological capabilities to the direct support of naval forces fighting in Vietnam. His request became the basis for the Vietnam Laboratory Assistance Program (VLAP). Under VLAP, in-house labs provided scientists and engineers to the Navy Research and Development Unit-Vietnam (NRDU-V) and to the 3rd Marine Amphibious Force (III MAF) to establish a direct line of communication from the operating forces in Vietnam to the laboratories. This arrangement brought R&D expertise to bear on immediate operational problems very quickly. In April 1967 NWL was one of the first laboratories to become involved with VLAP, sending gun specialist Donald H. George to Vietnam to augment NRDU-V. Over the next four years, NWL sent six more representatives into the country and participated in twenty-nine formal VLAP projects for both NRDU-V and III MAF. Likewise, NWL maintained an on-call small arms and armor expert in Saigon. Among the VLAP projects successfully carried out by NWL personnel was the increased ballistic protection, crash survivability, and improved machine gun door mount for the UH-1B "Huey" Gunship; the development of gunshields and armor for PCF river patrol boats; and an investigation into possible HERO effects on Claymore mines.[56]

VLAP was so successful that in November 1969 new DNL Joel Lawson Jr. extended the quick reaction services to fleet units and allied navies under the Navy Science Assistance Program (NSAP). The initial impetus for the program expansion came from the requests from a number of naval units for a science advisor who knew the stateside laboratory capabilities and could obtain solutions to shipboard technical problems. Likewise, the so-called Nixon Doctrine, which stated in principle that the United States could no longer assume the primary defense role of each country in the free world, called for laboratory R&D assistance to foreign navies so that they could better defend themselves. To fulfill both missions, Lawson established NSAP on a trial basis and sent NWL's Barney Smith to Korea in March 1970 for a three-month tour under the program. There, Smith worked directly as science advisor for the commander of Naval Forces-Korea, Rear Admiral George Steele, and helped the South Korean Navy develop its own self-sustaining R&D capability. After Smith returned in June, DNL Lawson made the NSAP official. While VLAP ended when American involvement in the Vietnam War formally ended in January 1973, NSAP thrived, expanded, and remains in operation to the present day.[57]

One other rotating team of scientists and naval officers called the Navy Laboratory Analysis Augmentation Group (NLAAG) supported the commander of Naval Forces-Vietnam, Vice Admiral Elmo R. Zumwalt Jr., directly as part of his staff in Saigon. As Dahlgren scientist and former NLAAG representative Wayne L. Harman recalled, Zumwalt was quite analytical and realized that the usual military staff functions might not give him all that he needed to run the naval war in Vietnam. Consequently, beginning in late 1968, the Navy's laboratories began sending volunteers familiar with operations analysis to help him under the NLAAG program. NWL sent Fred S. Willis as its first NLAAG representative in March 1969, who was succeeded by Harman in June. Herb Lacayo of the Naval Radiological Defense Laboratory was already in Vietnam when he transferred to NWL in August and briefly joined Harman as an NWL representative before returning home in September. Harman remained in Vietnam until December, when his six-month tour ended and he too returned to NWL. Before Zumwalt ended the program in 1970 (when he returned home to become CNO), NLAAG produced a number of working papers, research notes, and quick reaction analyses, most of which concerned the Navy and Coast Guard's joint coastal and "brownwater" interdiction and control campaigns against the Viet Cong, respectively called Operations MARKET TIME and SEA LORDS. NLAAG also studied such things as Vietnamese Marine Corps desertion rates, drowning accidents, mine countermeasures, anti-swimmer explosives, and the Vietnamese "junk" force, thereby assisting Zumwalt significantly as he successfully closed Vietnam's 1,200-mile coastline to enemy waterborne resupply and reinforcements and took the naval war into the heart of the Mekong Delta.[58]

Beyond VLAP, NSAP, and NLAAG, Dahlgren engineers helped the Air Force improve one of the more spectacular ground support and interdiction platforms to come out of the Vietnam War, the AC-130 Gunship. Called "Puff the Magic Dragon" by soldiers on the ground, the AC-130 sported a devastating array of 7.62-mm and 20-mm rotary machine guns and 40-mm cannon. It could literally rain death and destruction down upon enemy troop concentrations and supply convoys moving down the Ho Chi Minh Trail. Earlier versions of the gunship, such as the AC-47 and AC-119, had used the lighter weapons, but in April 1971 NWL installed the largest gun ever successfully fired on an American aircraft. The 105-mm Howitzer became standard on the final gunship model, the AC-130 Spectre, still in service today. Naval weapons engineers still chuckle at the irony that Dahlgren put the "gun" in Air Force gunships.[59]

Vietnam also wrought significant changes in HERO. Up until 1968, it was strictly a safety and reliability program for the Navy's electro-explosive devices. In 1969 and 1970, Dahlgren's management finally expanded HERO's mandate to include the assessment of solid-state devices based on the recognition of a pattern of electronic failure and interference that had been observed on aircraft, missiles, and ground support equipment through ground plane testing and evaluation. As part of the mission expansion, two HERO Department engineers, Robert L. "Bob" Hudson and Frank Rose, consolidated a vast amount of missile test data and briefed twenty-four senior flag officers within the Pentagon on missile vulnerability to high-power microwaves. The ramifications of their briefing were so startling that the Pentagon, NAVSEA, and NAVAIR "flags" all endorsed a Quick Response Capability (QRC) test for all Navy missile systems to determine the severity of the problem and to redesign, rebuild, or retrofit where required. The flag officers also tasked Rose with determining the feasibility of a high-powered microwave weapon.[60]

The missile testing program was dubbed P-19 for the nineteen weeks in which the Dahlgren team had to complete its work. Rear Admiral Julian Lake of the Naval Electronics Systems Command became the overall test coordinator and was Hudson and Rose's boss throughout the testing period. The two engineers thought that meeting the P-19 deadline was "hard enough" just for testing all the Navy's missile systems, but after Navy officials briefed Secretary of Defense Melvin R. Laird on their initial findings, their jobs became even harder when he directed the Army and Air Force to coordinate their own testing efforts with those of the Navy.[61]

Despite the added stress of inter-service coordination, the Dahlgren team completed the QRC testing within the required nineteen-week period. The results showed that much more work was necessary to insure that the military's weapons would survive and function in a "stockpile to target" scenario, and led to the complete revamping of specifications, standards, and testing for future weapons systems. During P-19, Hudson and Rose also validated the need for the nation's first high-power microwave anechoic chamber to test missile vulnerability in the high intensity electromagnetic environments which American missiles would encounter at launch and in the final approach to hostile targets. Additionally, Rose's secondary microwave weapon effort did establish the feasibility of some of the program's aspects, but his findings were not sufficient to warrant a full-blown weapon development effort.[62]

In 1975 Hudson left his position as Testing, Planning, and Reporting Branch head to become the new head of the Special Projects Branch within the HERO Division. There, he teamed with Sir Reginald "Reggie" Gray, an imminent British scientist who had been brought aboard previously by Jim Payne, to bolster the HERO program. Gray needed help with a new classified initiative to look at the pros and cons of new composite materials then being integrated into commercial and military aircraft. The data generated by the research branch quickly garnered national attention within the aircraft industries. DOD classified the data and authorized further research and testing under a new tri-service project called HAVE NAME, which the department assigned to the Joint Logistics Commanders for oversight. The Navy tapped Hudson to serve as its principal representative within the newly formed HAVE NAME Joint Technical Coordinating Group. His access to a broad range of platform vulnerabilities and suppression concepts allowed the Dahlgren team to begin a number of related "special programs" that would ultimately have unforeseen yet profound ramifications for both Dahlgren and the nation's warfighting capabilities in the distant future.[63]

NWL's EMR expertise, as showcased during FR-69, led to its involvement with Project EMPASS (Electromagnetic Performance of Aircraft and Ship Systems), which lay in the arcane realm of electronic warfare and intelligence. EMPASS started after the Navy, based on its early Vietnam experience, became worried that its ship and aircraft electronic systems suffered from "unknown deficiencies" and were unnecessarily vulnerable to enemy electronic intelligence and countermeasures (EI/ECM). OPNAV Intelligence and the Naval Electronic Systems Command (NAVELEX) determined that the fleet needed an aircraft that could identify potential problems so that they could be either fixed or minimized. Consequently, in January 1967, NAVELEX tasked NWL with developing an airborne system that could assess the fleets' electromagnetic (EM) capabilities, defenses, and vulnerabilities in an operational environment.[64]

To carry out this mission, NWL's engineers, led by Dave Colby and Jim Mills, outfitted a Navy NP-2H aircraft with commercially available electronic hardware that could collect, measure, and analyze the specific characteristics of EM emissions generated during several planned exercises with the First, Second, and Sixth fleets. Based at Dahlgren's airfield, the NP-2H subsequently flew and collected and analyzed a broad range of EM data from a wide variety of fleet units. The NP-2H flights confirmed the Navy's worst fears and convinced the service that it needed an even more advanced aircraft to conduct more strenuous EM emission measurements and analyses on its ships and aircraft.[65]

In 1969 NWL started work on a second generation EMPASS system by selecting the necessary military and commercial electronic hardware, designing and building special interface units, and integrating it all into a functioning EM performance evaluation system. NWL acquired an EP-3A aircraft in 1971 and extensively modified it to accept the upgraded EMPASS system. The new EP-3A EMPASS aircraft could collect radio frequency (RF) data on ships under way and electronically characterize such things as transmitted power and antenna patterns relating to EI and ECM as well as HULTEC (hull-to-emitter correlation) vulnerabilities, which could help an enemy conducting ocean surveillance to locate and track particular vessels. EMPASS development at Dahlgren continued until July 1975, when the EP-3A was transferred to Air Test and Evaluation Squadron One (VX-1) at the Patuxent River Naval Air Test Center and placed under fleet command. Although Dahlgren relinquished control of the EMPASS aircraft, the program contributed heavily to other signals intelligence and ECM work during the 1970s such as Integrated Cover and Deception (ICAD), Anti-Radiation Missile Countermeasures, and the Navy's advanced SLQ-32 electronic warfare suite that became a standard feature aboard the AEGIS-class cruisers and other warships.[66]

NWL also applied its expertise to problems even more exotic than EMPASS. One of these involved nuclear effects testing. The 1963 Partial Test Ban Treaty had restricted nuclear tests to underground detonations only, leaving the armed services and DOD's Defense Atomic Support Agency (DASA) unable to gather data except through expensive and then imperfect underground "tunnel" tests. As a result, the armed services and DASA scrambled to build facilities that could safely simulate nuclear-scale explosive forces. In 1966 DASA, in cooperation with the Navy, constructed a "conical shock tube" at Dahlgren under Operation CONSHOT to simulate 20-kiloton nuclear blast environments above ground using non-nuclear materials. The tube was effectively the world's largest cannon at 2,600 feet long with a maximum diameter of 25 feet. To operate it, DASA and NWL scientists exploded a thousand pounds of TNT in the tube's detonation chamber, which generated shock waves that traveled at ten times the speed of sound, with a blast concentration of 20 kilotons, entirely within the tube. The tube was capable of absorbing the full explosive force so completely that only a light breeze could be felt outside it. Since no radioactive fallout accompanied the shots, the tube represented a major advance in the field of nuclear research safety. Furthermore, its convenient location at Dahlgren and efficient design made blast tests much more economical for DASA and the Navy compared to doing them with actual nuclear devices at the underground test site in Nevada.[67]

In conjunction with nuclear weapons, biological and chemical warfare (BW/CW) agents emerged as potential threats during Vietnam. As a result, a biological/chemical devices test chamber comprising a full-scale mock-up of a shipboard magazine was constructed at NWL in which experiments and tests using toxic materials were safely performed. Dahlgren's work in this area proved so important that in early 1970 it was designated lead laboratory for Navy work in BW/CW. Although confined by Presidential directive to defensive countermeasures, it is a mission area in which Dahlgren still excels.[68]

By May 1967, fully one-third of NWL's total effort was devoted to the immediate problems of Vietnam, and as the war grew in intensity, the volume of gun work grew in proportion. Rapidly maturing as a weapons laboratory, Dahlgren seized the long-awaited opportunity to formally move into gun-based weapons systems development. The occasion for this new thrust was the first "Naval Gunnery Conclave," organized and hosted by NWL in 1969. The conference had been conceived by Armament Division Chief Engineer Carl Wingo and Armament Officer Commander (and physical chemistry Ph.D.) Jim Kirschke, who wanted to know if naval gunnery still had a future. An in-house study group led by Wingo concluded that "naval guns were not worth pursuing unless we could improve the intelligence in the bullet," so Wingo and Kirschke decided to bring together naval gunnery experts from all over the country to discuss, among other things, the feasibility of "intelligent" projectiles.[69]

Despite Wingo and Kirschke's fears that "the whole world would start giving us the big laugh," the conclave decided that technology had advanced to the point at which a guided projectile was certainly possible, and endorsed its development as a new weapons system. The Armament Division took up the task, using laser-based guidance technology developed by Texas Instruments and working with 8-inch rounds and finally 5-inch rounds in the early 1970s. After the division successfully demonstrated laser guidance in 8-inch projectiles, the Picatinny Arsenal in New Jersey gave Dahlgren a $650,000 contract to design a 155-mm guided projectile for the U.S. Army. The Armament Division completed and patented the resulting design, and the Army awarded Martin Marietta a production contract using this design as a baseline. The final product ultimately became the Army's M712 COPPERHEAD 155-mm laser-guided projectile, which was capable of penetrating armor and hardened targets at ranges up to ten miles away. One of Wingo's younger colleagues, chemist-turned-engineer Thomas "Tommy"

Tschirn, who had come to Dahlgren in 1968, later laughed about the project team's enthusiasm and ingenuity: "If we had a bunch of older folks who had a set attitude about how the world worked, we'd have probably said it's stupid, it can't be done. I think we were too stupid to know it couldn't be done, so we did it."[70]

Another idea that emerged from the conclave concerned the increase of the range, as well as accuracy, of naval gunnery through the development of extended-range, rocket-assisted guided projectiles (RAPs). Although Dahlgren had undertaken earlier studies of RAPs based on German research during World War II, a private contractor and ballistics engineer named Dr. Gerald "Jerry" V. Bull argued strongly for a revitalized effort through the incorporation of guided technology. A Canadian citizen, Bull was president of the fledgling Space Research Corporation (SRC), which he had formed in early 1969 as an outgrowth of McGill University's Space Research Institute in Vermont. Bull was also a maverick, working for both the Canadian government and the U.S. Army in the early 1960s. Under Project HARP (High Altitude Research Project), he had shot fin-stabilized electronic probes to extremely high altitudes from modified 5-inch and 16-inch guns at his Barbados test range. Now he sought desperately to break into the Navy gun and projectile business, and attended the conclave expressly to push his ideas about extending the range of naval gunnery and to seek contracts for SRC.[71]

Impressed by Bull's ability to launch electronics from a gun, Kirschke and Wingo invited him to join them. During SRC's first effort with NWL, called Project FLARE, he launched an 8-inch projectile carrying a spin-stabilized infrared sensor. While in flight, the sensor successfully detected target flares floating on the ocean and telemetered angular measurement data between them and the in-flight projectile back to Bull's command console. Four subsequent contracts enabled Bull to freely pursue his concept of extended-range, rocket-assisted guided projectiles. Unfortunately, SRC's follow-up performance was marginal, as its rocket motor designs proved too impractical and costly for Navy use, and also because Bull had been forced to gain range by sacrificing projectile size and weight.[72]

NWL's professional relationship with SRC ended abruptly on 14 August 1970 when the final contract was canceled only a month after it was issued. The ostensible reason given for the cancellation was a lack of funding, but in reality Wingo had detected a duplication of effort among SRC, Dahlgren, and Texas Instruments, which was much further along in its laser guided projectile research and therefore more attractive as a contractor. Moreover, Bull's inability to cope with byzantine Navy contracting regulations had resulted in the return of several of SRC's unsolicited proposals and some hard feelings among his project managers. Jim Colvard believed that SRC had "difficulty in recognizing that as a contractor rather than an in-house laboratory they must operate on a competitive basis." The personal friction between the upstart SRC ballisticians and Dahlgren's veterans in both the laboratories and on the gun line heightened tensions further, and a parting of the ways therefore seemed perfectly in order.[73]

The separation turned ugly when SRC sued NWL, accusing Dahlgren engineers of stealing proprietary information and harboring an "NIH" (Not Invented Here) attitude, in which ideas and proposals from outside NWL were either derided or dismissed out of hand. Bull's Washington lawyer took the issue into the stratosphere of the defense establishment, involving not only CNM Kidd and Navy Secretary John Warner but also Defense Secretary Melvin Laird and various other high level military officers. At the storm's height in March 1972, Bull, perhaps realizing that SRC was being hurt as much as Dahlgren, reached out to Barney Smith to make peace. After some difficult negotiations, colored by further angry broadsides up and down the command chain, NWL finally agreed to settle with SRC out of court for $100,000, without admitting culpability in the affair.[74]

Smith had never held Bull's ideas in high regard. When CNM Kidd asked for his opinion on the Canadian's original proposal, Smith's answer appropriated one of the admiral's own metaphors: "When you can't hit a bull in the ass with a shovel, it doesn't make much sense to extend the handle and reduce the scoop." Smith meant that range was a false issue since gunners had to clearly see and target an enemy before successfully hitting him. Improved fire control, then, was the real issue rather than extended range, especially since smaller projectiles required greater accuracy. Since "K" Laboratory had already enjoyed phenomenal success in handling digital fire control for the Navy's strategic missile force, why could NWL not extend that experience into the gunnery realm? The idea was a good one since mating fire control to gunnery as part of a complete weapons systems package fit DOD's total ships systems engineering concept perfectly and was a mission that Dahlgren was uniquely qualified to handle.[75]

Smith also recognized that sensors, the "eyes and ears of the fleet," represented the field of the future. In modern naval surface warfare, detecting and engaging an enemy before he could reciprocate was of paramount importance to the Navy. HERO had already given Dahlgren the tools to successfully conduct R&D in electromagnetic warfare, a capability best demonstrated in early 1968 when NWL built a replacement for the captured spy ship USS *Pueblo* (AGER-2) in only ninety days. Further, NWL had developed the capability to monitor with electro-optical devices not only the movements of friendly naval forces but also those of the enemy. Smith therefore saw no reason why sensors should not become a primary task for NWL under a greater surface warfare mission.[76]

Consequently, in 1968 Smith moved to bring gun fire control, electromagnetic warfare, and electronic sensors formally under NWL aegis as part of a more ambitious effort to have Dahlgren recognized as the lead laboratory for Surface Warfare. In an internal report prepared under the auspices of a "Mission Analysis Panel" and entitled *Recommendations for the Development of the Naval Weapons Laboratory: Mission, Organization, Program*, Smith argued, among other things, that Surface Warfare was an area of vital interest to the Navy and, except for the antiaircraft defense problem, had lagged sorely behind other warfare areas in terms of technological support. The need for an infusion of modern technology in surface warfare was not only recognized at higher echelons of the Navy and Marine Corps, but also by the President's Science Advisory Committee. Since no other laboratory had assumed a leadership role in this area, Dahlgren was the obvious facility for the mission. He thus concluded that "We should concentrate initially in the area of gunnery, further develop our capability in the areas of fire control and guidance, and eventually expand into the broad area covering all surface launched weapons."[77]

Meanwhile, Smith was aware that the shotgun wedding with White Oak had only been deferred and would likely happen sooner or later. Although he opposed centers on principle, Smith moved to ensure that Dahlgren would be in the best possible position to retain a dominant status after the inevitable happened and the honeymoon was over. To further this goal, he believed that NWL had to begin manifesting "many of the attributes of a center," specifically by focusing its efforts on important and visible responsibilities to the Navy and by stressing assignments related to the overall mission, ideally surface warfare. Likewise, all of NWL's organization components had to mutually support each other and focus on the prime responsibilities of the laboratory as a whole. In short, Dahlgren had to become its own center in everything but name.[78]

In July 1968 Smith and Hasler formally requested from CNM a change in mission and major tasks, in which Dahlgren would "conduct a program of analysis, research, development, test, evaluation, systems integration and fleet engineering support in surface warfare." Under the new mission, NWL would also handle all tactical and strategic warfare analyses and geoballistics projects, and also become the lead laboratory for electromagnetic warfare. Although most of NWL's customers had no objections, the naval RDT&E

bureaucracy was slow to approve the step, partly because of the continuing Navy laboratory mergers and consolidations, and partly because of White Oak interference. In the meantime, Smith embarked on a complete reorganization of NWL to keep his own troops happy, and incidentally to complete his revolution.[79]

Favoring an extremely loose management structure, Smith generally had little use for organizational charts, and even less for reorganizations. He had left Lyddane's old system in place when he became technical director in 1964. As the first few years passed, and after granting his program managers a great deal of managerial freedom to launch projects that made sense for the Navy, he found that new organizations began to coalesce around them in an "evolutionary way." In 1968, Smith determined that the time was right for a real, meaningful reorganization when a number of his bright young engineers, who had come to Dahlgren with a "fighter pilot attitude," as Jim Colvard characterized it, complained to him that "we're disorganized, and that we ought to get organized." Smith agreed, and allowed them to do the work of studying the present organization and proposing a new one. To his delight, the resulting new management structure formalized what had already taken shape. The old tri-laboratory system was abandoned in favor of a systems-oriented departmental scheme. "K" Laboratory became "K" Department (Warfare Analysis) and retained responsibility for strategic and tactical warfare analysis, geoballistics, satellite geodesy, and digital fire control and targeting computations for strategic systems. The old "T" and "W" Laboratories were both disbanded, and their functions were distributed among the new "G" (Surface Warfare), "F" (Advanced Systems), and "T" (Test and Evaluation) Departments. "G" Department was charged with a key mission, that of surface warfare systems, which included weapons systems, gun fire control, warhead R&D, and armor. "F" Department took responsibility for, among other things, HERO, electromagnetic warfare, sensors, chemical warfare, lasers, and materials research. "T" Department took over the old proof and acceptance mission and handled range operations, while "E" Department became responsible for base systems maintenance and shop services. Dahlgren's administration was divided between the military command staff in "C" Department and the technical civilian management in "D" Department, while a host of other administrative units, such as "A" (Comptroller), "W" (Public Works), and "H" (Medical) Departments, were created or reorganized to manage on-base services.[80]

The new organization was up and running by the summer of 1969, and Smith was content that "the revolution had taken place in an evolutionary way." The reorganization gave Dahlgren an added flexibility, permitting quick reactions to fleet emergencies, the most important function of a military laboratory. Official recognition of this flexibility came in 1970 through Dahlgren's participation as the Navy representative laboratory in the experimental, service-wide DDR&E Project REFLEX, which removed personnel controls and allowed technical directors to match staff levels to budgets.[81]

Dahlgren's thrusts into weapons systems development, gun fire control, electromagnetic warfare, and sensors were finally rewarded in 1972, when the Navy formally assigned the surface warfare mission to NWL. This was Smith's crowning achievement, since his tenure was drawing to a close. In 1973 the Navy awarded him the Distinguished Service Medal for outstanding accomplishments as Dahlgren's technical director. To encourage future "rebels" to make exceptional scientific or technical contributions despite unusual odds or significant bureaucratic opposition, the Navy also created the annual Bernard Smith Award. Smith interpreted these honors as "gentle notices by the fates" that it was time to go, so in June, after twenty-five years of service to the Navy, he retired.[82]

Smith's revolution had not only brought Dahlgren to the forefront of weapons systems development in the Navy but had also led to its ascension as the lead laboratory for surface warfare. By 1973, as the Vietnam War ended, NWL had become a Navy center in all but name, its budget rising from $24 million in 1964 to more than $100 million at Smith's retirement. Dahlgren was thus well positioned to maintain its identity and administrative control once the merger with perennial rival White Oak was finally resolved. Additionally, Dahlgren's new excellence in surface warfare systems and sensors would play a vital role in restructuring a badly bruised Navy around a new total ship system concept called AEGIS. This new breed of warship would greatly influence fleet doctrine in the post-Vietnam period and later during the defense buildup of the 1980s, when the U.S. Navy aggressively confronted the Soviet naval challenge and found important new missions in the increasingly turbulent Middle East.[83]

Chapter

6

On the Surface, 1973-1987

Barney Smith's revolution had left Dahlgren a much larger and far more dynamic Navy facility than it had been when he arrived in 1964. The changes in organization, management style, and culture that he had overseen during his nine-year tenure as technical director had fully prepared Dahlgren for the difficult post-Vietnam years. The accomplishments of Smith's revolution endured after his protégé James E. Colvard took Dahlgren's helm in 1973, just as the U.S. Navy began adapting, through systems engineering, to a new era of warfare that re-emphasized surface combat in American naval doctrine, which had been shaped almost exclusively by the carrier and submariner communities since the end of World War II.

THE CLASSICAL MANNER

Jim Colvard, like Barney Smith, was a China Lake alumnus. While working as a division director at China Lake, he realized that his future lay in administration, rather than in science or engineering. His superiors at China Lake praised his work, yet seemed to offer no avenue for advancement. Colvard realized that the senior positions at China Lake were tightly held by the same men who had run the station when he had first arrived eleven years before—he seemed locked in middle-management limbo.[1]

Early in 1969 Barney Smith threw Colvard a lifeline. Smith had heard about Colvard's fine technical work and his skills as an administrator. Always on the lookout for new blood as part of his ongoing campaign to "revolutionize" Dahlgren, Smith called Colvard and asked if he was interested in accepting a position at NWL. Colvard hesitated. While working for the Bureau of Naval Weapons a few years before, he had visited Dahlgren in an effort to sell a fully funded digital fire control program to the laboratory. Management had turned down his offer of free money, arguing that Dahlgren was too busy testing ammunition, and Colvard had left with a bad impression, exclaiming on the way out, "What a stick-in-the-mud outfit!" Accordingly, Colvard's initial response was, "Why in the hell should I leave the best laboratory in the Navy to come to a backwater outfit like Dahlgren?" Smith persisted, "Well, you ought to come and look at the place; we've made some changes." So on his next trip to Washington, Colvard drove down to Dahlgren out of courtesy, expecting only to pay his respects to Smith and depart. He got lost on the way, but once he arrived, his opinion began to change.[2]

At the headquarters building, Colvard reviewed the personnel records and was pleased to find that NWL had been hiring bright young people from good schools and was steadily increasing its capabilities and product lines. He was particularly impressed with the enthusiastic attitude that permeated NWL's departments and also noted with approval that Smith had imposed China Lake's system of management rotation on Dahlgren. After some quick reflection, Colvard decided, "What the hell, I've been in the desert long enough!" and accepted the new challenge. He began as head of the Electronics Warfare Department ("F" Department), his area of specialization at China Lake, but under the rotation system he soon moved to "G" Department to head up the work in "guns and bullets."[3]

Colvard, like Smith, was a rebel. Raised in Robbinsville, in western North Carolina, he graduated in 1958 from the small but excellent Berea College in Kentucky, with honors in physics. He spent a year employed at the Johns Hopkins Applied Physics Lab working on the design of the AN/BRN3 satellite tracking system, and then seven years managing programs at China Lake. Frank, outspoken, and able to negotiate the bureaucracy to get things done, Colvard was Smith's kind of manager.[4]

One episode early in Colvard's tenure at Dahlgren demonstrated his irreverent attitude and his ability to deliver for the Navy. While working as head of "G" Department, he received a call from a bright and energetic researcher at White Oak named Applebaum. White Oak had heard that Dahlgren was developing a guided projectile and was taking bids for fuze design work from the Army's Harry Diamond Laboratory, the Navy's China Lake, and White Oak. Applebaum told Colvard that this was not possible.

"Why not?" Colvard asked.

"Well, we have the charter to build fuzes for the Navy," Applebaum replied correctly.

"You can take that charter and dispose of it in the classical manner," said Colvard. "If you've got the best idea, great; if you haven't got the best idea, forget it."

Taken aback by Colvard's bluntness, Applebaum asked to speak with Barney Smith. Colvard got Smith on the line and complained, "This guy tells me I've got to do business with him just because he's got a charter." Smith replied, "Tell him to dispose of it in the classical manner." Smith and Colvard spoke the same language.

Colvard remembered that he was not necessarily angry with Applebaum, but he simply did not believe that because someone wrote down that "you'll do so and so," it made you capable of doing it. It helped, he always said, to know what you are doing. As it happened, that particular piece of fuze work was given to the Harry Diamond Lab, and its technical staff did a fine job. Colvard thought the episode illustrated a more serious problem at White Oak—what he called "mission mania," or "achievement by charter." By contrast, he believed, young researchers at Dahlgren had a competitive "fighter pilot" attitude. They saw a problem and competed to get the work to solve it.[5]

Colvard, like Smith, believed that leadership and management were crucial to a successful laboratory. The leadership would help identify and define problems. Once that theoretical work of defining the nature of the problem was done, the technical work of finding a solution would follow. The hard part was getting the problem defined and acquiring the funding to solve it. Researcher-managers who were aboard at Dahlgren under Smith's leadership had the go-getter attitude that enabled them to acquire new work that had never been "chartered" to Dahlgren. During the period, as the laboratories attempted to establish their roles under the McNamara reforms, all of them tried to develop mission statements that clearly separated and defined the laboratories to avoid duplication. But a little duplication and a little competition, Smith and Colvard believed, were good for the Navy.[6]

Smith and Colvard shared much more than a disdain for charters. Colvard, as a young North Carolina mountaineer, was culturally quite different from Smith, but they identified with and respected one another. Neither was the son of privilege. Colvard, like Smith, had worked for a living and had succeeded on the basis of talent, intellect, and ambition, and not because of family contacts or prestigious academic credentials. Colvard's education at Berea College in Kentucky, like Smith's at Reed College in Oregon, was at a small liberal arts school. Both schools prided themselves on cooperative work programs, highly qualified teachers, and a very bright student body. But as undergraduate colleges, neither school offered a network of well-placed doctoral graduates who could help provide contacts, access, and influence for alumni. While Cal Tech and MIT each had their old-boy networks, Berea and Reed did not.

The two men shared more than parallels in background and education. Both were good at identifying talent in others and were highly skeptical of the power of organization charts. Both believed that the Navy would be mistaken to rely on specialized, or "monopolistic," facilities, because without the driving motivation of competition, organizations and individuals would become intellectually lazy and complacent. Colvard fully supported Smith's ideas about management rotation, shared his respect for risk takers, and also hoped to keep the organization flexible enough to set up new project teams that could take on new work as opportunities arose. Smith, looking back on his management style in 1979, noted that "The death knell of innovation is the rigid control imposed by those who have found the 'right way.'" Colvard agreed.[7]

In 1972, after Colvard had served two years as head of "G" Department, Smith made him assistant technical director. When Smith retired in 1973, both Colvard and Charles Bernard were in the running to replace him. It was a close decision, but the Navy selected Colvard. Bernard was disappointed, but Colvard believed that he showed "a lot of guts" in taking the news graciously. As it turned out, Chuck Bernard and Jim Colvard remained good friends and continued to work well together as Colvard took the helm at Dahlgren.[8]

As the new technical director, Colvard extended the rotation system to include lower level managers as well. He maintained the active recruiting program and sought young leaders who would aggressively target projects as if they were fighter pilots in battle. His efforts confirmed Smith's revolution and extended the trend into the future. In later years Rob Gates credited Barney Smith, Jim Colvard, and their successors in the Technical Director position as representing "a long series of people with good vision and the knowledge and experience and willingness to go fight for things we needed." The leadership and advocacy by this handful of individuals explain not only the institution's survival through the decades of change following the Vietnam War, but its response, adaptation, and growth to become even stronger and more crucial to the Navy.[9]

A SHOTGUN WEDDING

Colvard had no sooner become technical director than a challenging managerial mandate landed squarely in his lap. Soon after taking office in August 1973, the new ASN (R&D) Dr. David Potter spoke at an honors luncheon at Dahlgren. Afterward, he drew Colvard aside to the bar at the station's club and said that he had decided to merge Dahlgren and White Oak into a new warfare center. He also wanted to make Colvard the center's technical director, if he was willing. If not, then he could continue at Dahlgren under whomever Potter selected as the new center's technical director. Like Smith before him, Colvard harbored strong feelings against consolidation. He believed that if a merger were necessary it would be more logical to merge White Oak with the laboratory at Carderock. Both were engaged in similar scientific research, and the two were only twenty minutes apart (when the beltway traffic was light), rather than the seventy-five minutes it took to travel between Dahlgren and White Oak. Moreover, as he later recalled, there was less than 10 percent duplication in the work done at Dahlgren and White Oak, and merging them would not "save any money or be more efficient because we had 3,000 people at each place, and it would cost just as much to heat and light the facilities and administer them as a center as it would for two separate laboratories." It nevertheless became quite clear to Colvard that Potter's decision was final, and with both Smith and Hartmann retired, the deferred shotgun wedding between Dahlgren and White Oak was back on again, this time for good. Faced with the choice of either leading the new center or being reduced to a subordinate status, Colvard accepted Potter's proposition, adding that if it had to be done, he would rather do it himself, because he did not have a parochial interest, because he understood the problem, and because if it could be done at all, he believed that he should be the one to do it. Potter was pleased, replying to Colvard that "You're just cocky enough to take a crack at it."[10]

Potter's decision was not taken lightly. He was interested in realigning the R&D community and improving its image as well as adjusting the shore establishment to reflect post-Vietnam fleet cutbacks. Additionally, Colvard believed that the consolidation choice in favor of Dahlgren and

White Oak rather than White Oak and Carderock was politically motivated. Congressmen in New Jersey and elsewhere had successfully lobbied against closing facilities in their states, and the Dahlgren/White Oak merger could be accomplished without layoffs or instigating a political backlash. However, there were deeper reasons behind the merger.[11]

In 1961 the Bureau of Naval Weapons had listed WOL as the lead laboratory for underwater ordnance and anti-submarine warfare, but many other laboratories acquired aspects of work for those areas by the end of the 1960s. Consequently, White Oak had failed to associate itself at a critical juncture with one of the Navy's strong blue-suit "unions": air, surface, and submarine. Instead, it had picked up a variety of minor missions such as fuzes, anti-mine warfare, underwater demolition team (UDT) weaponry, and small craft armament, which had no strong blue-suit sponsors and simply did not match the emerging major warfare areas from Chief of Naval Material Isaac Kidd's perspective. Therefore, in the laboratory consolidations envisioned under McNamara's reforms, defending the role and mission of WOL became increasingly difficult. By contrast, though, and through Smith's untiring efforts, Dahlgren had become solidly identified with the surface warfare mission.[12]

Furthermore, White Oak's stagnant management attitude had rendered the laboratory inefficient and resistant to change. Excessive layers of management and an overabundance of GS "super" grades among the laboratory's staff had compounded the problem, and in the view of many in the Navy Department, White Oak's aging senior managers, all at the terminal stages of their careers, had lost touch with the mainstream of the Navy. So ironically, instead of saving money or gaining efficiency, Potter's real objective was to save White Oak and its first-class technical staff through consolidation in the hope that Dahlgren's younger and more dynamic management team would infuse that lab's energetic management environment into White Oak.[13]

With Colvard aboard, Potter directed Kidd to make the merger happen. Needing documentary support and statistical evidence to present to SecNav in favor of consolidation, the CNM's first move was to commission a study by an ad hoc committee chaired by Barney Smith, who remained a private consultant to NAVMAT under a personal services contract, to examine "the economic and programmatic merits" of consolidation. Not surprisingly, the resulting *NOL-NWL Consolidation Study* of 12 October 1973 found few incentives to merge and recommended an examination of other possible consolidation schemes. It also suggested that if personnel and operating cost reductions were the primary aims of ASN (R&D) and CNM, then closing either one laboratory or the other was preferable to forcing a consolidation "without conviction of its merits" since recent experience had shown that lab mergers had not in fact achieved notable manpower and funding reductions but had required in some instances <u>increases</u> before consolidation could be effected.[14]

Frustrated with the findings, Potter asked the Naval Ordnance Systems Command (NAVORD) to study the impact of creating a Surface Warfare Center of 4,000 personnel by transferring WOL's surface warfare programs and staff to Dahlgren and moving the underwater programs elsewhere, while leaving the aeroballistic and Defense Nuclear Agency (DNA) Casio Facilities at White Oak as an annex. The study, completed in a week and entitled NAVORD Concept of Operation Under a NOL-NWL Consolidation, was submitted on 29 October 1973 and tacitly endorsed Potter's plan by only analyzing the process and impact of consolidation without questioning its assumptions. Other studies were subsequently generated by one side or the other that cast some doubt upon the benefits of merging the two laboratories, but the die had already been cast.[15]

Once all the studies were completed and the bureaucratic obligations fulfilled, Kidd made the formal request on 26 July 1974 for the consolidation of the Dahlgren Naval Weapons Laboratory and the White Oak Laboratory into the new Naval Surface Weapons Center (NSWC). The proposed effective date for the merger was 1 September, with "complete consolidation" by July 1975. Navy Secretary J. William Middendorf II approved the merger, and NSWC was born. Colvard was appointed technical director of the new NSWC and served in that capacity until 1980, when he was promoted to Deputy Chief of Naval Material, the senior civilian position at NAVMAT. White Oak's commanding officer, Captain Robert Williamson, who was senior to Dahlgren's Captain Conrad J. Rorie, became NSWC's first commanding officer, but after Williamson's retirement in March 1975, Rorie moved up in succession, giving the center's top management a thoroughly Dahlgren face.[16]

"CENTERIZING" THE CENTER

Many at White Oak feared that the "barbarian from North Carolina" would not support their research. But Colvard took charge by not taking charge. He recognized that White Oak's two previous technical directors, Drs. Ralph Bennett and Gregory Hartmann, had been competent scientists professionally and excellent men personally. Both were micro-managers, though, and their department heads had always deferred to the senior scientist as the technical decision-maker, even for matters involving a few thousand dollars.[17]

Colvard knew that a different approach was in order. He met with the department heads and explained that he did not regard himself as a brilliant scientist, and that in a large-scale "corporation" like the merged laboratories, nobody could possibly know the technical details of all of the disciplines represented anyway. He explained that power should be knowledge-based, and that since no one could be an expert in such widely separated fields as underwater hydrodynamics and fundamental chemical synthesis, power had to be diffused. The result was traumatic, but not in the way that White Oak researchers had anticipated. They not only retained authority over their work, they now had responsibility for it. All of a sudden, the senior managers had to make all their own decisions and were responsible if the decisions turned out badly.[18]

Colvard emphasized that department heads had to play a dual role: first as advocates of their component of the organization, and secondly as the corporate decision-makers who would take responsibility for allocating resources. Gradually, the trauma of decision-making wore off, since Colvard would publicly support the decisions, even if he had private reservations. If he erred, he thought in retrospect, it was in "under-management" rather than in over-management. He preferred to let ideas and initiatives of individuals find expression rather than constantly try to keep a lid on them. If mistakes were made, wrong pathways of research were pursued, or ideas did not work out, Colvard himself took responsibility. He was criticized for under-managing, but his under-management, he claimed, was intentional.[19]

After the unification of the two centers, the now joint NSWC continued some of the orphan programs from White Oak. Mine warfare, for example, remained an NSWC specialty, but by the end of the decade it still was a neglected and underfunded piece of the broader naval picture. Admiral Albert Monger, commander of the Mine Warfare Command, served as keynote speaker at a White Oak conference in 1979. He admitted the "sorry state" of mine warfare, and he regarded a major part of his mission as educating others on the importance of mines and countermine measures. He had no budget for mine development, yet had to assist the fleet commands in their mine requirements. Monger noted that "competing for the limited budget dollars with other Navy weapons is one of our most difficult problems."[20]

Other administrative problems soon surfaced in the newly formed NSWC, particularly with "stabilized rates," which required budgeting two years in advance. Management found that stabilized rates led to "disincentivizing" people. The reason was that when a budget was carefully constructed and then work was done at a cost under the estimate, the "profit" was simply taken back by the program officers and reassigned, often to another laboratory, or even to the Army. Thus, when people had been urged to be lean and mean and to operate an efficient organization, the reward was that the funding was transferred to a poorly managed operation. It was disheartening, to say the least.[21]

As previously described, Potter wanted Colvard to instill the dynamic management attitude that existed at Dahlgren as well as the tradition of management rotation into the staff at White Oak. Three years into the experience, Colvard felt he had partially achieved the goal of creating a new management environment. Both Captain Rorie and Colvard worked to preserve the separate identities of the two laboratories as much as possible. They recognized that people would identify with local units by nature and tried to restrict the number of individuals who would be required to travel the seventy-two miles between White Oak and Dahlgren. In addition to Rorie and Colvard, only Dan Shields and Len Klein, the associate technical directors at the two facilities, had to do any commuting between the facilities. The concept was to formulate common policy and common objectives and permit local execution.[22]

Nevertheless, tension surfaced over the question of whether White Oak had been absorbed by Dahlgren or vice-versa. The headquarters location would naturally be perceived by outsiders as the senior partner in the merger, just as in corporate mergers. Perhaps with this sensitivity in mind, the Naval Surface Weapons Center's postal mailing address for the first two years was White Oak, Silver Spring, Maryland.[23]

However, Rorie and his successor, Captain Paul Anderson, both advocated shifting the official headquarters to Dahlgren. For one thing, the Dahlgren location had some 3,500 employees, while the White Oak facility had about 1,000. For another, by 1976 Dahlgren had been designated by the Secretary of the Navy as the lead laboratory for the proposed new AEGIS Combat System, which, Rorie argued, would require more focus by the commanding officer. Additionally, more than 160 military personnel were stationed at Dahlgren, compared to only 10 at White Oak; the work at Dahlgren was more "fleet-interactive," demanding more of the commander's time; and less time traveling between the sites would mean more time available to the more crucial duties at Dahlgren. No one mentioned that the commandant's house at Dahlgren, built under such controversy in 1919 and 1920, still remained one of the most attractive homes for officers of any rank in all of the services. A "flap" ensued over the headquarters location throughout 1976, but the Navy finally resolved the issue in Dahlgren's favor. Consequently, in October 1977, Rorie's successor, Captain Anderson, moved into the spacious mansion overlooking the golf course and the Potomac River.[24]

When Anderson came aboard, he and Colvard agreed to hold a management retreat to map the future of the combined laboratories. Anderson believed the two labs suffered from a lack of "horizontal communication," which led to some duplication of effort. He later recalled asking a White Oak researcher involved with fuzes how he interfaced with the researchers at Dahlgren working on other aspects of fuzes. The researcher replied, "I don't think Dahlgren has [any work] about fuzes." The fact that he had never heard of colleagues working on the same system seemed symptomatic, and Anderson set out to correct the situation.[25]

Colvard agreed that the issue needed to be addressed. Meeting in Coolfont, West Virginia, the Board of Directors of NSWC decided that the nine technical departments should be restructured as seven, each with units at both sites. The seven proposed new divisions were Research and Technology, Engineering, Underwater Systems, Weapons Systems, Electronic Systems, Combat Systems, and Strategic Systems. These departments were established, and an eighth department, "Protection Systems," was subsequently added, responsible for systems safety, magnetic silencing, and electromagnetic/nuclear effects. Anderson hoped that creating the departments with divisions and branches at both facilities would help integrate and complete the merger.[26]

However, an Inspector General's report in the summer of 1977 criticized the separate management structure of the laboratories and mandated Anderson and Colvard to produce a more workable and unified management scheme. Anderson and the Board of Directors therefore decided "to complete the merger" by eliminating the separate management structure of the Dahlgren and White Oak sites, and to establish for Operations a Deputy Commander and a Deputy Technical Director to serve in a line management capacity over all technical and support departments of the center. Additionally, for Weapons Evaluation, a deputy commander and associate technical director would be established to serve in a staff capacity to the commander and technical director. These two individuals would also be responsible for all technical staff functions dealing with fleet and Marine Corps liaison and intelligence in an effort to focus center efforts toward improved weapons assessment and planning. Although formal approval for the new organization might take some time and considerable groundwork needed to be done to make it happen, Anderson set 1 January 1978 as the stand-up date. Aware of the negative impact that the reorganization would likely have on staff morale, particularly in light of the recent headquarters "flap" and Colvard's and Captain Rorie's past efforts to preserve Dahlgren and White Oak's separate identities, Anderson emphasized that these actions were not taken to reduce personnel or to relocate employees from one site to another, but were solely in response to the Inspector General's mandate to produce a structure that would allow the NSWC to operate as a "totally coordinated and effective activity."[27]

At the community level, Anderson also acted to establish some sense of overall identity for the newly merged Dahlgren and White Oak within the greater weapons center. Citing the Inspector General's report, which argued that "the Navy needs a Surface Weapons Center even if we must occasionally sacrifice the image of the two laboratories," he accordingly abolished each laboratory's familiar and long-trusted news sheets (the *Lab Log* and the *Oak Leaf*, respectively) and arranged an internal competition to name a single, new in-house newspaper that would henceforth serve both locations. The winning entry was *"On the Surface,"* which carried the usual mix of human-interest stories about staff, news of major technical programs, and selected reprints of important speeches, interviews, and announcements by major figures in the Navy's RDT&E community. Anderson hoped that the glossier *On the Surface*, first published in May 1978, would advance the cause of "centerizing" the NSWC by helping its personnel "focus on the totality of our work for the Surface Navy" while simultaneously serving as a continuing vehicle for both internal and external communication among the center's management, employees, dependents, customers, and DOD and congressional patrons.[28]

At the second management retreat after Anderson's arrival, held in Reston, Virginia, Colvard, Anderson, and the Board of Directors reviewed the initiatives they had put in place. These included a focus on combat systems integration efforts and increased involvement in the antisubmarine warfare area to offset declining work in the torpedo area. One bright spot was the coordination of "re-entry work" in the strategic systems department between White Oak and Dahlgren. Colvard believed that the integration of the two centers was progressing well by 1978. Even so, he remained concerned that the period of detente and the cuts in naval expenditures made it hard to attract new personnel. He argued that "the character of the R&D organization is driven by our ability to provide the training ground for young engineers which has to involve hands-on experience. With fewer new products, it's more difficult to maintain that environment." Colvard continued to believe that Dahlgren's greatest strength was its personnel, and in the face of hiring freezes, the lack of hands-on experience would make it difficult to keep up their caliber.[29]

For many of the mid-level managers who worked under Colvard, the experience was gratifying. Charles Eugene "Gene" Gallaher, who joined Dahlgren out of college in 1968, the year before Colvard arrived, later recalled just how Smith and Colvard both fostered innovation. "Our technical directors could have said to us, 'Hey, don't work on that, it's outside your area of responsibility.' But unless they saw us wasting money and duplicating effort in ways we should not be doing, there was a lot of freedom there to come up with good ideas for the benefit of the Navy and of the nation." The contributions of Dahlgren to improving specific weapons like TOMAHAWK, first employed in combat during the Gulf War in 1991, could be traced back, Gallaher firmly believed, to the creative work environment at Dahlgren established more than a decade earlier. The history of work on ballistic missiles, on TOMAHAWK, and on other systems bears out his sentiment.[30]

GUIDANCE BY THE STARS

"K" Department (Strategic and Strike Systems) at Dahlgren played a central role in upgrading and modernizing the Navy's leg of the strategic nuclear deterrent—the fleet ballistic missile (FBM), or, as it was later called, the submarine-launched ballistic missile (SLBM). The work at Dahlgren on the guidance systems for the SLBM remains a little-told story, because like much of the earlier work on sensitive weapons system, many of the details are classified. However, the broad outline of the program and the nature of its problems did find their way into the open literature. As Rob Gates, who devoted more than a decade to missile guidance engineering, remembered, the five branches working on SLBM had a total of about 75 people when he joined the team in 1971. By 1980 the work was done by two divisions with a total of about 300 people. The single branch he had entered with about 14 people had grown to about 42 by 1982.[31]

With its computer capability enhanced by the arrival of new CDC and Cray machines, and intimate understanding of the POLARIS and POSEIDON targeting issues, Dahlgren was ready when the Defense Department began developing the next generation of SLBMs during the 1970s. Called TRIDENT, the new SLBMs required significant increases in both range and accuracy during the 1970s and, consequently, "K" Department had its work cut out for it.

The study phase that led to TRIDENT began in 1966, with a report that recommended the "undersea long range missile system," or ULMS in military acronym parlance. ULMS would require a larger submarine, and the projected schedule for new submarines meant that ULMS would not be introduced until the 1980s. Dahlgren helped the Special Projects Office (renamed the Strategic Systems Project Office (SSPO) in 1968) and the Chief of Naval Operations staff to define the basic requirements for ULMS. The Secretary of Defense endorsed ULMS in September 1971, with the requirement that it be capable of a 4,000-nautical-mile range, with the ULMS I to be deployed in POSEIDON class submarines and the ULMS II to be deployed in the next class. In 1972 ULMS was renamed TRIDENT.[32]

Flight testing for the TRIDENT I (C-4) SLBM began in January 1977, and the first deployment of a TRIDENT I missile was aboard a converted *Benjamin Franklin* class POSEIDON boat, the USS *Francis Scott Key* (SSBN-657), in October 1979. Although the Navy deployed the first submarine designed solely for TRIDENT I, the USS *Ohio* (SSBN-726), in September 1982, the service had already begun work in October 1980 on the TRIDENT II (D-5) missile, which could be similarly carried on *Ohio* class submarines without major modification. Both TRIDENT I and II had ranges over 4,000 nautical miles, but TRIDENT II represented an improvement in payload capacity, and the *Ohio* class submarines were quieter, more capable, and stealthier than the earlier *Layfayette* and *Benjamin Franklin* classes of submarines. Congress had authorized the *Ohio* as the lead boat in her class in 1974, and her keel was laid in April 1976 at the Electric Boat Division of General Dynamics Corporation in Groton, Connecticut. The *Ohio* was launched almost exactly three years later, and she underwent a successful series of sea trials in the summer of 1981 before the Navy commissioned her on 11 November of that year. Nine more SSBNs of her class were subsequently commissioned between 1982 and 1989, and thus TRIDENT became the mainstay of the SLBM fleet through the heightened or "Second Cold War" that lasted from about 1978 through the mid- to late 1980s.[33]

TRIDENT SLBMs (I and II) were particularly lethal because of their Dahlgren-developed, "stellar aided" inertial guidance systems. This involved the missile taking a star sighting before releasing the re-entry vehicles or weapons. A specific star would be located and an onboard computer would correct for statistically known errors to estimate the position, velocity, and orientation of the missile and then send the data to the guidance computer. Dahlgren engineers analyzed the potential accuracy of the system, provided the computations to select the optimum star for accuracy, and developed the

"W-matrix" system to compute the error likely between launch point and target point, as well as an operational star catalog.

Each of these developments took Dahlgren into new areas. Since the position of a star in the sky is a function of time of day, launch point, and trajectory conditions, all changing as the submarine moves, the computation had to be made at the moment of launch. The operational star catalog had to include stars that exceeded a minimum brightness and were relatively constant in their brightness, and that also were well separated from other nearby stars that might be misidentified. Furthermore, the position of each star in the catalog had to be clearly predictable. Existing star catalogs developed by astronomers did not meet all of these specific requirements, and in fact, when the catalogs already in use were compared, many discrepancies appeared. Dahlgren specialists developed a thorough knowledge of these astronomical issues and developed a *Dahlgren General Catalog* from which a subset of stars was selected to provide the operational catalog for TRIDENT I.[34]

The accuracy of the TRIDENT missiles was improved by another Dahlgren development. Compensation for the oblate (rather than truly spherical) shape of the Earth and its effect on gravity, as well as local variations in the force of gravity, required a fast and very accurate trajectory model that could execute on the TRIDENT I fire control computer. A new computer with a new operating system and a new programming language were developed at Dahlgren to allow the TRIDENT missiles to make the complex calculations required to maintain accuracy at the longer 4,000-nautical-mile range.[35]

Under Admiral Levering Smith, the SSPO asked for an Improved Accuracy Program (IAP) for the SLBMs, and Dahlgren participated in that program. One approach that reflected the changed engineering concepts of the 1970s and 1980s was to take a systems-engineering look at accuracy. Rather than simply controlling subsystem errors, the whole system was evaluated. Among the many concepts studied during the IAP were a new error-weighting matrix, innovations in gravity modeling, the use of a system which coupled the then-incomplete Global Positioning Satellite (GPS) location system with some ground-station points of reference, and calculations of variation of gravity at altitude. In the end, the Navy selected for implementation only a subset of the entire range of accuracy improvement concepts that were investigated, with those developed by 1981 to be scheduled for inclusion in the SLBM force in 1989.[36]

These and other aspects of the IAP allowed a change in the strategic function of the SLBM. With a Circular Error Probable reduced to hundreds of feet, rather than thousands as with the early POLARIS, SLBMs could now be conceived as accurate, hard-target weapons. No longer just part of a second-strike capability, the SLBM could be used in a counter-force strategy designed to hit the enemy's own nuclear forces before they were launched. In effect, the improvement to TRIDENT represented a core part of the Second Cold War—an improvement in the ability of the United States to execute a successful first strike in a nuclear exchange that would almost negate the Soviet capability for a deterring second strike.[37]

THE SOVIET CHALLENGE

In the 1960s and 1970s, the celebrated commander-in-chief of the Soviet navy, Admiral Sergei Georgievich Gorshkov, adopted policies that many American naval strategists regarded as quite obscure in purpose. The Soviet navy expanded tremendously and apparently exchanged its old coastal defense policy for a more aggressive strategy, disconcertingly similar to Alfred Thayer Mahan's dictum of global sea power and force projection. Unsure of Gorshkov's intentions and the impact that a much stronger Soviet fleet would have on any naval war with the Soviet Union, CNOs and CNMs through the 1970s struggled to devise weapons systems that would offset the potential threat. In what naval historian George Baer has characterized as a decade of "disarray," those questions and their tentative answers shaped the direction of research at Dahlgren.[38]

Apparently, in the wake of the October 1962 Cuban Missile Crisis, Gorshkov sought more flexibility in his naval forces, hoping to have an independent sea-control force, rather than the simple blunt instrument of nuclear weapons parity. The blue-water, surface-combat force that the Soviets began to build, however, presented some new challenges. It was unclear whether the Soviet naval doctrine had become offensive or defensive, whether it had simply increased its strategic hitting power, or whether it was intended to allow for widespread intervention and projection of power around the world. When Gorshkov and his colleagues wrote and spoke about sea control, were they describing an intended future development or announcing what they had already achieved? In 1970 the Soviets conducted their first global naval exercise, known as Okean-70. It was impressive, displaying a modernized fleet of surface combatants, with a total of over 200 ships and submarines participating. Another Soviet exercise in 1975 included 220 ships as well as long-range Soviet Backfire bombers. That exercise, some of it conducted in the outer Atlantic, seemed designed to practice cutting Atlantic lines of resupply to Europe in case of World War III. Whatever its intentions, it was clear that the newly expanded and modernized Soviet fleet could present new classes of threats against the ships of the U.S. Navy if conflict erupted.[39]

Admiral Elmo Zumwalt, CNO from 1970 to 1974, began an effort to reconfigure the fleet, both in mission and in ship design. As to strategy for the ships, Zumwalt argued for sea control, focused on specific areas, such as international choke points and local engagements. In a 1970 plan called Project 60, Zumwalt called for several classes of less expensive ships. Eventually, only one of the types of ships that he suggested was built, the *Oliver Hazard Perry* class frigates that carried antisubmarine helicopters and HARPOON guided missiles. Zumwalt argued against spending too much on large, expensive aircraft carriers and was enthusiastic in his support of cruise missiles. Among those he backed was the existing HARPOON, with a range of thirty-five miles from a ship or submarine and a range of about one hundred miles from an airplane.[40]

Zumwalt's tendency to push through his reforms without consulting concerned individuals made him unpopular. And during the 1970s, the U.S. Navy was somewhat adrift, not only because of this resentment but also because of the more widespread malaise brought on by the national crises of the early 1970s, including the Watergate scandal, President Richard Nixon's resignation, and the evacuation of Vietnam. In the post-Vietnam era of the mid-1970s, defense budgets, especially appropriations for new naval weaponry, were tightened, and Zumwalt got very little of what he asked for. Indeed, only half his plan to scrap older vessels and replace them with smaller and cheaper ships was achieved—the scrapping half. As a consequence, by 1974 the number of commissioned ships in the fleet had dropped from 976 to 495. The Ford and Carter administrations further reduced the Navy's participation in national defense planning, as strategic decisions seemed increasingly to be governed by fiscal concerns. Instead of relying on an arms buildup to contain the Soviets, the administrations of the 1970s relied on diplomacy and arms-control agreements. Although naval officers saw the Soviet naval buildup as a set of threats, the political leadership at the presidential and DOD level desired detente, negotiation, and maintaining the status quo. Naval analysts concluded in 1971 that, whatever the Soviet intentions, the capability of fast, low-flying, and highly destructive missiles required "tactics and hardware to meet the new threat."[41]

Dahlgren's work continued to focus on improving strategic missiles to make them more accurate, developing guns and missiles suitable to the smaller frigate class, and extending the power of the ship to allow it to fire farther offshore and hit targets inland. Dahlgren also began research on systems to defend against new classes of threats represented by anti-ship missiles. The new classes of weapons systems for surface combatant ships considered in the 1970s and brought to fruition in the 1980s would no longer be just guns but complex coordinated systems that brought together advances in radar, computers, electronics, and missile propulsion.

SOME CHANGES ON THE SURFACE

Despite the Navy's decline in the 1970s, Colvard oversaw an increase in workload from about $100 million per year to more than $300 million by 1980, and the work at Dahlgren and White Oak shifted with the changing winds of naval surface weapons policy through the years of drift and into the revived Cold War during the administration of Jimmy Carter (1977-1981) and the first administration of Ronald Reagan (1981-1985). In the face of the decline in size of the fleet and the uncertain nature of the Soviet threat, naval surface weapons work took several simultaneous directions. One was the development of the TOMAHAWK land-attack cruise missile that increased the Navy's ability to conduct standoff attacks on land targets without risking ships to the hazards of close approach to coastal defense. Developed to protect Navy ships from increased threats from enemy missile and air attacks, the AEGIS system and the SPY-1 radar, the PHALANX close-in weapon system (CIWS), and later the STANDARD surface-to-air missile all led to 1970s and early 1980s work at Dahlgren. Other concerns that stemmed from the growing surface and submarine fleet of the Soviet Union, including better detection of the presence of enemy ships and protection against nuclear, biological, and chemical effects, also produced several research agenda items through the period.[42]

"F" Department research into electromagnetic vulnerability (EMV), under way since 1968, was particularly important, since the Navy was keenly interested in protecting its electronic equipment against any electromagnetic pulse (EMP) that would result from a high altitude nuclear detonation.
Dr. Vincent Puglielli worked with a test chamber first proposed in the 1960s and resurrected in 1973. By the late 1970s, a team under Puglielli used the chamber to develop devices to measure a system's susceptibility to electromagnetic bursts. The Electromagnetic Systems Division continued to conduct ship and aircraft evaluations of EMP, EMV, and HERO and provided direct assistance to the fleet on problems related to shielding against radiation damage to equipment and accidental detonation of ordnance.[43]

In another effort, one related to gun testing, the Navy, under direction from CNO Zumwalt, investigated light guns developed overseas and finally decided on the OTO Melara 76-mm, 62-caliber gun for installation aboard the *Perry* class frigates, as well as aboard experimental smaller hydrofoil ships like the *Pegasus* (PHM 1). "G" Department conducted the acceptance tests for the compact Mk 75 gun mount. The guns, originally designed and built by the OTO Melara firm in La Spezia, Italy, were manufactured for the Navy by FMC, Northern Ordnance Division, based in Minneapolis, after the incorporation of design modifications requested by a Foreign Ordnance Review team, with Dahlgren participation.[44]

Computer research likewise evolved with the changing nature of naval warfare in the 1970s. One inexpensive means of expanding the computer capability at Dahlgren came through the purchase of a small, QM-1 computer designed and manufactured by Nanodata Corporation. Using a program called Emulation Aid System, or EASY, developed by researcher Chuck Flink in "K" Department, programmers used the QM-1 to emulate the workings of other computers, either one-of-a-kind proposed development computers or existing standard types. Among others, the QM-1, using EASY, emulated the AN-UYK-7 standard Navy computer and the fleet-installed older Mk 148 computer. Since parts for the MK 148 were scarce and maintenance costly, it was possible to test capabilities and validate the running of programs in the QM-1. In addition, the new computers being developed for the TRIDENT system could be emulated and evaluated with the Dahlgren QM-1 machine.[45]

Through the mid-1970s, the Naval Research Advisory Committee advisory board on ordnance was concerned that the growth of AEGIS and the work on guidance for the Fleet Ballistic Missiles at Dahlgren were occupying "almost half of the technical base" with work "in large computer systems, software related, fleet support tasks." The board worried that such concentration could lead to neglect of gun development and "change the basic character" of Dahlgren. Taking that concern to heart, Colvard and his successors continued to keep alive gun work and maintain technical capabilities beyond the computer work of AEGIS and SLBM.[46]

By the end of Colvard's tenure in 1980, NSWC had worked on a wide variety of combat systems and weapons that were scheduled to come into the surface fleet over the next decade. In addition to AEGIS, Dahlgren specialists continued their research and development of guided munitions, especially laser-guided projectiles, and also worked on PHALANX, the SEA-FIRE Electro-optical Fire Control System, and a variety of other surface combat systems software, swimmer weapons, and gun systems. These included the gun fire control systems Mk 92, Mk 86, and Mk 68. NSWC had established a role for itself as the Navy's lead laboratory for both Anti-Ship Missile Defense systems and for Anti-Radiation Missile Countermeasures, and was very active in the research and development of anti-air warhead technology. NSWC's "G" Department also fielded the Continuous Rod (CR) warhead for both the STANDARD and PHOENIX missiles and, as targets evolved, developed controlled fragmenting warheads, which the Navy subsequently deployed as the WAU-17/B for the SPARROW missile, the WDU-29/B for the PHOENIX missile, and the Mk 115 Mod 0 for the STANDARD missile. Additionally, the TOMAHAWK missile penetrating warhead utilized a modified Dahlgren BULLPUP B warhead initially developed for the BULLPUP missile, while the PENGUIN missile used a Dahlgren BULLPUP A warhead.[47]

Although neither Smith nor Colvard ever claimed that it was enough to have a mission assigned or a role defined, by 1980 the NSWC did retain or acquire the role of lead laboratory on twenty technical missions, some less well known. In addition to those just mentioned, NSWC had responsibility for Arctic anti-submarine warfare, undersea mines, metal matrix composites, nuclear weapons effects, combat systems engineering and analysis, high-angle threat, warfare gaming systems, and electromagnetic analysis measurements.[48]

Supported by field activities at Solomons, Maryland, Ft. Monroe, Virginia, and Fort Lauderdale, Florida, the NSWC had a wide range of facilities in addition to the gun line. These included a chemical and biological defense complex, a fleet ballistic missile disk pack production facility, and a magnetic structure facility. Between White Oak and Dahlgren the facilities list was extensive and included a wind-tunnel complex, a hydroballistics tank, explosives testing laboratories, nuclear weapons effects simulators, and laboratories for chemistry, plastics, and metallurgy. On the Eastern Shore of Virginia (the Delmarva Peninsula), thirty-five miles south of Salisbury, Maryland, NSWC developed a cooperative relationship with NASA at Wallops Island, Virginia, in 1980. There, the Navy's AEGIS Combat Systems Center tested and evaluated radar signatures, chaff dispersion, anti-radiation missile countermeasure techniques, and recovery systems for projectiles, in support of the AEGIS ship system. The facility, right on the shore and exposed to maritime weather conditions, allowed simulation of conditions experienced at sea.[49]

When Colvard left in 1980, he was replaced by Ronald S. Vaughn, who had previously been a senior member of the Center Management Group at the Naval Air Development Center at Warminster, Pennsylvania, and had served as the director of the Sensors and Avionics Technology Directorate. Vaughn and another engineering manager named Lemmuel Hill, who had headed the Physics Research Department at White Oak before rotating to head the Weapon Systems Department at Dahlgren, interviewed for the position, but Hill later recalled that he had not done well in the interview. While Vaughn served as technical director, Hill was selected to head the Navy's 6.2 (or Program 6—RDT&E, Budget Element 2—Exploratory Development) program in Washington, then was invited back aboard NSWC to become technical director when Vaughan left NSWC in 1983. When Hill took over, he tried to split his week more or less evenly between the two sites, with two days at White Oak, two days at Dahlgren, and one day downtown. It fit his particular "druthers" to emulate the Colvard manner of management, that is, to select good people and leave the management to them. Indeed, Lemmuel Hill was cut from exactly the same cloth as both Smith and Colvard. He firmly believed in the system of management rotation, and that the department head had to be a "general manager." For this reason, there was no sense having a person qualified only as a technical expert at the helm of such a group, but rather the head had to have the broader skills of an administrator. Likewise, he warmly supported the "dual executive" system of joint military and civilian leadership.[50]

Like his predecessors, Hill sought to make sure the NSWC retained a solid record of developing new systems, and he regarded several areas as key in the 1980s: electronic warfare, strategic systems, combat swimmer/ underwater demolitions work, mines, and guided projectiles. He later admitted that he had been accused of suppressing AEGIS because he believed, as the NRAC advisory board did, that one program should not become the entire reason for being for a whole organization. In another move, he responded to the Navy's demand for "mission purification" by transferring some responsibility for TOMAHAWK's in-flight software to China Lake, expecting reciprocity in mission alignment.[51]

Hill learned, however, that sometimes the policy of management rotation had to be carefully applied, especially when it affected a major, high-profile program. When Hill rotated the well-respected Thomas A. Clare out as head of the AEGIS Laboratory and replaced him with the equally competent Paul Wessel, Rear Admiral Wayne Meyer, the project's powerful military chief, called Hill into his office for a chat. During an ensuing forty-five-minute verbal barrage, Meyer sternly informed Hill that he could not make such a change without consulting him. Hill's discomfort level was increased further by the chair that Meyer had him sitting in, which had deliberately shortened front legs, a trick Meyer used to intimidate his subordinates and visitors. Withering under Meyer's assault, Hill remembered that he "did one of those out of body things" and, mentally removing himself from the awkward chair, he quickly developed a healthy respect for Meyer's point of view. Finally, at the end of his tirade, Meyer let Hill off the hook, and said, "Well, we'll see how it goes."[52]

A force of nature in his own right, Meyer's leadership abilities had become legendary. Born in Brunswick, Missouri, on 21 April 1926, Meyer was one of the Navy's most technically educated men, holding a bachelor's degree in electrical engineering (communications preradar option) from the University of Kansas and a master's degree in Aeronautics and Astronautics from MIT. He had attended the Army/Navy/Air Force Officer's Guided Missile School in 1950, and later taught for three years the technical and employment aspects of Special (Atomic) Weapons.[53]

By the early 1960s, Meyer had earned the reputation as one of the Navy's premier guided missile officers. In 1963 he was ordered into the Special Navy Task Force for Surface Missile Systems, commanded by Admiral Eli T. Reich, and worked on modernization and fire control for the TERRIER missile system. Two years later, he was "converted" into an ordnance engineer and later served as the Second Chief Engineer at the Surface Missile System Engineering Station at Port Hueneme, California. In December 1969 the Navy selected him to become the Weapons Systems Manager for the new Advanced Surface Missile System, subsequently renamed AEGIS.[54]

Meyer not only possessed a finely honed intellect but a towering public presence as well. Hill especially remembered Meyer's address to a large, mixed audience of officers and civilians at the dedication of the facility at Wallops Island. Hill found himself hanging on every word, and then realized that the huge after-lunch group was absolutely quiet, not a teaspoon rattling. He looked around and discovered that even the busboys and serving staff were frozen in position, holding their trays, listening spellbound to Meyer's exposition of the role of the Navy in defending American principles. Hill, like many others, believed that Meyer's leadership and vision were key to AEGIS' success.[55]

SHIELD OF THE FLEET

Despite some diversification, the largest and most visible work of NSWC through the 1970s was the development of the AEGIS system. Named for the mythological shield of Zeus, AEGIS had formally started in December 1969 with Defense Capability Plan (DCP) 16, the operational implementation of a Navy study led by retired BUORD chief Admiral Withington that called for a dedicated warship system capable of defeating missile threats to the fleet. The decision to build AEGIS stemmed from a disturbing development in naval warfare that echoed back to World War II. This was the successful employment of anti-ship cruise missiles against a target on 21 October 1967, when Soviet-built *Komar*-class missile boats of the Egyptian Navy sunk the Israeli destroyer *Eilat* off Port Said with three SS-N-2 STYX missiles, the latest in Soviet technology.

Many senior officers in the U.S. Navy were haunted by the Japanese *kamikaze* attacks of the late Pacific War. Like other officers in the Navy's air defense community, Meyer understood that suicidal Japanese *kamikazes*, rather than German U-boats, had been the most lethal force to confront the United States on the high seas during the war, and in fact had almost brought the Navy to its knees. Consequently, the *Eilat's* sinking held frightening implications. The STYX missiles, as Meyer observed, did not have men in them but were relentless "robots," giving the *kamikaze* a modern day flavor. In global sea power terms, Meyer had no doubt that Gorshkov was then building the entire Soviet navy around the *kamikaze* concept, as evidenced by the Egyptians' successful use of Soviet technology and presumably tactics. Later, Meyer insisted that "there is not a single warship, of any note, or an airplane of any note, or boat of any note, and even shore batteries, in the Soviet Union that aren't armed with the *kamikaze*. It is the common weapon of choice. We call them anti-ship cruise missiles . . . the whole AEGIS fleet was built and formed around that dominant threat."[56]

Consequently, DCP 16 laid down for AEGIS three basic functions and five performance requirements, later called *cornerstones*. The three basic functions are: Detection (first finding the target, and in terms whereby a weapon can engage it); Engagement (killing the target); and Control of the other two. The cornerstones consisted of Reaction Time (how much time is there allotted from detection until first motion to attack?); Countermeasures (resistance to enemy detection and attack); Firepower (how much "lead" is in the air at any one time?); Availability (what does the performance have to be from a manpower and reliability point of view?); and Coverage (who must the system protect?). These formed the architecture of the AEGIS system and in time would dominate American naval warfare.[57]

Navy planners wanted a totally integrated weapons system, from detection to kill, rather than simply joining individual radar and sonar systems and separately fired guns or missile launchers. The Radio Corporation of America (RCA) was awarded the prime contract for AEGIS, and Meyer was chosen to spearhead the project, not only because of his expertise in radar and guided missiles but also for his knowledge of systems engineering. Because of Dahlgren's outstanding reputation in the sensors field, the Combat Systems Division handled the program through an AEGIS Data Center at Dahlgren and the AEGIS Combat Systems Center at Wallops

Island. The Navy likewise established at Dahlgren in 1982 a high-tech AEGIS Computer Center for the development of AEGIS software and to provide facilities and computer engineering services for vessels equipped with the system.[58]

During the program, Captain Paul Anderson and Technical Director Colvard were both impressed with the systems thinking that Meyer applied to AEGIS development. Not only did incoming aircraft have to be spotted on a radar and that information linked to the fire control system, but Meyer soon discovered that other aspects of the ship had to be integrated for the AEGIS system to work to full advantage. This drew the team into engineering questions such as loading arrangements, placement of equipment, linkage of computer information, and integration of navigation and electrical power systems into the complete system. In effect, AEGIS could not work unless the whole ship—not just the radar target acquisition system—was fully integrated. That outlook, with its implications for management as well as for engineering, soon had leaders at NSWC thinking in terms of the place of their particular subsystem research project in the larger system of the ship and of naval warfare more generally. Anderson, an avid convert to systems thinking, helped spread the gospel.[59]

For more than a decade, travelers on the New Jersey turnpike had been mystified by the RCA facility at Moorestown, the "ship in the cornfield" used to test AEGIS's radar. More than one-quarter of the NSWC employees working in the AEGIS division spent full time at the Moorestown site and at a nearby Production Test Center in Moorestown. A similar land-based ship was later constructed at Wallops Island to support the fleet in developing tactics and solving technical problems closer to the AEGIS Data and Computer Centers at Dahlgren. In 1982 the AEGIS computer library maintained at NSWC was a satellite facility of the Moorestown center. After 1983 the roles reversed, with NSWC becoming the main delivery point for computer programs to the ship and the RCA New Jersey test facility becoming the satellite.[60]

At the heart of the system was the four-megawatt AN/SPY-1A phased array radar system. Interfaced with the associated UYK-7 onboard computer system, AN/SPY-1A could track hundreds of targets simultaneously and guide eighteen missiles in the air at once, including four in the terminal phase of flight. An AN/SQS-53A bow-mounted sonar, interfaced with ASROC missiles and Mk 46 torpedoes, was incorporated in AEGIS to defend against submarine threats. As an "automatic battle system," as Lemmuel Hill later characterized it, AEGIS could identify and acquire hostile targets, aim its weapons, fire rapidly, and control more missiles with greater accuracy than any system previously built, all without human control should a captain ever throw the system's switch to full automatic mode, which was seldom ever done. Targets in the air, on the surface, and under water could be engaged by a number of different weapons systems, depending on the specific threat, including STANDARD (SM-2) surface-to-air missiles, ASROC anti-submarine missiles, Mk 46 torpedoes, HARPOON surface-to-surface missiles, two lightweight Mk 45 5-inch, 54-caliber guns, and two PHALANX close-in weapons systems (CIWS).[61]

AEGIS officially entered service on 23 January 1983 when the Navy commissioned the first guided-missile cruiser of its class, the USS *Ticonderoga* (CG-47). Built on the same hull as the *Spruance* (DD-963) class destroyers, *Ticonderoga* was 567 feet long, with a 55-foot beam and a 31-foot draught, and a top speed exceeding 30 knots. A crew of 33 officers, 27 chief petty officers, and approximately 324 enlisted men manned her, and she was propelled by four General Electric LM 2500 gas turbine engines, generating an impressive 86,000 sustained horsepower. Two Kamen Aerospace SH-2F Seasprite Light Airborne Multipurpose System (LAMPS) helicopters, capable of carrying two Mk 46 torpedoes, two Mk 11 depth charges, and a host of other attack weapons, rounded out *Ticonderoga's* complement. Twenty-six more of the cruisers were planned for the fleet, at an estimated cost of $1 billion each.[62]

To prepare and instruct select officers and sailors from the fleet and from Allied navies in the operation and maintenance of AEGIS, the Navy established the AEGIS Training Center (ATC) at NSWC on 9 November 1984. Beginning with a staff of eight enlisted personnel, ATC quickly grew to a full-time staff of 14 officers and 89 enlisted personnel by 1987. The curriculum included Combat Information Center (CIC) team training, operator courses, prospective commanding and executive officer courses, a Combat System officer course, and maintenance training. Courses ranged in length from one week to twenty-seven weeks, with the average class lasting eight hours a day, five days a week. An influx of new students, reflecting the growing number of AEGIS warships entering the fleet in the late 1980s, required a $2.7 million expansion of ATC in 1988 and an increase in the teaching staff from 142 to 160. Another $9 million expansion, with $150 million in equipment, was scheduled for completion in the spring of 1989 in anticipation of the new *Arleigh Burke* (DDG-51) AEGIS destroyers (the first entering service on 4 July 1991). On 16 August 1991 the training complex was officially named the Rear Admiral Wayne E. Meyer AEGIS Education Center, in honor of the man who had done more to bring AEGIS from the drawing board to the mainstay of the fleet. As Dahlgren's largest tenant, the AEGIS Training and Readiness Center (ATRC) continued to provide the U.S. surface fleet and

AEGIS-equipped allied navies with trained, high-quality personnel capable of operating, maintaining, and employing AEGIS against hostile forces in an increasingly dangerous world into the twenty-first century.[63]

AEGIS epitomized the conversion from an analog to a digital radar and integrated fire control system. In the days of analog technology, the amount of communication between the fire control system and the gun was quite limited. Gunners trained, elevated, and corrected for initial velocity and fire based on manual sights and range tables. With the flow of digital information directly from the fire control radar to the gun, all of that changed.

The irony, however, was that the technological revolution increased the threat dramatically. Cruise missiles flew at near-supersonic speeds and were very small. Therefore, the defensive system had to have several new features: an automatic mode to facilitate short reaction time; defense against multiple missiles; and the capability of dealing with an environment cluttered with electronic signals and jamming frequencies. Furthermore, the system had to be kept ready for conflict for many years without actually being used in a true defensive situation. By 1980, Rear Admiral Meyer estimated that there were more than one hundred people at Dahlgren involved in the AEGIS project in one way or another. Lemmuel Hill remembered that by the time he took over, about 1984, there were some five hundred people working on AEGIS.[64]

The need for such defensive systems had already been demonstrated in several conflicts around the world. In the 1982 Falklands/Malvinas Island War, Argentine EXOCET sea-skimming missiles were used effectively against British ships, sinking the HMS *Sheffield* and *Atlantic Conveyor* and damaging the HMS *Glamorgan*. Likewise, in March 1987, two Iraqi EXOCETs struck the American frigate USS *Stark* (FFG-31) in the Persian Gulf. One of the new *Oliver Hazard Perry* (FFG-7) class frigates, the *Stark* was equipped with an OTO Melara gun and with PHALANX to defend itself in an engagement. However, in the few seconds during which the approaching missiles were detected, no response could be effectively mounted. The *Stark* was severely damaged and thirty-seven American sailors lost their lives. The use of EXOCETs by the Iraqis and Chinese-built SILKWORM missiles by the Iranians in their war against each other in the 1980s revealed how little time was available for defenders to react. Such experiences underscored the fact that the missile "threat environment" was deadly and evolving, and the defensive systems designed to counter them had to rapidly evolve with upgrades and modifications as well.[65]

However, the AEGIS system revealed its share of bugs in the early years of its deployment. In 1983, while acting as a "de facto" control tower off the coast of Lebanon, the *Ticonderoga* reportedly failed to detect a small, incoming plane that was visible to the naked eye. During Operation ATTAIN DOCUMENT III, the third in a series of Freedom of Navigation exercises conducted 23-29 March 1986 in defiance of Libyan dictator Muammar Qadhafi's "Line of Death" in the Gulf of Sidra, the AEGIS cruiser USS *Yorktown* (CG-48) reportedly targeted and shot at a dense cloud on the water's surface that its SPY-1 radar had mistaken as an attacking aircraft or vessel. The most tragic failure occurred on 3 July 1988 during Operation EARNEST WILL in the Persian Gulf. During that operation, U.S. naval vessels were escorting Kuwaiti oil tankers that had been reflagged and registered as U.S. ships in order to protect the flow of Iraqi and Kuwaiti oil from attack by Iranian forces. In the confined Gulf waters, Captain William C. Rogers III, commanding the AEGIS cruiser USS *Vincennes*, decided not to operate the AEGIS system on fully automatic mode because the rules of engagement with Iranian forces required that he respond only to hostile actions. While Rogers's ship battled several gunboats that had fired on one of his helicopters and tracked an Iranian patrol boat that was on an intercept course, the AEGIS system detected an apparently hostile, incoming aircraft about six minutes away. After broadcasting several warning messages and with only a few remaining seconds to respond, Rogers fired two STANDARD missiles at the target, shooting it down. The target, tragically, turned out to be an airbus, Iranian Airlines Flight 655. All 290 passengers and crew aboard were killed.[66]

In the subsequent investigation into the incident, the AEGIS tapes were hand-delivered under chain-of-custody to Dahlgren and studied at the Wallops Island facility. After initial data reduction, technical representatives from Dahlgren, the AEGIS Program Office, and NAVSEA traveled to Bahrain to conduct further analyses aboard the *Vincennes*. Among other things, they looked closely at the performance and operation of its AEGIS Weapon System Mark 7, its AN/SPY-1A radar, its UPX-29 Identification of Friend or Foe (IFF) system, and the possible environmental effects on system performance. The team also attempted to reconstruct the command and decision (C&D) console operator actions and compare the tape data analysis with the operators' statements, as well as ensure that C&D doctrine had been followed.[67]

Under close scrutiny, and amid a media frenzy that erupted outside NSWC's front gates, the *Vincennes'* tapes had revealed AEGIS's limitations in target identification. Additionally, they showed that despite eyewitness observations aboard the *Vincennes* that reported the aircraft descending toward the ship in the last seconds, like a *kamikaze*, the airbus was actually ascending. Such information, in conjunction with the investigation team's on-site analyses in Bahrain, allowed a close reconstruction of the tragedy. No blame was attached to the officers and crew of the *Vincennes* as a result of the investigation, although the news media and some members of Congress believed the action had been unnecessarily hostile. While the chairman of the Joint Chiefs of Staff Admiral William J. Crowe admitted that AEGIS "has not solved all of our problems and it does not defy the laws of physics," in the sort of "hybrid not-war/not-peace" situations such as EARNEST WILL, in confined waters, simultaneous civilian aircraft flights and military operations could have led to an accidental shoot-down even without AEGIS. In any event, the computer tapes recorded by AEGIS and analyzed at Dahlgren, along with the on-site analysis conducted in Bahrain, allowed a precise reconstruction of the episode that could never have been accomplished from human memory alone.[68]

THE *IOWA* TRAGEDY

Dahlgren experts were called on not only to help with the *Vincennes* investigation, but also to help study the possible causes for the explosion aboard the battleship USS *Iowa* (BB-61) on 19 April 1989 that resulted in the death of forty-seven crew members. After studying the remnants of the destroyed turret aboard the ship, the official report, prepared under the direction of Rear Admiral Richard Milligan, concluded that one of the victims of the accident, Gunner's Mate Clayton Hartwig, had intentionally placed a detonator in the powder bags that were being loaded into the 16-inch gun. Foreign matter embedded in the ruins pointed to such a device, and only Hartwig could have had access to the gun, and a possible motive, for the self-destructive detonation. The episode and the report produced a lingering controversy, with the Federal Bureau of Investigation indicating it could neither support nor disagree with the Navy's conclusions.

After members of Congress and the press harshly criticized the report and its apparent rush to judgment against Hartwig, an independent investigation by Sandia National Laboratories concluded that the foreign matter, consisting of iron fibers found in the ruins, could have come from materials normally associated with the firing of the gun. Furthermore, Sandia identified other factors that may have contributed to the explosion, such as the use of aging powder and ramming the powder bags too forcefully into the gun. The latter error, the Sandia report concluded, appeared to result from improper training, inadequate supervision, and possibly defective equipment.[69]

In July, experiments at Dahlgren dramatically confirmed that over-ramming could create pressures that would cause premature detonation when full-scale drop tests resulted in a violent explosion of 16-inch powder bags. However, experts at Dahlgren and at Sandia continued to disagree as to whether the *Iowa* incident was an accident or an intentional act. Eventually, CNO Admiral Kelso issued a statement indicating there was no proof of an intentional act and that it was improper to leap to conclusions blaming a deceased individual for an accident.[70]

The lead Sandia scientist engaged in the research, Richard Schwoebel, has noted that the Dahlgren facility was named after the man who had brought scientific methodology to the Bureau of Ordnance after the *Peacemaker* accident of 1844. It seemed that the Navy had somehow come full circle, and that the laboratory named in Admiral Dahlgren's honor was investigating the very sort of tragedy that had initiated organized ordnance research more than a century and a half before.[71]

A NAVAL PHALANX

The *Eilat* sinking and rekindled worries about *kamikaze*-style warfare spawned other innovations beyond AEGIS. The most prominent of these was the PHALANX Close-In Weapon System (CIWS), which was developed in the 1970s after General Dynamics responded to a Navy Request for Proposal (RFP) for a point-defense gun-based system that could protect ships against the rapidly evolving anti-ship cruise missile threat.[72]

PHALANX, named for the ancient Greek military formation that presented enemies with an impenetrable wall of pikes, was more than just an anti-missile gun. The system was comprised of a search and track radar, a closed-loop fire control system, a rapid fire M-61A1 20-mm rotary cannon capable of firing 3,000 rounds per minute, and a specialized armor-piercing discarding sabot round made from depleted uranium (changed to tungsten after 1988). Moreover, PHALANX could also be integrated into existing combat systems, such as AEGIS, to enhance its impressive sensor and fire control capability. While General Dynamics' Pomona Division (sold to the Hughes Missile Systems Company in 1992) oversaw the project and worked to produce a prototype by 1975, NSWC's "G" Department conducted an extensive testing and analysis effort to quantify the systems performance of PHALANX against a variety of anti-ship cruise missile threats. The lethality for the 20-mm gun became a critical technical issue in PHALANX's development. Congress at one point canceled the program until its effectiveness could be adequately quantified. To resolve the issue,

"G" Department's scientists and engineers developed a unique analysis methodology that predicted the effect of multiple projectile impacts against an incoming missile. They subsequently conducted an extensive series of tests, including projectile tests against full-scale mockups of cruise missiles with 1,000- and 2,000-pound warheads, to validate their analysis.[73]

In 1977 PHALANX underwent operational tests and evaluation aboard the destroyer USS *Bigelow* (DD-942) and exceeded its specifications. System production began in August 1979, and the first system was deployed aboard the aircraft carrier USS *Coral Sea* (CV-43) in 1980, with ammunition deliveries arriving in 1981. Ultimately, over 750 PHALANX systems were built for the U.S. Navy and several allied navies in the succeeding years.[74]

The spectrum of aerial threats faced by the PHALANX system widened through the 1980s, with the development of low-observable, low-altitude, and high-speed anti-ship missiles as well as slow-speed, propeller-driven aircraft (such as the Cessna or a helicopter). The changing threats led to system improvements and the need to test and evaluate each system change. Dahlgren assumed responsibility for PHALANX as lead laboratory and technical direction agent. "G" Department maintained range facilities for testing and evaluating the modified PHALANX systems as they appeared.[75]

A PHALANX Development Steering Group identified several critical threats, including low-radar cross-section weapons and sea-skimming missiles with consequent reduced detection range and problems of lowered elevation of the weapon. Maneuvering missiles increased the fire control prediction errors and decreased the probability of hits. Active Electronic Countermeasures could also deny detection or introduce false targets to confuse the system. All the environmental threats and risks had to be examined, and Dahlgren took task leadership on characterizing the effect of the ammunition against high-speed targets, and also assumed responsibility for planning by developing a PHALANX threat guide with detailed descriptions of evolving weapons that might be used against U.S. ships. Much of the work on evaluating the effect of the PHALANX against anti-ship missiles could be done by computer simulation. By the mid- and late 1980s, the PHALANX tasks at Dahlgren had a budget of over $3.6 million a year.[76]

The *Stark* had been equipped with a PHALANX CIWS, but Navy technicians in Bahrain reported that the system had not been working properly when the EXOCETs hit the frigate in 1987. "F" Department technicians, who began their analysis of the critical computer components and software associated with the *Stark's* SLQ-32 electronic warfare suite within 24-36 hours of the strike, confirmed this initial report and ultimately discovered a number of key shortcomings in both the ship's defense systems and anti-cruise missile warfare procedures. Despite the discovery of faulty sensors, the Navy relieved Stark commander Captain Glenn Brindel of duty, left his recent promotion unconfirmed, and later forced him into retirement at the rank of commander for poor combat-oriented leadership and failing to properly prepare his crew for the mission. While losing faith in Brindel, the Navy did maintain its confidence in PHALANX and maintained the system as an integral part of its warships' anti-cruise missile defenses.[77]

Following the first Gulf War in 1991, naval analysts concluded that a weapons system was necessary to protect ships against emerging *littoral* (shallow water/amphibious/special operations and small craft) threats. Swedish-built BOGHAMMAR and Boston Whaler type speedboats, which the Iranian Revolutionary Guards had deployed during the Tanker War of the 1980s, had proven especially effective in swarming attacks against larger ships (the *Vincennes* had been fighting several such vessels when it accidentally engaged Flight 655). Accordingly, the Navy initiated the Advanced Minor-Caliber Gun System (AMCGS) program to develop a system to meet the fast attack boat threat.[78]

Simultaneously, the Stabilized Weapons Platform System (SWPS) program was studying a system to counter low-performance, close-range aerial threats. In late 1991, just as the first post-Cold War budget cuts loomed on the horizon, the Navy undertook a study to determine if a single weapons system could satisfy both AMCGS and SWPS requirements. The resulting AMCGS/SWPS Cost and Operational Effectiveness Analysis revealed that with only minor modification and without any further development, the existing PHALANX design could handle both the traditional aerial anti-ship threat as well as the new surface threat. Consequently, "G" Department, in tandem with Hughes Missile Systems (subsequently merged in 1997 with Raytheon to form the Raytheon Missile Systems Company), enhanced PHALANX with an advanced search and track radar system integrated with a stabilized, forward-looking infrared (FLIR) electro-optical fire control sensor. Tests against full-scale model patrol boats proved with devastating effect the potential lethality of PHALANX against littoral threats, and the Navy subsequently deployed the upgraded Block 1B PHALANX system for the first time aboard the frigate USS *Underwood* (FFG-36) in 1999.[79]

FLYING BOMBS REDUX

One of the most significant new weapons to enter service in the 1980s was the BGM-109 TOMAHAWK, a sea-launched cruise missile (SLCM) that could be used as an anti-ship missile or for nuclear or conventional strikes against distant land targets. TOMAHAWK's ancestry can be traced directly back to Elmer Sperry's "Flying Bomb" experiments at Dahlgren in 1919. Although that project and Carl Norden's subsequent automatic pilot project faltered for a number of technological reasons and because of budget cuts, the Navy's interest in flying bombs was not entirely extinguished. While attending the Second London Conference in 1935, Chief of Naval Operations Admiral William H. Standley witnessed the successful use of British "Queen Bee" radio-controlled target drones. He returned to the United States in early 1936 thoroughly convinced that the American fleet needed realistic aerial targets for proper gunnery practice. On 20 July of that year, the Chief of the Bureau of Aeronautics, Rear Admiral Ernest J. King Jr., ordered Lieutenant Commander Delmer S. Fahrney to reactivate and supervise the radio-controlled aircraft program. Fahrney was a creative thinker and soon suggested the development of what he termed assault *drones*. Remotely controlled by airborne observers, these could attack seaborne targets from a distance without exposing the master aircraft to unnecessary risks.[80]

By August 1941, newly developed television cameras negated the need to maintain visual contact with drones altogether, and the program appeared to gain real momentum. Delays pushed back tests until February and March 1943, though, and the first Special Task Air Group (STAG 1), fielding four TDR-1 drone squadrons, did not see combat until 30 July 1944, when it successfully attacked the beached Japanese freighter *Yamazuki Maru* at Cape Esperance. Another successful strike on a beached anti-aircraft ship at Khili, South Bougainsville, on 27 September 1944 confirmed the viability of TDR technology, but the program was soon canceled because of the success of conventional carrier forces in bringing the Japanese fleet to heel.[81]

While the TDR-1 program had ultimately failed, a concurrent program would have far-reaching ramifications in the history of TOMAHAWK. Supervised by BUORD and the Massachusetts Institute of Technology (MIT), with the National Bureau of Standards in charge of the overall development, the ASM-N-2 "Bat" became the world's first fully automatic, radar-guided, "fire-and-forget" anti-ship missile. The Bat originated from RCA's 1941 DRAGON anti-ship "aerial torpedo" and the subsequent PELICAN anti-submarine program. The missile packed a 1,000-pound warhead and was carried by PB4Y-2 Privateer patrol bombers and other similar aircraft. It entered service in April 1945 and soon sank a number of Japanese ships off Borneo, including a destroyer from a range of twenty miles. Foreshadowing the first TOMAHAWK combat strikes decades later, several modified Bats destroyed Japanese-held bridges in Burma and elsewhere in the closing months of the war.[82]

After V-J Day, when air and atomic warfare reigned supreme in Allied defense thinking, the Navy maintained an interest in guided anti-ship missiles, and in the late 1940s and through the 1950s developed a series of first-generation sea-launched anti-ship missiles. The first of these was the LOON, a reversed engineered variant of the German *Vergeltungswaffe 1* (Revenge Weapon 1) "buzz bomb." The first LOON was launched (and crashed) on 7 January 1946, but it soon became the Navy's primary surface-to-surface missile in the early Cold War years. LOON was followed by RIGEL (1950) and the turbojet-powered REGULUS (1954), but the successor systems Regulus II (1955) and the anticipated TRITON both fell victim to POLARIS and the Navy's fervor for FBMs/SLBMs over cruise missiles as the fleet weapons of the future. Although the technology was rapidly becoming available for more advanced cruise missile systems, especially with the introduction of onboard guidance computers in the late 1950s, the Navy largely turned its back to the potential lethality of precision-guided anti-ship cruise missiles in surface combat and land attack situations.[83]

While the Navy's interest in cruise missiles diminished in the 1960s, the Soviets surged ahead in the technology in their drive to counter American aircraft carriers. They developed during the period a number of submarine-, surface-, and air-launched anti-ship cruise missiles, beginning with the surface-to-surface SS-N-2 STYX and air-launched AS-1 KENNEL, both of which entered service in 1956. Improved models started appearing annually, and by the late 1960s the Soviets had assembled a potent cruise missile arsenal and began exporting the weapons to client states such as Egypt, Syria, Vietnam, North Korea, and India, among others.

Appalled at the unexpected success of Soviet cruise missiles—as demonstrated by the *Eilat* sinking, and again on 4 December 1971 when three *Osa* class missile boats of the Indian Navy sank the Pakistani destroyer *Khaibar* and the minesweeper *Muhafiz*, and crippled another destroyer, the *Shajahan*, using STYX missiles—the U.S. Navy suddenly realized the technology's surface combat value after years of neglecting it. Consequently, the service pushed for a next-generation weapon comparable to the Soviet cruise missiles, which ultimately resulted in McDonnell-Douglas's AGM-84 HARPOON anti-ship missile, flying for the first time in December 1972 and entering service in 1977. HARPOON in turn paved the way for a larger, more ambitious cruise missile for the future.[84]

In January 1972 Secretary of Defense Melvin Laird ordered the development of the Strategic Cruise Missile, which was later called the Submarine-Launched Cruise Missile (SLCM). After seven years of development and competitive "fly-offs" against rival designs (including the Air Force's AGM-86B Air-Launched Cruise Missile (ALCM), manufactured by Boeing), General Dynamics' BGM-109 TOMAHAWK emerged as the clear winner for naval service. It was originally intended as a nuclear-capable submarine weapon, but the Navy expanded TOMAHAWK's scope to include surface ship deployment, resulting in yet another official designation change to *Sea*-Launched Cruise Missile. On 19 March 1980, the destroyer USS *Merrill* (DD-976) became the first surface vessel ever to launch a TOMAHAWK. In May 1983 a TOMAHAWK was launched from the *Iowa* class battleship USS *New Jersey* (BB-62) as a test event. This test was important for Dahlgren since the Navy wanted to bring its four *Iowa* battleships out of mothballs and equip them with Armored Box Launchers (ABLs) for TOMAHAWK. As Dahlgren scientist Wayne L. Harman recalled, the launchers' contractor was struggling to meet the battleship refit schedules, and the Joint Cruise Missile Program Office (JCMPO) was on the verge of seeking another contractor when Harman suggested that Dahlgren could do the work. In a couple of days, Harman and his staff drew up a proposal and submitted it to the JCMPO, which accepted Harman's proposal and assigned the ABL work to Dahlgren. This later led to Dahlgren's assignment to the TOMAHAWK Vertical Launch System (VLS) project, which encompassed the development of special, vertical missile launchers that were integrated within a warship's internal structure, as opposed to the bulky and exposed ABLs that occupied precious deck space.[85]

Eighteen feet long (20 with booster), 21 inches in diameter, and with a wingspan of nearly 9 feet, the 3,500-pound TOMAHAWK is a flying bomb in every sense of the word. It carries either a nuclear or 1,000-pound conventional warhead for use against seaborne or land-based targets, and can be launched either through a submarine's torpedo tubes or vertically from specially installed launching tubes. It can also be launched via VLS aboard surface warships and from attack aircraft and bombers. Using a booster engine to clear a launching platform before extending internally retracted wings and cruising at subsonic speeds (about 550 miles per hour) toward its target, TOMAHAWK employs a special *Terrain Contour Matching* (TERCOM) guidance system to achieve exceptional accuracy. TERCOM had in fact been developed in 1958 by the LTV-Electro Systems Company (later E-Systems), but the Navy's flagging interest in cruise missiles at the time had left the system largely unexploited until the TOMAHAWK and Air Force AGM-86 programs. TOMAHAWK was designed to penetrate Soviet territory using a combination of stealth, low-altitude flight, a pre-programmed flight path using TERCOM to avoid hostile detection systems, and large numbers of missiles to overwhelm the enemy ground defenses.[86]

Because of Dahlgren's well-known expertise in missile guidance and computer technology, and its emergence as the Navy's surface weapons center, JCMPO asked Dahlgren in 1979 to participate in the ALCM competition between Boeing and General Dynamics. As a result, the Space and Surface Systems Division (K10) took on the Guidance Software Independent Verification and Validation Task, with four of its engineers (including Harman) working in a quality control capacity. At roughly the same time, K10 began its involvement with weapons control system (WCS) software, helping JCMPO supervise the contractor's development efforts and doing phases IV and V of the project. As Harman remembered later, this experience with WCS software helped prepare K10 for the responsibility in 1982 for developing this software for the Vertical Launch Tomahawk program. Moreover, K10's work eventually led to the establishment of the Cruise Missile Weapon Systems Division (N40) in 1984 and a TOMAHAWK team of over one hundred people, dispersed throughout NSWC's technical departments.[87]

Work for TOMAHAWK was funded and directed out of the Naval Air Systems Command (NAVAIR) in the Cruise Missiles Project (CMP). CMP divided the work on the missile between the government and private sector, assigning the role of principal support laboratory to NSWC. In addition, NSWC developed and supported the computer software for the weapons control system's operating system. Other laboratories around the United States had other aspects of the work assigned to them by CMP, but Dahlgren, with its computer capabilities, housed the Software Support Activity for TOMAHAWK. Within NSWC the support for CMP was coordinated within "N" Department (the Combat Systems Department) in N40, the Cruise Missile Weapons Systems Division.[88]

Of the four divisions within "N" Department, the AEGIS ship combat system and TOMAHAWK weapons system were separate divisions. The other two divisions, devoted to assessment and design, dealt with a variety of weapons systems. By the mid- and late 1980s, AEGIS and TOMAHAWK work at Dahlgren reflected two of the most crucial ship and weapons system developments of the period.

N40 coordinated software development for the program and wrote simulations, maintained math libraries, and provided programs and disks for shipboard use. A Product Assurance Branch managed all tactical, support, and operating system software as well as data analysis for all kinds of tests of the TOMAHAWK system. A VLS for TOMAHAWK required separate programs and procedures, and N40 provided all the operating and testing software for that system as well. As TOMAHAWK was modified and upgraded, N40 served as the system integration agent, assuring that all subsystems worked properly in the modified models. Housed in Building 185, the program had computer capacity for simulating the TOMAHAWK performance on ships from battleship size to destroyer for the trainable launch system. For the vertical launch system, a separate facility was maintained in Building 1580, with the capability of simulating vertical launches on destroyers and cruisers. Such simulation testing allowed verification in computer facilities that the missile control system, radar, and associated ship computer systems could all interface successfully. By the end of the 1980s, the budget for N40 ran in the range of $18 million a year.[89]

Within N40, the work was structured in four separate branches: Systems Engineering, Software Engineering, Cruise Missile Analysis, and Product Assurance. However, the formal organization chart did not reflect the fact that much of the work was performed in a more flexible matrix structure, with team members from different branches working together on specific problems or projects. The importance of TOMAHAWK at NSWC was reflected, though, in the construction of a new TOMAHAWK Weapon Systems Development Laboratory at Dahlgren in 1989, which was designed to accommodate N40 and to conduct RDT&E on new cruise missile weapon system designs, support TOMAHAWK system integration, study future improvements to TOMAHAWK, and support the fleet during TOMAHAWK deployments.[90]

IT WORKED PERFECTLY . . . UNFORTUNATELY!

While "N" Department was engaged with both AEGIS and TOMAHAWK in the 1980s, "G" Department assumed the lead role in developing another weapons system, the STANDARD Missile. Succeeding the earlier TERRIER and TARTAR missiles of the 1950s and 1960s, the STANDARD missile was developed as a fleet area air-defense and ship self-defense system against airborne threats, whether low-level or high-altitude. Designed for integration into AEGIS-equipped ships, STANDARD had to be capable of being launched and guided to targets acquired by the new radars and fire control systems. At first launched from an angled or trainable launcher, a later version for installation on some ships was a vertical launch system (VLS) in which the missiles would be fired straight up from tubes and then angle over in flight to acquire the target.

NAVSEA designated NSWC the lead laboratory for the STANDARD missile in December 1986, and the appointment was confirmed in 1987 by the Space and Naval Warfare Systems Command (SPAWAR). NSWC led the field activities, worked with contractors involved in developing the missile, assured that the missile met the operational requirements, assessed technical progress, and coordinated the development team throughout the missile service life. Other labs and centers involved in the work included China Lake, Martin Marietta Aero and Naval Systems, the Applied Physics Laboratory of Johns Hopkins, and the General Dynamics plant in Pomona, California. The STANDARD Missile Program Office was located at Dahlgren, in the missile division of "G" Department. Areas of work included not only electronics but target vulnerability and system effectiveness, warhead design, batteries, materials, guidance, navigation and control, electromagnetic vulnerability, propulsion improvements, nuclear hardening, and missile system safety. "G" Department target vulnerability and effectiveness engineers continually evaluated emerging threats and developed requirements for improvements in the STANDARD missile. This work led to the development and deployment in 1991 of the Mk 125, DOD's first velocity-enhanced, aimed warhead with automatic aiming selection at target encounter.[91]

The program office had responsibility for interface issues with other systems, including the TARTAR missile and the AEGIS systems, as well as the modifications that established the VLS for the STANDARD missile. As the weapon's warhead, guidance and control system, aerodynamics, propulsion system, and storage system were modified, there were repeated evaluations for safety, interfacing with other systems, and vulnerability to radiation, electro-magnetic pulse, gamma, and neutron effects. More ordinary hazards, such as shock, shipping risks, and degeneration in storage had to be considered. Each risk had to be studied and possible remediation paths recommended.[92]

The missile that brought down the Iranian airbus in 1988 was an AEGIS-guided STANDARD Missile-2 launched from a Mk 26 rail launcher. Although the shootdown was an accident of warfare, it proved that the system itself performed as expected. As Dr. Thomas Amlie, former director of the Naval Weapons Center at China Lake, said, "It worked perfectly . . . unfortunately."[93]

THE SYSTEMS PHILOSOPHY

The three weapons systems, PHALANX, TOMAHAWK, and the STANDARD missile, all reflected the underlying change in approach that was exemplified by AEGIS. That is, in order for the new systems to work effectively in the new threat environment of rapid aircraft and missile warfare, they had to be linked electronically to other systems in the ship. But the implications of the systems approach that had evolved through the 1960s and 1970s were even greater. In engineering work it meant that designers, programmers, mechanical engineers, electrical engineers, and all others involved had to work together across discipline lines to ensure that the total system operated effectively, rather than concentrating only on optimizing their own subsystem.

This systems approach eventually extended into most of the departments at NSWC. Even in the old debate about the efficacy of guns versus missiles, gun designers began to use the new systems arguments. They contended that the number of munitions carried aboard a ship for guns could be several hundred that could be replenished at sea, compared to only dozens of missiles that could only be replenished by a visit back to a port. In addition, ordnance was cheaper than missiles. By such logic and considering the total system, guns continued to hold an advantage over missiles, or so the gun folk argued.[94]

The managerial methods instituted at Dahlgren by Barney Smith and carried forward by Jim Colvard were well adapted to the needs for cross-disciplinary work and systems integration. The rotation of managers, the culture of respect for the capabilities of the specialists at the bench by the administrators and managers, and the creation of matrix teams from across disciplines all meshed well with the needs of AEGIS, PHALANX, TOMAHAWK, and the STANDARD missile. At NAVSEA, the program officers understood and relied on the ability of NSWC to rapidly construct teams that included experienced physicists, mechanical engineers, testing technicians, and computer designers and programmers, all managed by administrators with respect for the capabilities of the technical and scientific personnel. The reputation of NSWC, with the "Dahlgren way" nurtured by Smith through the 1960s and by Colvard through the 1970s, had paid off.

Granted, this new systems approach represented a challenge to some naval traditions. For naval officers accustomed to receiving orders and giving orders, the new bottom-up decision-making processes could be disturbing. In the context of naval engagement at sea, having weapons that would fire not on command but by following computer protocols could be disturbing.

Yet, by the 1980s, as the new weapons systems began to find their way into the fleet, the revolution expanded, and naval officers began expanding their systems thinking—from integrating the weapon into the fleet to integrating the whole ship and fleet operation into a larger warfighting scenario engaging satellites, land forces, and Air Force operations. With Dahlgren's help, *Systems* would soon take on the new meaning, affecting not just the weapons but also the whole approach to fighting a war.[95]

7
Chapter

A New World Order, 1987-1995

Dahlgren had spearheaded a revolution in naval surface warfare technology by pioneering systems engineering and a completely new approach to ship, sensor, and weapon design and development. Through AEGIS and its associated subsystems, Dahlgren had helped the U.S. Navy counter the Soviets' challenge to American naval dominance throughout the world. Unbeknownst to politicians and defense planners, however, the Soviet Union was verging on economic and military collapse. When it came, the Navy and its RDT&E establishment faced a seismic shift in political, financial, and military priorities within the U.S. government that led to base closures and realignments as well as deep civilian and military personnel reductions. While dealing with these difficult political realities, the Navy—and Dahlgren—also confronted completely new, unconventional threats to national security in the so-called "New World Order."

SPAWAR

The advent of total systems engineering in the 1970s and 1980s not only changed how the U.S. Navy conceptualized warfare but how it managed RDT&E. A number of Reagan-era studies and initiatives, including the Packard and Grace Commission Reports of 1983 (which studied bureaucracy reduction and "privatization" of government services, respectively) and the 1986 Goldwater-Nichols Department of Defense Reorganization Act, had spurred Navy Secretary John Lehman to undertake an enormous reorganization of the laboratory establishment. To eliminate a reporting layer and to decentralize Navy acquisition management, he disestablished NAVMAT in April 1985. The Director of Naval Laboratories (DNL) and the labs (including Dahlgren), which had functioned under NAVMAT since 1966, were first transferred to the Chief of Naval Research, and then to the new Naval and Space Warfare Systems Command (SPAWAR) in 1986. SPAWAR was an expansion of the Naval Electronics Systems Command, which assumed responsibility for the development of integrated space and weapon systems. As a management scheme, in the opinion of Technical Director Tom Clare, it was a decade ahead of its time. Responsible for developing an "overall systems architecture" for the Navy, SPAWAR was to coordinate all strike, surface, air, and subsurface technologies and to translate mission changes and warfighting strategies into a coordinated plan for future technological development. It essentially envisioned taking the systems engineering approach developed during AEGIS beyond the ship and even the fleet to the entire warfare arena, including space. SPAWAR, in short, had the potential to propel both naval research and warfare into the "Buck Rogers" universe, imagined by the old Bureau of Ordnance men during World War II.[1]

ANOTHER NAME CHANGE

In early 1987, SPAWAR recommended that the Naval Surface Weapons Center at Dahlgren be designated the principal R&D center for platform-level combat systems to better focus its work toward the fleet's surface warfare needs. Captain Carl A. Anderson took the opportunity to request a name change for the installation. Writing to the CNO on 22 May 1987, via the commander of SPAWAR, Anderson noted that "Since 1974, both NSWC and the Navy have undergone a revolution in their approach to surface warfare. With the advent of many systems such as AEGIS, we no longer think in terms of individual weapons but rather in terms of an integrated weapons system together with its platform." Therefore, he continued, "the concept of the battle force is changing the character of these systems and the manner in which they are employed." Since NSWC had transcended its surface weapons and components mission to lead the surface ship engineering, integration, and analysis fields, he argued that the current name "has not kept pace with our role in surface warfare." Changing the name from Naval Surface Weapons Center to Naval Surface <u>Warfare</u> Center, he believed, would "align our title with our mission as the principal R&D Center for surface warfare."[2]

Anderson's seemingly innocuous request sparked some controversy at Dahlgren. Engineer John E. Holmes asked in a memo forwarded to both Anderson and Lemmuel Hill, "Does the Captain know that it will take a lot more than OP-09's (CNO) say-so?" He noted that "past such efforts have taken a lot of work," and that CNO "ended up reviewing the missions of all centers, got the ASN's office involved, and required a reissue of NAVMAT 5450 [the activity's authorization order]." Holmes hoped that a simple name change might "squeak through a lot of this," but he advised his superiors to "be ready for more work" since "it's really a mission change."[3]

From "K" Department, an alarmed Dave Colby wrote Anderson and Hill, asking "What happened to our mission as the 'Principal Navy RDT&E Center for . . . Ordnance, Mines, and Strategic Support'?" He pointed out that over the years the NSWC name had not kept pace with a number of important mission areas, yet no one had considered that a handicap. If it was a problem now, Colby suggested, generic names such as "Riverside Center" and "Beltway Center" would better allow Dahlgren to host many different laboratories and commands. He warned though that "by over-emphasizing the platform-only center mission" with the new surface warfare oriented name, then "we may re-energize the perceived need to break up the present NSWC arrangement at both Dahlgren and White Oak."[4]

Despite these concerns, Anderson pushed through the name change. On 17 August 1987, the CNO approved the captain's request and reissued OPNAVNOTICE 5450 announcing the new name. The order only changed the name of the headquarters site at Dahlgren, but the CNO noted that White Oak and the NSWC detachment at Fort Lauderdale, Florida, would have to be changed as well. Consequently, the Naval Surface Weapons Center formally became the Naval Surface Warfare Center, and Dahlgren's weapons laboratory legacy was eclipsed by the rising new legacy of surface warfare systems.[5]

A "STRATEGIC PERSPECTIVE"

In May 1988, after the name change controversy had cooled, Hill and Anderson distributed throughout NSWC a paper entitled *A Strategic Perspective on the Future of the Naval Surface Warfare Center,* which carefully described the Center's purpose and operating philosophy and outlined its future goals within the context of the recent changes in Navy RDT&E management. Noting that NSWC was at a crucial point in its history, and that the complex RDT&E environment in which the Center had operated was still changing unpredictably, Hill and Anderson announced that they were charting a course that would build upon Dahlgren and White Oak's past strengths, develop new capabilities, and assure the Center's continued importance to the Navy. The Center would continue to act as the Navy's "technical conscience," but its most significant management challenge, in their opinion, lay in recognizing "the difference between what is good for the Navy in the short run and what is best for the Navy in the long run, and to act in accordance with the long-term view." This would require the Center's management to be "actively and directly engaged in advancing the state of the Navy's technical know-how, across the entire RDT&E spectrum."[6]

Hill and Anderson were also concerned about the impact of "privatization" on Navy RDT&E and the NSWC. They argued that as a part of the American naval family, NSWC was solely dedicated to serving the Navy and the nation, as opposed to private contractors, whose motivations for maximum profit would naturally conflict with the Navy's desire to minimize costs. NSWC, therefore, could act as the guardian of the Navy's long-term technical interests and a check against contractor excesses. This demanded, in language reminiscent of Dr. L. T. E. Thompson, the ability to make sound technical judgments, supported by the best available scientific and engineering capabilities, and the "professional integrity to challenge the positions of others when such challenge is warranted—even if that means taking unpopular positions."[7]

Addressing the question of how the "NSWC of the future" might be different from that of 1988, Hill and Anderson could only opine that it was "unrealistic to presume that today's programs, today's organizational structure, or even today's mission will continue indefinitely into the future." However, by working within and supporting the Navy's Warfare Systems Architecture and Engineering concept for warfighting, they envisioned a center that "both maintains the strength of our diversity and focuses that strength more cohesively, particularly on the needs of the Surface Navy." To shape this "center of the future," Hill and Anderson wanted to hold it at approximately its current employment level (some 5,000 total workers) and to limit the extent to which NSWC contracted out its technical responsibilities. As reflected in the new name, they expected that the Center's future balance of work would be weighted toward systems and components that directly support surface warfare, including the prosecution of anti-air, anti-submarine, anti-surface, strike, and electronic warfare from a surface ship perspective.[8]

"WE KNEW YOU'D HANDLE THAT!"

Hill and Anderson's *Strategic Perspective* was prophetic, but it fell to Hill's successor, Dr. Thomas A. Clare, to deal with a rapidly changing and increasingly difficult RDT&E environment. A New York native, Clare was an aerospace engineer who had come to Dahlgren in 1970 after earning his Ph.D. from Notre Dame. Clare was another product of the Barney Smith school of government lab management, having learned Smith's unorthodox management philosophy in a particularly intimidating way at the very start.[9]

Clare's dissertation had concerned the rolling motion of missiles, and NAVAIR expressed an interest in funding $75,000 for further research. Thus, even before Clare showed up for his first day of work, his branch head told him to gather his dissertation slides and viewgraphs and take them to Washington, D.C., to present his research to the head of NAVAIR's Science and Technology Directorate. As if this pressure was not enough, Clare was further informed that he could bring $75,000 to Dahlgren, depending on his performance. Clare dutifully made the journey to Main Navy and found his way to the designated conference room. When the NAVAIR officials filed into the room at the appointed hour, he was stunned to find no one from Dahlgren among them. Alone, he gave his presentation, and as he later recalled, "I was sweating bullets, and I was very angry and very nervous." Clare did well, though, and after a few questions the NAVAIR officials said, "That's really good stuff! We're going to fund that!"[10]

Although he got the $75,000, Clare brooded over his apparent abandonment during the seventy-five-mile drive back down to Dahlgren, getting angrier by the mile. By the time he arrived, he had worked himself into such a fury that he charged into his boss's office demanding to know, "Where the hell were you?"

The bemused branch head replied with a simple question, "Well, how did it go?"

Simmering, Clare answered, "It went fine."

The branch head then asked, "Did you get the funding?"

"Yes," Clare replied.

"Well, congratulations! We knew you'd handle that."

"What?"

Through that experience, Clare quickly learned, from day one, that at Dahlgren even the most junior engineers were expected to develop their own proposals, make their own presentations, do their own fund raising, and manage their own projects. He took this to heart and became one of Dahlgren's most successful managers, rotating among the various departments in the early 1970s and serving a stint from 1975 to 1976 as the Science Advisor to the commander of the Atlantic Naval Surface Forces in Norfolk, Virginia. Returning to Dahlgren, he headed the AEGIS Ship Combat Systems Division until he was reassigned to the Electronics Systems Department in 1979. Clare entered the government's Senior Executive Service in 1980 and became the head of the Combat Systems Department, where AEGIS was headquartered. In 1983 he rotated to "K" Department, where he successfully tackled some serious problems in the TOMAHAWK program and got it back on track. His systems engineering expertise took him to SPAWAR in 1986, where he helped start the warfare systems architecture and engineering business for the Navy. When Clare became NSWC's new technical director on 27 February 1989, he was, perhaps, the one person most suited for leading the station safely through the tough years ahead.[11]

AT THE PINNACLE

With NSWC reaching its Cold War pinnacle, Clare inherited an enormous corporate entity. With a total funding of $720.7 million in fiscal year 1990, the Center employed, at both Dahlgren and White Oak, 5,119 civilians, including 2,640 engineers and scientists, while the military complement numbered 33 officers and 80 enlisted personnel. NAVSEA remained NSWC's chief customer, funding approximately 45 percent of the Center's work, and eight departments (Engineering and Information Systems, Electronics Systems, Weapons Systems, Protection Systems, Strategic Systems, Combat Systems, Research and Technology, and Underwater Systems) produced a vast range of high-tech products for the fleet. NSWC also directed major field test facilities and activities at Fort Lauderdale, Florida, Fort Monroe, Virginia, and Wallops Island, Virginia.[12]

At the Dahlgren site, the list of independent military tenants had grown to include the Naval Space Surveillance Command (NAVSPASUR), the AEGIS Training Center, and the Naval Space Command (NAVSPACOM), which Secretary of the Navy John Lehman commissioned on 1 October 1983. A $62 million per year operation in 1989, NAVSPACOM consolidated several space-based activities, including NAVSPASUR, into a single command. With forty-four military and fifty-three civilian personnel working at the NAVSPACOM headquarters, the command managed Navy satellite systems and provided space systems support to the U.S. fleet and Marine forces. NAVSPACOM also managed the Navy's new Over-the-Horizon Radar system, a ground-based system that could provide over-the-horizon air and surface radar coverage capable of detecting ships and aircraft at ranges exceeding 1,000 nautical miles. When the U.S. Space Command was established in September 1985, NAVSPACOM became the naval component in the unified command and supported the unified commander's efforts to pull together all the satellite capabilities and resources of the separate armed services.[13]

Within NAVSPACOM, NAVSPASUR continued its original mission of tracking over 7,000 space objects and their orbits, but it also acted as a single integrated "sensor" component of the Space Surveillance Network (SSN) and reported, via the commander of NAVSPACOM, to the U.S. Space Command Space Surveillance Center in Cheyenne Mountain, Colorado. In December 1984, NAVSPASUR began functioning as the Alternate Space Surveillance Center (ASSC) and exercised backup command and control of SSN. When the commander of NAVSPACOM was assigned control of the Tactical Event Reporting System in October 1985, he delegated the day-to-day administration and operations to NAVSPASUR, which established the necessary communications and processing capabilities and was operational by June 1986. On 21 November of that year, the Commander-in-Chief of the U.S. Space Command designated NAVSPASUR as the Alternate Space Defense Operations Center (ASPADOC) and charged it with monitoring all space events and informing all U.S. system operators of potential impacts to their satellite systems.[14]

NAVSPASUR had further become a major communications center, processing over 700,000 messages annually for the Navy and NAVSPACOM. It was so efficient that in 1983 all General Service Telecommunications were placed under NAVSPASUR control. Later, it began processing all messages for NSWC, ATC, and all surrounding commands in the Dahlgren area.[15]

THE PEACE DIVIDEND

Although the prospects for NSWC and its tenants appeared excellent as Clare took charge at Dahlgren, a series of dramatic international events began unfolding, which would bring contraction, realignment, and turmoil to the Navy's RDT&E establishment. In February 1989, Soviet forces withdrew from Afghanistan following a disastrous eight-year guerrilla war with U.S.-backed Muslim mujahadeen fighters. That summer, as the Soviet economy nosedived and unrest spread throughout the Communist bloc, Premier Mikhail Gorbachev began pulling troops out of Eastern Europe. In November the Berlin Wall fell, and in 1990 the former satellite countries, in quick succession, overturned forty-five years of Soviet rule through free elections. The peaceful liberation of Eastern Europe climaxed with the dissolution of the Warsaw Pact and the Soviet Union itself in 1991.[16]

As the ships of the former Soviet fleet were scrapped, sold, or abandoned in their berths, the U.S. Navy was left with neither a serious adversary nor justification for its size, budget, and large shore establishment. True, Operation DESERT STORM showcased the Navy's TOMAHAWK and guided munitions strikes, but that conflict was largely an air and land war; the Navy acted largely in a support capacity by sealing off the Kuwaiti coast, enforcing the U.N. embargo against Iraq, and keeping the Persian Gulf oil lanes open. If DESERT STORM was a model for future naval operations, then the 600-vessel fleet of the Reagan era was hardly sustainable. The shoe dropped even before the last remnant of the Iraqi Republican Guard had finished retreating toward Baghdad. On March 6 President George H. W. Bush boasted that a "New World Order" had emerged, in which it was time "to turn away from the temptation to protect unneeded weapons systems and obsolete bases" and to "rise above the parochial and the pork barrel, to do what is necessary." Bush's oblique rhetoric translated into what pundits and politicians were already calling the "peace dividend"—vast sums of money extracted from defense budgets and reinvested into domestic programs to boost the sputtering American economy. Basking in DESERT STORM's triumph, Bush failed to foresee that the "New World Order" would ultimately degenerate into a new world of disorder, and accordingly pushed ahead with his planned defense drawdowns. With the Soviet fleet decommissioned and dismantled, and no other challengers on the horizon, the U.S. Navy braced itself for the decommissionings, base closures, and "realignments" that suddenly loomed ahead.[17]

IDENTITY CRISIS

As the Soviet bloc fell apart in 1990, Clare came to grips with the fallout from the Reagan-era reforms in Navy RDT&E management and the crisis that had already engulfed the warfare centers. Coming on the heels of NAVMAT's disestablishment (and also several embarrassing military procurement scandals), the Goldwater-Nichols Act had enormous ramifications for NSWC. The legislation formally removed the acquisition process from the military and placed it firmly under civilian control within each service branch. As part of its compliance with Goldwater-Nichols, the Navy created Program Executive Officers (PEOs) who supervised groups of program managers and controlled all program funding. The systems commands, such as NAVSEA, were directed to act as resource managers that enabled and supported the PEOs, providing contracting, financial management, administrative, and logistics skills. At the same time, the long-range 6.2 Exploratory Development funds, which drove the technology base business and had been controlled by the SYSCOMs, were transferred to the Office of Naval Research. This left NSWC's biggest customer, NAVSEA, without exploratory development funding and largely out of the R&D business.[18]

Although he believed that the PEO/NAVSEA arrangement made sense, Clare remained very concerned about the Navy's apparent shift in RDT&E philosophy. Like his predecessors, Clare had risen through the ranks at Dahlgren believing that the Navy wanted organizations such as NSWC to strike a balance between its short- and long-term programs and to plan for the future, making the necessary investments in people, equipment, facilities, and capabilities. Under the Navy's new RDT&E management scheme, however, the planning horizon appeared disconcertingly close.[19]

Clare also worried about the general confusion within the Navy about the centers' fundamental purpose. Following discussions with DNL Gerald R. Schiefer, Clare observed in a white paper entitled *Identity of the Navy R&D Centers: Known or Unknown?* that Navy officials held widely differing opinions and attitudes about why the centers exist, what they should do, who they should do it for, and how their performance should be evaluated. Unfortunately, he argued, all of the conflicting views eventually bore upon the R&D community in one way or another, and the resulting internal conflict had produced an "institutional identity crisis" for the Navy's R&D centers.[20]

To resolve the "identity crisis," Clare proffered his own ideas as to what the centers' purposes should be, which incidentally echoed those outlined in Hill and Anderson's *Strategic Perspective:* 1) help the Navy be a smart buyer, 2) be the Navy's technical conscience, and 3) provide corporate memory in science and engineering to give the Navy a continuity of experience relating to its missions over time. Within this framework, he believed that there were two additional features of the Navy centers and their internal operations that were central to their capacity to be of value to the Navy. The first was the centers' direct execution of research and development activities, and the second was the breadth of center experience across a broad range of technical efforts, including a close relationship with the fleet. Clare suggested that these should serve as a basis for a serious examination of the centers' intended future roles.[21]

Clare added that current international developments, coupled with domestic economic difficulties, were raising profound questions about national security policy, strategy, military missions, and future force levels. These demanded to be addressed and resolved through public discussion, in Congress, and within the executive branch. If the Navy continued to be a part of the nation's defense, and if the R&D centers should remain vital, long-term contributors to the service's strength, then there could be no doubt that conscious attention to the Navy R&D centers must be an integral part of the contemporary debate. The time to act, he argued, was now since a number of ongoing DOD studies were examining Navy shore activities, and because the current management environment was receptive to change and improvement. He therefore urged DNL and SPAWAR to take the opportunity to champion the development of a long-term corporate perspective of the value and purpose of the Navy's R&D centers. The result, he concluded, would be both a stronger Navy and stronger centers.[22]

DNL Schiefer responded to Clare on 9 February 1990 with an e-mail copied to all the other center TDs and COs. He agreed with Clare but insisted that the issue was even broader, and that it was well known that the Navy's technical community did not have an overarching "vision" or "aim." Schiefer informed the COs and TDs that his office had thoroughly considered Clare's points and that he was correct in the timing. It was either now or never, and Schiefer thought that policymakers were already taking the proper steps in establishing "Vision" and "Identity for the DOD R&D community with an emphasis on what the Navy labs are best at." He promised that one of the prime efforts in the various studies concerned education on what the labs had done, what they can do, and why they have to play such a major role in the foreseeable future. However, Schiefer had to remain "close-mouthed" on the rest of his various "activities" for the time being.[23]

BRAC

What Schiefer could not discuss, even with the centers' TDs and COs, were the politically sensitive issues of military facility closures, consolidation, and realignment. President Bush had signaled his intent to shrink the military establishment as early as February 1989 when he directed Secretary of Defense Richard B. "Dick" Cheney to develop a plan to implement the Packard Commission findings and review overall defense management. By January 1990, when Clare was crafting his paper, the Office of the Secretary of Defense had developed a set of criteria with which to judge facilities for possible closure or consolidation. Special DOD study teams were already considering options, ranging from complete closure through reduction, privatization, partial elimination, and conversion, to conversion to government-owned, contractor-operated facilities, and selective establishment of lead laboratories.[24]

As part of the greater DOD effort, one of DNL Schiefer's mysterious activities during this period was heading up a panel that examined Navy facilities for possible consolidation. Initially, few consolidations appeared practical, and few economies appeared available in the near term, but there were some possibilities. It was possible, for example, to align several field laboratories with related R&D activities into one field organization and place it under a single headquarters organization. This idea of creating "megacenters," in which related smaller facilities would be clustered around existing Navy RDT&E centers, gained early momentum because of the potential for eliminating redundancies while reinforcing strong technical cores for the Navy's warfare areas.[25]

After Schiefer submitted a follow-on Phase II study that looked at consolidation and closure from a broader perspective, Gerald A. Cann, the Assistant Secretary of the Navy for Research, Development, and Acquisition (ASN(RD&A)), forwarded an extensive memo to the Undersecretary of Defense that outlined the Navy's evaluation of how its RDT&E facilities should be realigned in light of the nation's changing domestic and international priorities. Among other things, Cann described how realigning the Navy's RDT&E community would lead to the creation of four full-spectrum RDT&E megacenters characterized by the warfare arenas in which their products would be employed. These centers would encompass Naval Air Warfare, Naval Surface Warfare, Naval Undersea Warfare, and Naval C³I (Command, Control, Communications, and Intelligence). A fifth facility, called the "Corporate Laboratory," would comprise only the Naval Research Laboratory.[26]

The numerous DOD studies and recommendations concerning military reduction culminated in the Defense Base Closure and Realignment Act of 1990, which President Bush signed into law on 5 November 1990. The act established the Defense Base Closure and Realignment Commission (DBCRC), which would recommend to the President and Congress which facilities should be closed and which consolidated. The Fiscal Year 1991 Defense Authorization Act subsequently required each service to develop a base realignment and closure (BRAC) procedure. To comply, Secretary of the Navy Henry Lawrence Garrett III established a Base Structures Committee (BSC) that would make recommendations to the Secretary of Defense, who would then forward them on to DBCRC. To keep members of Congress from protecting inefficient installations and to shield the lawmakers from their constituents' wrath, the DBCRC's final recommendations had to be accepted or rejected in their entirety by both the President and Congress. Though the procedure seemed draconian to those most affected, BRAC represented a long-needed reform and seemed to be the only fair way to cash in the "peace dividend."[27]

In December 1990, Navy Secretary Garrett ordered the BSC to review all Navy and Marine Corps installations and recommend which should be realigned or closed. Four months later, in April 1991, the BSC forwarded a target list of ninety-four bases and facilities. The BSC followed Schiefer and Cann's plan for realigning the Navy's RDT&E installations into four megacenters: the Naval Air Warfare Center (NAWC), the Naval Command, Control, and Ocean Surveillance Center (NCCOSC), the Naval Surface Warfare Center (NSWC), and the Naval Undersea Warfare Center (NUWC). Additionally, the BSC recommended ten facilities for closure, consolidating their functions into other facilities, and another seventeen for realignment, some with partial relocation to other facilities. As the dual components of the existing NSWC, Dahlgren and White Oak were both included on the target list for realignment.[28]

Although the BSC's recommendations were only preliminary, Democratic Congresswoman Patricia Schroeder released the target list during a joint Armed Services Subcommittee hearing on 24 April. The very next day, the *Washington Post* reported in bold headlines that the Navy had put ninety-four bases on a "hit list." At Dahlgren, the rumor mill creaked back to life, as fears about possible layoffs or outright closure gripped the station to an extent not seen since the 1950s. The Fredericksburg *Free Lance-Star* contacted NSWC for a response. Commanding Officer Captain Robert P. Fuscaldo issued the prepared statement "At this time, NSWC does not anticipate any major impact on our employees." A Dahlgren spokeswoman did not sound as confident, though. She told the newspaper that although "it's so vague right now," the Navy "may be moving functions in and moving functions out." She added that "we're not expecting any major impact right now," but "five or ten years from now, who's to say . . . ?"[29]

Secretary Cheney's decision in early May to extend a freeze on new military construction projects and to review more than two hundred projects for possible cancellation intensified the closure rumors. Among the projects listed for possible elimination was a $1 million child development center at Dahlgren that Congress had approved as part of the 1990 federal budget. Coming on the heels of the "hit list," the news about the preschool cast doubt on Dahlgren's prospects for BRAC survival.[30]

Quietly assured by SPAWAR that none of the NSWC facilities on the target list would be closed, Dahlgren's senior management moved quickly to dispel the rumors swirling throughout the station. On 10 May 1991, Captain Fuscaldo and Clare issued a joint memorandum officially informing all hands that Secretary of Defense Cheney had announced his recommendations for defense base closures and realignments and that these would impact NSWC. Even if the station was not closed, Fuscaldo and Clare wrote, it was inescapable that the Navy had to draw down its acquisition workforce by 20 percent over a five-year period. That would be a huge change even without BRAC.[31]

Lacking concrete answers, Fuscaldo and Clare could only commiserate and wait for policies currently being developed at higher levels within the Navy, which were not immediately clear. Fuscaldo and Clare understood that the uncertainty was stressful for NSWC employees and their families, so they planned a variety of internal programs to address employee concerns and prepare the Center for consolidation and drawdown. The goal was to provide timely, useful, and accurate information for everyone and help to those destined for "personal and career transition."[32]

Despite SPAWAR's back-channel insistence that Dahlgren would not be closed, Clare was worried. He was acutely aware that there was nothing physically but the Potomac River to keep Dahlgren where it was, and with gun proving no longer the station's primary *raison d'être*, even that was of diminished importance. He was afraid that the BRAC committee would embrace the view that civilian employees could be moved about freely like chess pieces. Indeed, the BRAC process hearkened back to the McNamara days, when decisions were made according to cold, hard figures rather than through any criteria that involved the human element. Throughout the BRAC years, NSWC managers and their staffs often worked around the clock to compile volumes of facts and figures to satisfy the seemingly endless "data calls" from the BRAC committee. Everything from White Oak's hypervelocity wind tunnel to Dahlgren's Main Battery and golf course was quantified to ensure efficiency and objectivity in BRAC. Unfortunately, though, as Clare later lamented, "people were certainly not considered the most precious resource during that process."[33]

As the gloom of BRAC engulfed NSWC, Clare became consumed with keeping the organization alive. He fully understood that the Center's intellectual capacity—its people—was its strength, but this alone would not suffice. However, in his estimation, he did hold one trump card. NSWC was the only Navy lab that excelled in complex, large-scale systems management, something that industry could not do because of private sector competition. In Clare's mind, a system was all about relationships. To cite his example, a hundred parts could sit in a room, but if there is no relationship among those parts, then no system exists. Clare further explained, "They can be the best radars and the best missiles and the best launchers and the best computer programs and the best displays and the best ship," but "if they're not interrelated, consciously and intentionally, in a consistent fashion, with a common objective, they're not a system." Since contracting corporations have an innate tendency to work for their own individual benefit rather than toward a common mission, particularly by withholding proprietary secrets and erecting barriers against competitors, good systems engineering was inhibited in the marketplace.[34]

Beginning with AEGIS, NSWC had successfully been managing large systems for some time. Clare and the Board of Directors therefore concluded that because of the systems-inhibiting competition in industry, "the government needed to play an inherent role in the leadership and management of systems engineering for the Navy." As a survival strategy, then, they decided that NSWC should stake its identity on being the Navy's systems management leader. It would be the Department of the Navy's Warfare System Engineer, thereby "making the whole greater than the sum of the parts."[35]

Clare and the rest of NSWC's senior management essentially "bet the ranch on competition in private industry," but the merger boom of the early 1990s made it a tough sell— even to skeptical employees, who inevitably asked, "Don't you read the papers? Lockheed Martin's buying up everybody. There's not going to be any more competition!" Clare was willing to bet that the trend would not continue. "If the competition in the defense industry goes away," he admitted, "then we need to find another job because our jobs will not be needed." "And people accepted that," said Clare. "They didn't like it," and "they were worried about it a little because they saw things being scarfed up, but they accepted it."[36]

NSWC's gamble on being the Navy's Warfare System Engineer became a rallying point and, as Clare remembered, "the cornerstone of everything we did during the BRACs in the nineties." It also worked. On 1 October 1991, Congress accepted the DBCRC's recommendations, which reduced DOD's original "hit list" significantly. Each of the four megacenters was created by realigning, closing, or consolidating four to eleven related centers, laboratories, or stations, for a total number of thirty-four facilities involved in the reorganization. Since "Naval Surface Warfare Center" was one of the titles that the Navy wanted for its new surface warfare megacenter, it therefore became the name of the new headquarters in Washington, D.C., while the old NSWC became the Dahlgren Division of the NSWC. Ironically, White Oak and the Panama City Coastal Systems Center were aligned alongside Dahlgren Laboratory within the new Division, some twenty-five years after ASN(R&D) Robert A. Frosch had attempted the very same arrangement for his prospective Naval Ordnance Center. However, White Oak, which had then attempted to establish its dominance over the other two facilities, took a serious hit during BRAC '91. Most of its technical programs were realigned to other Navy facilities, and only 650 personnel from the Research Department, the underwater warhead development program, and the support staff remained at the site. The laboratory itself was perhaps spared at this time only because of its unique facilities and Clare and Scott's lobbying, and through the efforts of Maryland Congresswoman Constance A. Morella.[37]

The new Division formally "stood up" on 2 January 1992, and Clare became its first "Executive Director." Captain Norman S. Scott, who had been Dahlgren's CO since June 1991, became the Division's overall CO. Now responsible for an additional laboratory in far away Florida as well as at White Oak, Clare and Scott found their management problems multiplied as they reorganized the Division's management structure in August to integrate Panama City into the organization. Panama City's realignment proved troublesome. Formerly known as the Mine Development Laboratory (MDL) and the smallest of the Navy's labs, the Panama City Coastal Systems Station enjoyed a world-class reputation as the Navy's primary RDT&E activity for mines and countermeasures, special Sea Air & Land (SEAL) commando warfare, amphibious warfare, diving, and other naval missions that involved coastal regions. Like Dahlgren, Panama City had operated rather autonomously for some fifty years and had developed its own unique RDT&E culture with an informal management style. Moreover, the personalities at Panama City were largely different from the personalities at Dahlgren, resulting in a serious culture clash. Unfortunately, this became a significant challenge for both Clare and Scott as the lab was suddenly thrust under Dahlgren's equally headstrong directorate.[38]

In the beginning, Clare and Scott did not handle the Panama City realignment well, and Clare later characterized the whole process as "pretty rough." He already had reservations about expansion in general and thought that Dahlgren and White Oak were enough. When the folks at Panama City demonstrated some early resentment about taking orders from Dahlgren, Clare and Scott had to "go down and talk with people." As Scott later remembered, "We got a little heavy-handed at times," and "there were probably some things said that alienated folks in Panama City." However, Clare and Scott believed that "in the end it worked out and we were able to pull it together."[39]

THE DEATH OF WHITE OAK

The Panama City merger was difficult, but White Oak's closure was much harder for Clare and Scott. The lab had just barely survived BRAC '91, and it had become increasingly apparent that its days were numbered. One factor that worked against White Oak, oddly enough, was its deer population. Over the years, the urban sprawl of Silver Spring, Maryland, had completely engulfed the formerly rural laboratory, which still maintained a sizable enclosed natural preserve. Within the preserve, and without any natural predators, the deer grew unchecked in size and numbers. When animal rights activists cut holes in White Oak's fences to free the deer, the animals became not only an annoyance to backyard gardeners but also a hazard to Silver Spring motorists. Those same activists howled in protest when the government tried to thin the population by opening White Oak to deer hunters. This sparked a backlash against the Navy, as many locals blamed the laboratory for the deer problem rather than the activists.[40]

White Oak's potential danger to the surrounding Silver Spring community was seriously illustrated on 28 June 1992 when an earth-covered, reinforced concrete magazine containing 4,500 pounds of high explosives detonated, although no one was injured and the explosion caused only minor blast damage. When the ensuing investigation determined that the most likely cause of the explosion was the spontaneous ignition of unstable and improperly stored explosives, remedial action was swift in coming. All formulation and testing of explosives were transferred to NSWC-Indian Head Division and the Navy revoked the explosives operations certifications of all personnel found culpable in the incident.[41]

The explosion sealed White Oak's fate. With all of its explosives work now at Indian Head, only the materials research work and its special facilities remained, leaving White Oak unsustainable from a cost perspective unless another large tenant moved in. During the second round of base closures in 1993 (BRAC '93), the DBCRC recommended the disestablishment of the White Oak detachment and the relocation of its functions, personnel, equipment, and support to Dahlgren. White Oak was given one last ray of hope, though, as the commission also directed NAVSEA to move out of its leased space in Crystal City near Arlington and relocate to the White Oak site. NAVSEA, however, soon expressed a desire to move to the Washington Navy Yard instead, and during the third round of closures in 1995 (BRAC '95), the DBCRC consented. On 28 February 1995, Secretary of Defense William Perry announced DOD's recommendations to close or realign 146 more bases. White Oak was finally marked for closure, leaving the Silver Spring community stunned, the more so since it had been preparing for the arrival of NAVSEA and 4,000 accompanying jobs. Opposition to the plan by Maryland lawmakers was immediate and fierce. Democratic Congressman Albert R. Wynn declared "We will be fighting this every step of the way," while Democratic Maryland Senator Barbara A. Mikulski announced that she would "fight tooth and nail for White Oak . . . I'm in my camouflage and fatigues and all set to go." This time, though, there was no reprieve. Maryland's congressional delegation proved powerless to change the decision, as the 1990 DBCR Act had intended, and both Congress and President William J. Clinton signed the recommendations into law. Thus rang the death knell for White Oak.[42]

On 31 July 1997, the laboratory formally closed. Its Magnetic Silencing Group was transferred to NSWC-Carderock Division, and the nuclear weapons effects facilities were dismantled and distributed among the NRL, NAWC-Patuxent River, Aberdeen Army Research Laboratory, and Arnold Development Center at Tullahoma, Tennessee. The Hydroballistic Tank, the Mine Tank, and acoustic testing facilities were all abandoned in place, while the Air Force took control of the hypersonic wind tunnel, considered a national asset and used to support NASA's space shuttle program and for missile defense research. The site itself was transferred to the General Services Administration for further government use and renamed the Federal Research and Development Center.[43]

White Oak's closure was traumatic for all concerned. Those hurt most were former White Oak employees who had to choose between moving to Dahlgren, Indian Head, or another military facility even farther away, or just leaving government service altogether. As Clare had foreseen, most chose to leave the Navy's civil service and enter the private sector rather than tear up their roots and move away. Clare later estimated that during the BRACs only 20 percent to 25 percent of affected workers chose to relocate, with the rest separated from the Navy forever. Captain Scott keenly felt the service's loss. He later recalled ruefully, "It was almost sad to see us break [White Oak] up . . . they were world-class people," and "you wouldn't find any more of them."[44]

A TOXIC PROBLEM

In the midst of the BRACs, Clare and Scott also had to wrestle with myriad other problems arising out of the political climate of the times. Environmental issues in particular demanded much of Scott's attention. Throughout its history, Dahlgren's environmental health had taken a back seat to its primary function as an RDT&E station. Consequently, over the years, weapons testing and byproducts related to the station's numerous projects had accrued an enormous environmental mess. Unexploded shells, hardware and casings, scrap metal, asbestos pipe wrappings, and batteries were buried at one known site on the station from the 1940s into the 1980s. At an old chemical burn site, frequently used in the 1960s and 1970s, small amounts of decontaminated chemical warfare agent solutions were incinerated using fuel oil or gasoline. Even worse, polychlorinated biphenyls (PCBs) had contaminated a 1950s transformer drainage area, while poisonous mercury had seeped into a fifteen-acre manmade "Hideaway" pond along a marshy drainage area flowing into Gambo Creek. Other heavy metals likewise found their way into Dahlgren's soil and groundwater from a variety of old landfills and service areas.[45]

The Navy had identified several sources of hazardous materials at Dahlgren as early as 1983, and by 1987 the station was under mounting pressure by environmentalists and local residents to make its ordnance testing methods more compatible with national and state environmental objectives. In 1989 Scott's predecessor, Captain Robert P. Fuscaldo, "conducted an all-out campaign" to monitor and eliminate hazardous substances on the station. As part of his cleanup effort, Fuscaldo consolidated station safety and environmental responsibilities— previously spread among three departments—under one officer who reported directly to him.[46]

Fuscaldo's environmental efforts ironically brought the station into conflict with the Environmental Protection Agency (EPA) in September 1990, specifically over its sewage treatment system. At issue was an electro-plating operation that discharged effluent into the station's privately owned sewage system until 1981 and again from 1985 until May 1990. Although it would have been perfectly legal to dump the effluent into the Potomac, Dahlgren had been routing it into the treatment system to avoid polluting the river. As it happened, the standards of Virginia Department of Waste Management (VDWM), which governed sewage systems, were much higher than those of the state agency that oversaw discharges into waterways. The VDWM accordingly cited Dahlgren for improperly analyzing and mishandling hazardous wastes since the effluent contained nickel that contaminated the system and the sludge emanating from it. Dahlgren officials denied that the sludge was hazardous and argued that it contained less nickel than is acceptable for sludge currently used as farm fertilizer. The EPA intervened at the VDWM's request and issued Dahlgren a notice of noncompliance for violations of federal hazardous waste regulations. After the EPA demanded that the system be shut down permanently, Dahlgren representatives and EPA officials met to discuss the sludge, but no agreement was forthcoming. The issue was unresolved when Scott came aboard in March 1991. When he moved into the commandant's mansion, he found that the Public Works Department had used some of the dried sludge for fertilizing the flowers around the house. The EPA naturally took exception to his new flowerbed.[47]

The environmental issue at Dahlgren crystalized further on 7 February 1992 when the NSWCDD was proposed for the EPA's National Priorities List (NPL) as a "Superfund" site. The Superfund program had been enacted by Congress in 1980 to identify and clean up the country's worst polluted sites, and getting onto the NPL represented a dubious distinction. However, in coordination with the Virginia Department of Environmental Quality (VDEQ), the EPA found that NSWCDD's problem sites met its Superfund criteria and therefore formally added Dahlgren to the NPL on 14 October 1992.[48]

Scott directed NSWCDD personnel to cooperate with EPA and VDEQ in the ensuing cleanup efforts, which were necessarily painstaking because of all the unexploded ordnance buried throughout Dahlgren. In the following years, the Dahlgren toxic site list grew to include seventy-five sites, eleven of which were of sufficient concern to become top priorities in the effort. Between 1980 and 2004, Dahlgren spent $46 million on environmental remediation, with forty-three projects completed but work continuing only at six other locations because of the diminishing Superfund appropriations by Congress. Interestingly, one unintended consequence of the environmental trouble was that should BRAC ever earmark Dahlgren for closure, it could cost DOD hundreds of millions or even billions of dollars to clean up the entire reservation before its reversion to private or state ownership. In pragmatic terms, the very cost of the cleanup could convince the Navy to keep Dahlgren open and under its jurisdiction.[49]

ORGANIZED CHAOS

While Dahlgren's military leadership came to terms with the EPA, Clare and his senior managers struggled to deal with the organized chaos and plummeting morale among the station's civilian workforce. Although BRAC and personnel reductions were the primary sources of tension that permeated supervisor-employee relationships up and down the line, a series of seemingly self-defeating internal Navy initiatives only stoked the fires of discontent. The imposition of *Total Quality Leadership* (TQL) management principles upon the laboratories was particularly vexing. TQL was a quality management approach based on the philosophy of the statistician, physicist, and business guru William Edwards Deming (1900-1993), who had contributed to Japan's post-World War II economic recovery. Deming's theory was based heavily on the scientific method and work psychology. The Navy Department had become enamored of it in 1984 when the service sought to improve the performance of its logistical organizations. After initial tests resulted in quality improvement at the North Island Naval Aviation Depot, the Navy began extending the approach, originally called "Total Quality Management," throughout the SYSCOMs, to other aviation depots, shipyards, supply centers, headquarters, and field activities, including NSWC. The CNO had changed the name to TQL in 1990 to emphasize the crucial role of leaders in promoting quality improvement, and by 1993 senior managers at the laboratories were fully expected to assimilate Deming's teachings into their organizational cultures.[50]

Dahlgren's management was not nearly as smitten with TQL as the Navy would have wished. "G" Department Head Paul Credle fired an early salvo against TQL in January 1994 when he bluntly told the Board of Directors that "the TQL flight vehicle is never going to leave the ground in the context of the Deming definition." Key to Deming's philosophy was organizational stability, something sorely lacking at NSWCDD because of BRAC and downsizing. Additionally, he believed that TQL in a military organization might not be achievable even when stable, since "it takes a real silver-tongued devil to empower someone who is still saluting the boss." Credle conceded that the process had taught them how to "better treat people and customers, and how to not implement TQL," but he would "submit [that] we grandfather the latter as our goal in the first place, declare a victory, turn out the lights, and stop spending precious overhead dollars pursuing the 'Holy Grail' known as TQL."[51]

An outpouring of condemnation followed in May, after Clare circulated a draft memorandum on perceived deterioration in NSWCDD supervisor-employee relationships. When Clare suggested redefining the relationship between Dahlgren's management and employees in accordance with TQL principles, per Navy Department policy, the response was overwhelmingly negative. Gene Lutman in "N" Department commented that he "did not see the point of all this perceived need to change the management structure to cope with the new-world order." He had already felt "the negative effects of this movement on morale" and thought that "this will be more widespread if we push forward." Group Leader Michael W. Masters responded, "With all due respect to Dr. Deming, practices that work in a manufacturing environment will not work where creativity and initiative are required. We are not a widget factory; we are a scientific and engineering center with a long history of service and accomplishment." Another colleague was not so nice: "Don't waste too much time worrying about TQL; we won't implement it, it is too hard, and there ain't no glory in it. . . . Deming's dead, let the evil he did be buried with him."[52]

The dissatisfaction among NSWCDD supervisors and employees was exacerbated by the hiring, promotion, and overhead expenditure freeze that DOD imposed as part of the defense drawdown. Raises, awards, and performance bonuses for higher Government Service professional levels were included in the freeze, and the results were predictable. As early as December 1992, Clare and Scott warned NSWC headquarters that "the high grade freeze has been extremely painful for the Dahlgren Division. Not only is it hurting the morale of our senior level and experienced people, but some of our very best GS-12s and GM-13s are leaving." Nothing was done, however, and during the TQL/Employee-Supervisor uproar in May 1994, more high-ranking people announced their intentions to leave. One division manager told Clare outright that he "was going to look elsewhere for employment." Plainly put, he felt that "the environment is not one that makes me want to come to work" and that the price of staying at Dahlgren was "now too high." The manager prophesied that "others struggling with quality of life concerns will be making the same choice."[53]

The exodus was cataclysmic for Dahlgren. Scott later recalled that a "lot of brain trust was lost," since under the "freeze" mandate, Clare was prohibited from hiring replacements. Middle management accordingly moved up, leaving no one in the middle to supervise the junior engineers at the bottom. Much later, as Robert Gates remembered, a colleague named James O'Brasky led a study called Project EZRA to determine the extent of the damage. O'Brasky compiled an impressive body of data concerning managers' length of service and "where people were, when people were going to retire, and where the people were who ought to be in the pipeline." What he found was sobering. His analysis revealed that by 2005 it would be theoretically possible for all the department and division heads to be retired, as well as a good number of branch heads and most of the senior program managers. While Captain Scott optimistically estimated that it would take Dahlgren from eight to ten years to recover from this generation "chasm" as he called it, Project EZRA suggested that it could extend well into the twenty-first century before a new generation of senior managers reached full maturity.[54]

One symptom of Dahlgren's management turmoil was the "unsatisfactory rating" given during the Inspector General's Naval Occupational Safety and Health (NAVOSH) oversight inspection conducted from 2-6 August 1993. Specifically, the Inspector General found deficiencies in sixteen of twenty-six NAVOSH administrative programs, including six repeat offenses from a previous inspection in 1990, and 168 standards violations in the workplace. The Navy had cracked down hard on NAVOSH following the *Iowa* explosion and a string of other deadly accidents that resulted in the service's worldwide forty-eight-hour stand-down and a congressional investigation in November 1989. Dahlgren had always prided itself on its commitment to on-base safety, and NAVOSH compliance had never been a problem in the past.[55]

However, the station's failure in this instance came as a grave shock. Embarrassed and facing the prospect of direct intervention by NAVSEA and CNO (Logistics), Captain Scott informed Dahlgren's senior management that "we need to get our arms around this and get on with life." To that end, he established an oversight board that would meet monthly to supervise corrective actions. Scott believed that one factor in the failure related to staffing deficiencies and turnover in both management and experienced personnel. He complained to Rear Admiral "Skip" McGinley, his superior at NSWC, that this "management discontinuity disrupts and sometimes eliminates the voice needed to convey critical safety issues to Command." Furthermore, the "system" made hiring experienced personnel "almost impossible" because of the high-grade freeze, with the result that it sometimes took over a year to fill key positions in the Safety Division.[56]

Scott quickly put a "get-well" program in place, and by November, seventy-nine of the workplace violations had been resolved and fifty-five more entered into a "Deficiency Abatement Program" for further tracking until corrected. The fix only applied to a single symptom of a much greater problem, though, specifically that the NSWCDD organization, through no fault of its own, was increasingly overburdened and understaffed in an enduring, hostile fiscal environment. The problem cascaded as numerous disaffected personnel left for the private sector because of "Reductions-in-Force" and hiring and overhead freezes that promised longer hours without any prospects for performance rewards or position advancement. There was no quick fix for this problem, since it was the direct result of higher government policy. Clare and Scott did, however, hold the organization together during the tumultuous BRAC years through superior leadership reinforced by internal and external diplomacy. Despite the adversity, NSWCDD continued to fulfill its mission where it mattered most, by providing critical technical and systems-based products to the Navy's warfighters in the New World Order, the ugly nature of which had already been glimpsed in 1991.[57]

INTO THE GULF

On 2 August 1990, Iraqi dictator Saddam Hussein's large, modernized army overran oil-rich Kuwait at the northwestern end of the Persian Gulf. Saddam then threatened to wheel south and seize Saudi Arabia's oil fields, which produced over a quarter of the world's crude oil supply. Faced with losing Saudi oil to the Iraqis, U.S. President George H. W. Bush responded by building up American forces in the Persian Gulf region and organizing a broad coalition of nations to contain the Iraqis and protect Saudi Arabia. The mobilization and defense operation was named DESERT SHIELD. On 16 January 1991, Bush authorized the transformation of DESERT SHIELD into DESERT STORM to forcefully evict the Iraqis from Kuwait. A six-week air campaign then devastated the Iraqi military and civilian infrastructure, followed by a four-day ground war that completely swept Saddam's vaunted army out of Kuwait.[58]

The Navy's role in the war was largely littoral, since Iraq had no real naval force with which to contend. Its mission, therefore, involved securing the confined Persian Gulf oil lanes and focusing on anti-mine, SEAL/UDT, amphibious, and coastal operations in the northern gulf. These included high-profile shore bombardments and Marine landing exercises near the Kuwaiti coast that were designed to hoodwink the Iraqis into believing that a massive World War II-style seaborne invasion was imminent. While its sea control mission was necessarily limited by circumstances and geography, the Navy did participate in the air war by striking deep inland targets with guided munitions from carrier-based attack aircraft and with TOMAHAWK

---

cruise missiles, which were televised zooming along the Baghdad city grid and homing in on their targets, usually with remarkable precision.[59]

Dahlgren supported naval operations during the Persian Gulf War on a number of levels, doing surveys of potential electromagnetic hazards both ashore and in the fleet as well as developing new computer software and upgrades for AEGIS. To prepare the fleet for combat operations, AEGIS engineers likewise deployed to the Red Sea to study SPY-1A radar performance in a desert environment. In the electronic warfare arena, "F" Department personnel responded in late August 1990 to an urgent request to upgrade AN/SLQ-32V electronic warfare threat libraries for Saudi ships, which provided the latest naval and electronic intelligence for the Persian Gulf region. By 4 September, "F" Department had developed, tested, produced, and delivered updated library tapes to the Navy International Program Office, less than a week after receiving authority to respond. Additionally, "F" Department personnel enhanced the AN/SLQ-32V's electromagnetic countermeasures capability and delivered the necessary software to American vessels in the Persian Gulf. "F" Department also obtained during the conflict major components from Iraqi "Cluster Psalter" electronic jamming equipment, which represented a major threat to naval operations. Analysis of the equipment allowed NSWC to evaluate the capability of Navy electronic warfare systems, especially the AN/SLQ-32V electronic warfare system, against Iraqi jamming.[60]

"G" Department engineers supported safety certification of ammunition and load-out for the battleships USS *Missouri* (BB-63) and USS *Wisconsin* (BB-64), and also developed weapons strategies to defeat the BOGHAMMAR threat and Iraqi SCUD surface-to-surface missiles. At White Oak, mine and counter-mine specialists supported the fabrication of Mk 57 Bomblets (for use in a special Mine Neutralization System), trained the crew of the minesweeper USS *Avenger* (MCM-1), and expedited the initial load-out and resupply of that vessel. The specialists also measured the acoustic and magnetic signatures of a Benthos Mini-Rover Remotely Operated Vehicle to determine the signature levels and safe standoff distances from several different mine types. Finally, they analyzed and passed along threat intelligence on the Italian MP 80 ground mine.[61]

"K" Department also contributed to the war effort by supporting TOMAHAWK missile operations. Although as department head Sheila Young later explained, "Every TOMAHAWK shot is shot using our software," mission planning was not conducted at Dahlgren since TOMAHAWK was an operational system. "K" Department did stand ready to immediately respond to fleet problems, if necessary, through its capability to analyze data from stray "war shots" and to conduct flight simulations to determine what might have gone wrong. Fortunately, the fleet did not call on "K" Department very much for assistance because TOMAHAWK worked well.[62]

After DESERT STORM ended, "K" Department undertook a "lessons-learned" analysis to see if the system could be improved. One issue that came under close scrutiny was the cumbersome mission planning and targeting process. While AEGIS cruisers, battleships, and submarines could launch TOMAHAWKs, they were incapable of programming the missiles. This had to be done either on aircraft carriers or ashore, with the pre-generated flight path and targeting data then transmitted electronically to launching platforms. "K" Department believed that the process could be streamlined considerably and, funded by the Office of Naval Research, started looking for simpler ways of doing TOMAHAWK mission planning aboard host warships. The research continued through the 1990s. Test flights proved that the onboard planning concept was valid, and development promptly started on a wholly new missile system called TACTICAL TOMAHAWK, which incorporated the new technology and was scheduled for deployment in 2004. "Now, it doesn't have all the bells and whistles," Young observed, "and it doesn't let you do some of the real detailed targeteering that can be done on some of these other platforms, but it's perfectly adequate for most missions."[63]

During the 1980s, Saddam had used chemical weapons against both the Iranian army and his own restless, minority Kurd civilians, killing thousands altogether. Saddam's scientists were also believed to be researching biological agents and nuclear weapons to add to his arsenal of mass destruction. It therefore was feared during the DESERT SHIELD phase of the campaign that the dictator would unleash these so-called "Weapons of Mass Destruction" (WMD) upon coalition forces once combat began.[64]

The Navy was particularly concerned about the possibility of a chemical/ biological attack against its vessels and personnel, which presented a special set of problems in sea environments. As division head James F. Horton informed the Fredericksburg *Free Lance-Star*, the Army can move out of a contaminated area during a land attack, but "in Middle Eastern waters, ships have fewer chances to move out of target range—and if a ship is attacked, those on it can't get away." Consequently, the service turned to NSWCDD's "H" Department (Protection Systems) for critical support since Dahlgren had been the lead laboratory for chemical and biological warfare defense since the late 1960s.[65]

During DESERT SHIELD, "H" Department conducted an extensive research and testing program to determine the persistence of chemical agents on shipboard deck surfaces under a broad range of environmental conditions. The resulting data was integrated with existing NSWC test data that determined how well wash-down systems cleaned chemical agents from contaminated decks. "H" Department compiled all of the information into a Chemical Hazard Assessment Guide (CHAG), which the Navy distributed to its ship captains for consultation in chemical warfare situations.[66]

Anticipating Iraqi chemical attacks from shore batteries, "H" Department likewise developed a computerized Chemical Warfare Naval Simulation Model during DESERT SHIELD. The model simulated the impact of chemical agent attacks on ships and allowed commanders to determine the "threat line" from the attacks by tracking vapor, liquid, and solids from impacting munitions based on particular agent types and meteorological conditions. "H" Department also assisted in the installation and evaluation of Collective Protection Systems (CPS) aboard ships heading to the gulf, beginning with the construction ship USS *Gunston Hall* (LSD-44). CPS freed crewmen from wearing cumbersome protective clothing and masks within chemical, biological, and radiation containments on naval vessels.[67]

"H" Department provided decontamination and casualty handling training to fleet physicians and corpsmen, and collaborated with Canadian and British forces on chemical and biological warfare hazard assessment. The department also provided special protective clothing, detection and monitoring equipment, and chemical and biological warfare training to Marines, SEAL commando units, and Navy civilians deployed in the gulf. Fortunately for the coalition, none of these measures proved necessary since Saddam did not employ his WMDs during DESERT STORM.[68]

A QUESTION OF "JOINTNESS"

By all accounts, several unsung heroes at Dahlgren contributed significantly to the success of American warfighters during this period. First and foremost was Robert L. "Bob" Hudson, a 1960 graduate of Randolph-Macon College who had cut his teeth in the Hazards of Electromagnetic Radiation to Ordnance (HERO) program in the 1960s and had served as the head of the Research Branch of the Electromagnetic Vulnerability Division from 1971 to 1975. Winning the Bernard Smith Award in 1975 for his Electromagnetic Vulnerability work, Hudson had become head of the Special Projects Branch of the Electromagnetic Effects Division that same year. Hudson's management and electrical engineering skills had also led to his appointment as the Navy's principal member for the Joint Technical Coordinating Group (JTCG) classified tri-service materials assessment program called HAVE NAME. The program, conducted under the authority of the Joint Logistics Commanders, had given Hudson vital experience working in a "joint" environment with Army and Air Force counterparts.[69]

"Jointness" in American military thought appeared as early as 1903 when the Secretaries of War and the Navy established a Joint Army-Navy Board to devise broad policies for both services and to do joint contingency planning. The initial concept spelled out the responsibilities of both services during amphibious operations, such as those later conducted during World War II. In 1942 the board's strategic planning role was assumed by the Joint Chiefs of Staff (JCS), comprised of the uniformed chiefs of the Army, Navy, and Army Air Forces, and the president's chief of staff. The JCS advised the country's civilian leadership on strategic matters and also controlled joint operational commands, while the commanders of those organizations generally commanded all service elements assigned to their respective theaters of operation. The theory was refined and revised after defense unification in the late 1940s, and by the 1980s "jointness" meant fully integrating the military services and intelligence agencies at all levels to combine capabilities, achieve greater efficiency, avoid duplication and incompatibility, and confront unconventional threats with a greater variety of options than otherwise available. It was very similar to a systems approach since ideally the whole would be greater than the sum of its parts. However, the very nature of "jointness" required the services to relinquish—at least partially—their cherished independence, traditions, and identities. As a result, "jointness" became politically volatile among the services, with some within the Navy remaining quite hostile into the late 1990s.[70]

The need for joint operations became abundantly clear during the Iranian hostage crisis of 1979-81. Because of his ongoing Navy special programs and his JTCG/HAVE NAME experience, Hudson, along with division head Charles E. "Gene" Gallaher and chief scientist Sir Reginald Gray, were called to Washington, D.C. to brainstorm with Joint Staff planners for potential suppression and power projection options that might be used to free the hostages. They faced a daunting situation. The supreme Shi'ite leader of Iran, Ayatollah Ruhollah Khomeini, had announced that the hostages would be killed if any shots were fired in a rescue attempt. Additionally, the Carter administration desperately wanted to avoid casualties, enemy as well as friendly, and thus the Joint Chiefs of Staff were grappling with the tough question of "how to get the hostages out of Teheran without killing anybody." Since the government was interested in finding a non-lethal way of freeing the hostages, as Gallaher recalled, "we began in earnest to work to try to look at ways that we might be able to do that."[71]

The complexity and sensitivity of the problem dictated that all planning be held in tight secrecy and that the ultimate rescue attempt be rapidly executed with "minimal force." Rear Admiral Robert B. Fuller, an ex-Vietnam prisoner of war working within the J-3 Special Projects Division, briefed Joint Chiefs chairman General David C. Jones (USAF) on a number of classified minimal force ideas from Dahlgren that might be applied during the "leading edge" of the anticipated operation. Hudson was on hand during the briefing to support Fuller and to offer technical advice on the issue. Jones was impressed, and shortly afterward the Joint Staff told Hudson to rapidly develop his team's proposed ideas for modifying and equipping appropriate Air Force and Navy "platforms" that might be used in the rescue attempt. Accordingly, Hudson's technical group set up shop in the Pentagon, initially in the office den of the Assistant Secretary of Defense, Dr. William J. Perry, with all cover stories, planning sessions, test and evaluation activities, and rehearsals coordinated within the secure spaces of the Joint Staff.[72]

For the next ninety days, the Joint Staff, OSD, NRL, and private industry would work closely together with Hudson's Dahlgren group to qualify and deploy several systems deemed necessary to support a rescue operation. The audacious National plan that ultimately emerged from the Pentagon involved several Central Intelligence Agency (CIA) and Defense Intelligence Agency (DIA) operatives, all of the armed services, a combination of some forty-four Navy and Air Force rotary and fixed-wing aircraft, 100 Army Rangers, and 120 elite Army Delta Force counter-terrorist "operators." Initially, the DOD planning team had toyed with the idea of using long-range C-130 cargo planes outfitted with Dahlgren-supplied TARTAR and Zuni rocket motors to land in a soccer stadium across from the American Embassy, pick up the freed hostages, and then launch out of the stadium using the rocket boosters. This idea was scrapped in favor of using shorter-range, carrier-borne Navy RH-53D *Sea Stallion* helicopters, which would be refueled at a rendezvous point deep within Iran called "Desert One" before flying on to Teheran for the hostage extraction mission, called Operation EAGLE CLAW.[73]

As it happened, EAGLE CLAW was a debacle. Eight servicemen were killed and a number of others badly burned when one of the *Sea Stallions* collided with a fuel-bearing C-130 at Desert One. The mission had been troubled from the start. DOD and the Joint Chiefs of Staff had refused to coordinate with the State Department or any other civilian agency outside of the CIA to bring all of the country's considerable assets to bear on the problem. Likewise, no psychological warfare tools that could have favorably maneuvered Iran's new Islamic government and the hostage-takers into making tactical or strategic blunders were employed in the weeks and days leading up to EAGLE CLAW. More critically, mission command, communications, and control were not unified, and training had been compartmentalized among the services and conducted at scattered sites throughout the United States without any full-dress rehearsals. Delta Force commander Colonel Charlie Beckwith testified before the Senate Armed Services Committee soon after the botched rescue attempt. He told the senators that "In Iran we had an ad hoc affair. We went out, found bits and pieces, people and equipment, brought them together occasionally and then asked them to perform a highly complex mission. The parts all performed, but they didn't necessarily perform as a team. Nor did they have the same motivation."[74]

In EAGLE CLAW's aftermath, the Carter administration commissioned an inquiry headed by former CNO Admiral James L. Holloway to study the operation and suggest ways to prevent similar disasters in the future. The Holloway commission agreed with Beckwith's assessment and particularly criticized the lack of centralization and excessive secrecy among the armed forces, which together had hampered the operation from start to finish. An outraged Congress also held a series of hearings on the mission's failure but was slow to enact potentially remedial legislation.[75]

Similar problems later plagued the 1983 invasion of Grenada (Operation URGENT FURY) and the troubled U.S. peacekeeping mission in Beirut, Lebanon, which all but ended on 23 October 1983 when an Islamic suicide truck bomber plowed into the Marine barracks near the Beirut airport, killing 241 servicemen. Afterward, a number of prominent military officers, defense officials, politicians, and intellectuals began calling for legislative-mandated reform since it was apparent that DOD and the services were incapable of reforming themselves. One of the sharpest critics of DOD was historian Edward N. Luttwak, whose *The Pentagon and the Art of War: The Question of Military Reform* was published in 1984 and influenced a number of powerful Republican and Democratic lawmakers, including Senators Barry Goldwater, Samuel Nunn, William S. Cohen, and Representatives William Nichols and Les Aspin, among others. Despite stubborn resistance from DOD and the Joint Chiefs of Staff, Congress finally passed the Goldwater-Nichols Department of Defense Reorganization Act in 1986. It ended much of the independence of the individual armed services, centralized the authority of the chairman of the Joint Chiefs of Staff as a presidential advisor and decision-maker, and mandated a new emphasis on joint operations in future missions. Moreover, the act established promotion preferences and other incentives for officers who served in "joint," or "purple" (the notional color of the combined uniforms) assignments, which would effectively force the services' future leaders to learn to work together if they wanted to be promoted. The accompanying Cohen-Nunn Act reorganized and consolidated all Special Forces under a single United States Special Operations Command (USSOCOM), based in Tampa, Florida.[76]

While the reforms were positive and long overdue, it would take time for them to change the long-standing operational cultures of the individual armed services. By 1989, though, significant progress was evident, as demonstrated in the successful joint invasion of Panama during Operation JUST CAUSE, when the conventional armed services and special operations task forces, supported by civil affairs and psychological operations specialists, worked effectively together to overthrow dictator Manuel Noriega with minimal casualties and collateral damage. Despite this success, true joint interoperability still encountered stiff resistance from the more conservative elements of the U.S. military—especially in the Navy—even after the Gulf War, when it became obvious that joint operations were the wave of the future.[77]

While the United States struggled militarily and politically to deal with successive non-war incidents and operations in the Middle East, the Caribbean, and Central and South America during the early 1980s, Hudson realized that warfighting was evolving. Although the United States "still had a peer competitor called the Soviet Union," as Gene Gallaher recollected, Hudson observed that the world "was getting smaller and people were getting nastier," and fully "understood what was coming down the road" for the U.S. long before the Warsaw Pact collapsed. Therefore, after digesting the hard lessons of EAGLE CLAW, in which he saw a real need to improve the process and tools used by warfare planners during the target identification and mission planning stages, Hudson began looking at ways to bridge the gap between "showing the flag" around the world, which the Navy does very well, and conducting full-scale, conventional war, which the Navy also does very well.[78]

The problem that bedeviled him, though, was that no alternatives existed between the two extremes. The old naval warfare models, based on World War II paradigms, were becoming obsolete because the emergent threat was "asymmetric," in which it was impossible to tell the "bad guys" from the "bystanders," much like Vietnam but on a more dangerous international scale. Asymmetric warfare therefore required something besides always "going in with a gun or a missile," which could cause unnecessary civilian casualties as well as unacceptable political damage to the United States and its military and elected officials.[79]

Hudson ultimately developed a strategy that involved looking at potential enemies, including Third World and non-state "bad actors" outside of the Communist Bloc, from a systems engineering perspective and determining how they could be positively influenced early in conflicts by attacking their "centers of gravity" through either lethal or nonlethal means. Hudson developed a process that focused on providing the military with options across the entire "force escalation curve," from peace to war and back to peace, with many intermittent stages in between. To implement his "grand vision," which depended on joint doctrine, he began developing a new series of special programs, some highly classified, that helped the Navy move into a leadership position in the "joint" warfare arena through the 1980s. In the process, he became a joint warfare program manager and ultimately "masterminded" the creation of a wholly new department at Dahlgren, dedicated to providing the country with new options for fighting the conflicts of the future as he envisioned them.[80]

When Hudson moved up in 1981 to handle the broader programmatic aspects of his early joint warfare programs, his colleague Gene Gallaher replaced him as the head of the Special Projects Branch of "F" Department's Special Electronics Warfare Systems Division. A West Virginia native and an electronics engineer, Gallaher had come to Dahlgren in 1968 from the West Virginia Institute of Technology. After completing the one-year Junior Professional Development Program in 1969, Gallaher had accepted Hudson's invitation to join him in his new Electromagnetic Vulnerability project, which was concerned with the effects of radio frequency (RF) emissions from powerful aircraft and shipboard devices such as radars on smaller solid-state devices. From 1969 to 1972 Gallaher had worked under Hudson's guidance and developed an electromagnetic environment handbook and electromagnetic interference design guide for weapon systems. Afterward, he directed the dual development of a special anechoic chamber, which absorbed ambient external radio waves as well as internally generated emissions, and an RF generation complex, both of which were built to study and test solid-state components for susceptibility or compatibility within a wide electromagnetic effects environment (E^3). In 1974 Hudson brought him on board for HAVE NAME, and over the next seven years Gallaher gained valuable experience working inside a joint program.[81]

Working through the 1980s, Hudson, Gallaher, Reggie Gray, and the members of the Special Projects Branch looked for ways to confront the asymmetric threat from the standpoint of assessing it, countering it, influencing it, and ultimately defeating it at any point on the force escalation curve. Their efforts resulted in the development of "measured response options," or "operations other than war" in later parlance. These measured responses placed a heavy emphasis on nonlethal combat methods, on targeting centers of gravity rather than people, and on the questions of how to minimize the number of targets and with what weapons to best strike those targets. Through employment of measured response options, the Special Projects Branch anticipated that collateral damage and the anger of both hostile and friendly populations would be minimized to America's benefit during future asymmetric situations, such as combating terrorism, counter-drug operations, and humanitarian interventions.[82]

Hudson's special programs within OPNAV provided him with numerous opportunities to promote his idea of a special planning organization that would identify critical infrastructure targets and suppression mechanisms within a hostile country's military, political, economic, and social systems, and then rapidly analyze, process, and distribute the information to the Joint Staff and warfighters around the world. During a period between 1985 and 1988, he successively briefed most of the key flag officers within OPNAV and the systems commands and secured approval for the development of pieces of the capability as he went. In early 1987, Hudson finally sought permission to establish a limited prototype organization at Dahlgren and received support from all quarters except Naval Intelligence, which reasoned that "surely some [other] organization [DIA, CIA, and NSA] is [already] doing this." In view of this objection, the Navy undertook a six-month study of the issue and, as Hudson recalled, it found that "bits and pieces existed but there was no total orchestration capability and basically no effort to improve the mission planning/targeting process." As a result of the findings, the prototype infrastructure analysis capability was authorized soon afterward. "From that time on," said Hudson, "I cannot think of one flag officer or DOD official who did not endorse the idea of the new organization."[83]

A number of key naval officers supported Hudson in his effort to build the new organization. Dahlgren's commander Captain Fuscaldo, who appreciated the need for a joint culture at the station, was an especially strong advocate during the 1980s. Hudson remembered that "he provided an excellent atmosphere at Dahlgren for the development [of measured response options]" and also "ran interference to keep NAVSEA from interfering with 'joint activities.'" Likewise, Hudson later highlighted the contributions of Captain Pat Patrick, who provided "tremendous support" for the effort to build classified suppression mechanisms, as well as those of Captain Bill Evans (Executive Agent for Special Programs in OP-08 (Navy

Program Planning) and Commander Tom Wilson (OP-08 Executive Agent for Security and future director of the DIA), who pushed Hudson's special programs to "the front of the line in the review process" at the Pentagon and ensured that they were always staffed and funded. According to Hudson, Evans's assistance was particularly significant. He not only guarded the programs in the Pentagon but he also took every opportunity to participate in their development. Hudson recalled:

> "We could call Bill during the day and after getting out of the Pentagon he would come down to Dahlgren to critique the problem at hand and contribute to our vision of the future. Much of our original structure was developed on the back of Chi Chi Restaurant napkins when we were out briefing fleet CINCs . . . Even after being transferred from the Pentagon Evans continued to check in and contribute to the success of the program. He was truly one of the first joint visionaries." [84]

Hudson's vision for the future progressed even further toward reality at a conference in November 1989, in which the fleet commanders-in-chief discussed the need for a more systematic approach to joint warfare applications. They had been intrigued by Hudson's prototype infrastructure analysis organization, which could conceivably allow the Navy and DOD to assess a hostile country's centers of gravity and to develop sophisticated warfare options that addressed a range of measured responses against them. As Gallaher later commented, with such a capability they could "bring a piece of it down, bring the entire thing down, bring it down for just a few minutes or hours, or bring it down for many days, weeks, and months."[85]

The political and economic benefits of such a damage-minimizing warfighting tool could be tremendous, and as Gallaher recounted, the admirals asked themselves, "How do we grow that effort? How do we develop and mature those capabilities?" They decided that Hudson's prototype organization should be permanently institutionalized. Consequently, the director of Naval Warfare, Admiral Paul D. Miller, himself a joint warfare visionary, asked DNL Schiefer (then within the SPAWAR organization) in February 1990 to establish the new technical department and a program office at Dahlgren to nurture and expand Hudson's programs. According to Gallaher, Miller specifically wanted to give Hudson the room he needed to mature the capability so that he could offer it to the Joint Chiefs of Staff as a joint program by 1992. Miller's request was very unusual, perhaps the only such one ever made from an OPNAV flag officer to DNL, and Schiefer quickly agreed to accommodate it.[86]

The new department, christened "J" Department, formally "stood up" in May 1990 with sixty-two people. Gallaher, who wrote the new department's organizational documentation, chose the letter "J" as its designator because he wanted to communicate its jointness. Although many high-ranking officers and civilians within the Navy Department, JCS, and OSD generally shared Hudson's joint ideas, the issue was still extremely sensitive within the Navy's lower echelons. Consequently, more conservative elements within NAVSEA refused to allow Gallaher to use the word "joint" in the organizational name, and so he simply called it the Warfare Systems Department. The naming difficulty foreshadowed other troubles that Gallaher would encounter concerning the joint mission, and as he recollected, "from 1990 to 1996, I was in a kind of no-man's land, fighting to even be able to use the word joint."[87]

The new program office, named the Special Programs Office, was placed on the Dahlgren commander's staff as C07. Hudson was named program manager, and Admiral Miller selected Captain Robert "Bob" Tolhurst to serve as the Special Programs director. Gallaher remained in charge of the line organization. On 2 April 1991, the "J" Department position was approved as a Senior Executive Service (SES) position. Ted Williams, the former head of "F" Department, came over from Clare's staff on 8 April to become the first head of "J" Department. A member of the SES, Williams not only coordinated between the program office and the department and led it on a day-to-day basis, but he also provided "high cover" and "greased the skids" in the naval bureaucracy to ensure that the department received everything that it needed. Moreover, his diplomatic skills were invaluable during this time when the government had implemented the promotion and hiring freeze. As Gallaher recalled, Williams had to tactfully explain to other organizations, without revealing the classified nature of "J" Department's work, why "J" was allowed to hire while everyone else was "hard frozen and suffering."[88]

Ten years after the disastrous Operation EAGLE CLAW, "J" Department vindicated Hudson's measured response options and infrastructure analysis approach during DESERT STORM. Unlike in EAGLE CLAW, Hudson and his team at Dahlgren were able to rapidly provide U.S. joint forces with "high leverage" targets that were attacked at the outset of this latest conflict. Moreover, some of those targets were attacked by Air Force and Navy weapons systems that had been derived and matured from the same systems that had been originally developed during the ninety-day planning period leading up to the 1980 hostage rescue attempt. "J" Department also conducted classified analyses for Special Forces operations and provided additional precision-targeting support and classified suppression systems. Its success led the J-3 Director of Operations for the Joint Staff, General John J. Sheehan (USMC), to declare in 1994 that "Dahlgren's contribution was an unqualified success and consequently saved many lives." Sheehan further added that as "this methodology matures, it may one day prevent a war."[89]

Another of the department's important achievements was the short-notice development of a special Identification of Friend or Foe (IFF) system for ground vehicles following a number of tragic "friendly-fire" or "blue-on-blue" fratricide incidents during the air campaign that resulted in the deaths of thirty-five American and twenty-four British servicemen. Captain Norm Scott later recalled that the device "wasn't much bigger than a coffee cup, with an antenna that would go up and down, that would allow [friendly forces] to identify one another." It was, he continued, "very temporary in nature, not very expensive, but it worked and it prevented further blue-on-blue." This IFF system was especially crucial because of the political considerations attendant to coalition warfare, as witnessed in the contentious British investigation surrounding the deaths of nine of their servicemen and the wounding of eleven others by two United States Air Force ground attack aircraft.[90]

The war was a kind of "operational test and evaluation" opportunity in Gallaher's opinion, and in the process "J" Department developed a reputation in defense circles as a miracle worker, capable of taking fleet experience and applying it within a national setting to emergency "must-have-now" types of projects. Afterward, the Secretary of the Navy rewarded "J" Department's success by allowing it to grow throughout the DOD hiring freeze. From 1990 to 1993, "J" Department accordingly grew from the initial 62 people to 235, and its methodology and technical products improved proportionally, so much so that a steady stream of flag and general officers from all services began flowing into Pentagon and Dahlgren conference rooms for briefings on Dahlgren's new infrastructure analysis capabilities.[91]

However, success wrought major changes to "J" Department. During NSWCDD's October 1992 internal reorganization, Williams rotated out of "J" Department to head up "N" Department, and Gallaher moved up as acting head of "J" Department. NSWCDD posted the SES position in 1993, but it was not until June 1994 that the Office of Personnel Management and NAVSEA approved both the position and Gallaher's formal promotion as SES head of "J" Department.[92]

"J" Department's analysis capabilities rapidly matured, right on schedule within Admiral Miller's three-year time frame. Up to this point, "J" Department had necessarily been a Navy show. Senior Air Force and

Army commanders had not been briefed on Hudson's special programs until the late 1980s. Once the Air Force learned about the work at Dahlgren, it had commissioned numerous studies at RAND, Maxwell Air Force Base, and Langley Air Force Base that largely validated the Navy approach. DESERT STORM confirmed the Air Force's findings, and Hudson believed that the service was ready to develop a similar system when JCS, OSD, and the Navy began a series of discussions aimed at expanding "J" Department and formally transforming it into a joint command. In the fall of 1992, Hudson and Tolhurst personally briefed JCS Chairman Colin L. Powell on "J" Department's capabilities. Powell was impressed and also expressed his desire for the organization to go "joint."[93]

Although Powell's blessing sealed the deal, Hudson and Tolhurst soon "learned the true meaning of 'the devil's in the details.'" By law, the service branches had to establish their joint "positions" on the viability of the desired "Joint Warfare Analysis Center" (JWAC). Once that was done, each service's position statement had to be forwarded to each of the other service chiefs, who in turn forwarded their own respective positions to the J-3 Director of Operations, who then compiled and distributed all the cumulative information for consideration by both the JCS chairman and vice chairman. After receiving their written comments, the Joint Staff then produced a unified position paper for OSD's review. The position, once consolidated in JCS and OSD, was returned to the service chiefs for one last look before it was formally acted upon. Hudson chafed at the tedious process, and just when he "thought we could see light at the end of the tunnel," he was irritated further by unwelcome outside interference from DIA and CIA, which "weighed in to say that the JWAC function should belong to them." After some additional debate and review, Powell finally decided that the JWAC function should remain with the armed services.[94]

Admiral William A. Owens (OP-08), who would soon become vice chairman of the JCS, was frustrated too at the cumbersome process and determined to speed things up by making the program office and portions of "J" Department a separate naval command. This would allow Hudson and Tolhurst to begin the formation and staffing of an appropriate military command structure in 1992 while JCS finished its paper chase. Once the new organization was established, it would be easy to change its name when it was transferred to JCS. Therefore, in 1993 he directed NAVSEA and Dahlgren's commanding officer, Captain Norm Scott, to essentially divide "J" Department in half and to establish a new military tenant command at Dahlgren, called the Naval Warfare Analysis Center (NAVWAC). Hudson would become NAVWAC's executive director and Captain Tolhurst would serve as its military commander. The rest of "J" Department would remain under Gallaher as part of NSWCDD. This was quickly done over the winter, and by May 1994, now JCS Vice Chairman Owens finally elevated NAVWAC to the joint command level under JCS, and it was renamed the Joint Warfare Analysis Center (JWAC).[95]

The split between JWAC and "J" Department was executed without difficulty. JWAC took control of country analysis and all targeting nominations and planning functions against hostile adversaries, while "J" Department retained weapons analysis, new hardware development, and pre-existing hardware support. Some discussion ensued about where the Modeling and Analysis program for U.S. internal infrastructure should reside. Ultimately, it was transferred to "J" Department, where it continued to expand because of increasing tri-service and DOD interest that was prompted by new terrorist threats against America's civilian and military assets. The program soon became one of "J" Department's major "thrusts."[96]

Since JWAC was no longer a Navy organization, Hudson had to relinquish all of his Navy program management responsibilities to Gallaher and focus solely on joint warfare R&D for JCS. While Gallaher had lost half of his department to JWAC, he was pleased with the outcome. He later noted that "it's great when somebody comes to you at the three or four star level and says that the work you're doing is so important that we need to make it a command at the national level." And as far as Gallaher was concerned, it mostly happened because of Bob Hudson, who "will never get the credit for what he really did to support our warfighters" because of the classified nature of his work and his refusal to ever call attention to himself.[97]

GATHERING "HYENAS"

"J" Department's success and JWAC's birth in the midst of the BRACs and defense drawdowns proved fortunate. By 1995, the New World Order that President Bush had announced with great optimism was already promising to be even more dangerous than the old Cold War order, as more insidious types of threats began filling the vacuum left by the collapse of the Communist bloc. Gallaher perhaps said it best that "when that elephant died in 1989, then all the hyenas came out." The "hyenas" included not only ambitious regional strongmen armed with weapons of mass destruction, such as Saddam Hussein, but a new generation of militant Islamic fundamentalists who sought to impose their uncompromising brand of Islam throughout the Middle East and rest of the world. These radicals were emboldened by perceived American weakness due to the "peace dividend," an increasing desire to turn inward, and humanitarian missions gone awry, as happened in Somalia in 1993. The Somalia disaster particularly influenced the Saudi terrorist chieftain Osama bin Laden, who drew an ominous lesson from the American pullout following a bloody street brawl in Mogadishu between U.S. Army soldiers and Somali militiamen. Bin Laden later told ABC reporter John Miller that "[Our] youth were surprised at the low morale of the American soldiers. . . . After a few blows, they ran in defeat. . . . They forgot about being the world leader and the leader of the new world order. [They] left, dragging their corpses and their shameful defeat." Thus encouraged, bin Laden and his minions would soon step up the tempo of their global *jihad* against the United States.[98]

As the hyenas began gathering, America lowered its guard in the mid-1990s and reveled in a deceptive, self-absorbed peace. But Dahlgren quietly stood ready for an uncertain, dangerous future, thanks to the untiring efforts of leaders such as Captains Fuscaldo and Scott, and Tom Clare, Bob Hudson, Captain Bob Tolhurst, Ted Williams, Gene Gallaher, and others who held the organization together in the face of political and financial adversity within DOD and the Navy. Dahlgren's unique and powerful technological capabilities would once again rise to help defend the country, though, as the asymmetric threat would fester and grow in the Middle East throughout the late 1990s and ultimately reach American soil, culminating in the catastrophe of 11 September 2001.

On 21 March 1958, a display of SIDEWINDER (air to air), BULLPUP (air to ground), and TERRIER (surface to air) missiles on a truck for a parade showed the variety of missiles tested and analyzed at Dahlgren.

The Fleet Ballistic Missile (FBM) Program required a complete mathematical description of the Earth's gravitational field for accurate trajectory computation. This "geoid" display represented one of the first detailed global gravity field models developed at Dahlgren. Some of the laboratory's leading scientists who worked on the FBM guidance problem and pioneered the science of "geoballistics" included (from left to right) Walter P. Warner, Raymond H. Hughey Jr., David R. Brown Jr., and Dr. Charles J. Cohen. (Dahlgren Historic Photograph Collection)

Launched from the USS *Ethan Allen* (SSBN-608) on
23 October 1961 and guided by a Dahlgren-computed
presetting, this second generation POLARIS A2
missile begins its journey down the Atlantic Missile
Test Range. (Naval Historical Center)

Dahlgren's "K" Laboratory generated millions of these punched target cards for early POLARIS
fire control. Each card contained preset targeting data that a submarine weapons officer used to
manually program a POLARIS missile's guidance system. (This card is unclassified because it is
not punched.) (Courtesy of Dr. Robert V. Gates)

In 1964 Dahlgren's senior management opened the Computation and Analysis Building, located near the front gate, to help modernize the laboratory's image. (Dahlgren Historic Photograph Collection)

Bernard "Barney" Smith served as Dahlgren's technical director from 1964 to 1973. Because of the major organizational and cultural changes he fostered during his tenure, many of his junior colleagues believed that he "brought the Dahlgren Way back to Dahlgren." (Dahlgren Historic Photograph Collection)

Barney Smith's management rotation system at Dahlgren, depicted here as a musical chairs game, became notorious within Navy RDT&E. Left to right: Department Heads Chuck Bernard, Jim Colvard, Jim Mills, Ralph Niemann, and Dick Rossbacher, with Smith playing the "Dahlgren Boogie" on the piano. (Drawn by Paul Wasser at Dahlgren, 1973)

Completed in 1967, Dahlgren's half-mile-long conical shock tube allowed DASA and NWL scientists to simulate 20-kiloton nuclear explosions without using radioactive materials. (Dahlgren Historic Photograph Collection)

This 8-inch lightweight gun has just fired a semi-active laser guided projectile (inset image). Its installation and test firing aboard the USS *Hull* (DD-945) in April 1975 represented the first time that a major-caliber gun was successfully mated to a destroyer-sized warship. (Dahlgren Historic Photograph Collection)

Dahlgren's Main Battery remained active in major-caliber gun and ammunition testing during the 1980s. (Dahlgren Historic Photograph Collection)

A group of Dahlgren engineers pose with the CG-47, USS *Ticonderoga*, first ship in the new AEGIS class of guided missile cruisers in 1982, while it was still under construction in the shipyard in Pascagoula, Mississippi. The Dahlgren team was responsible for integration and test of the early AEGIS computer programs with the shipboard equipment. That team, and their successors, has continued this support in the 25 years since that time for all newly constructed AEGIS Cruisers and Destroyers. From left to right: CWO3 Joe McGlade (FCS); Ms. Nadine Blyn (WCS); Mr. Pete Dacri (FCS); Mr. Winston Langston NSWC Team Lead; Ms. Linda Clark (WCS); Ms. Dee Faccini (Colmer) (CM/Documentation); Mr. Bill Sealand (WCS/C&D); Mr. James Clark (FCS); Mr. Parminder Duli (SPY); Mr. Guy Rich (WCS); Mr. Ken McCullum (Documentation); Ms. Trish Hamburger (Smith) (C&D); Mr. Ron Schaffer (FCS); and Mr. Graham O'Neill (Sys Eng).

From 1987 through 1995, ship defense anti-air warfare radar systems installed at Dahlgren's Land Based Sensor Test Site supported a variety of programs to improve sensor performance for the detection of low altitude, low radar cross-section, fast incoming missiles. (Dahlgren Historic Photograph Collection)

The AEGIS class cruisers USS *Bunker Hill* (CG-52) and USS *Valley Forge* (CG-50) returning from deployment in support of the Global War on Terror. The warships' AN/SPY-1 phased array radar is prominently integrated into their forward superstructures. (U.S. Navy photo)

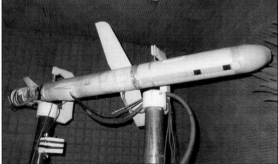

A TOMAHAWK cruise missile undergoes electromagnetic vulnerability testing in Dahlgren's anechoic chamber, 23 May 1991. (Dahlgren Historic Photograph Collection)

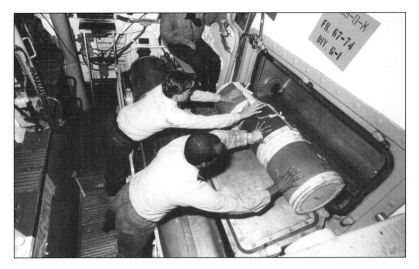

During the investigation into the *Iowa* explosion, a gun crew at Dahlgren demonstrates a proper bag charge ram by loading three powder bags into a 16-inch gun. (Dahlgren Historic Photograph Collection)

A Mk 45, 5-inch, 62-caliber gun fires down range, with the projectile just exiting the barrel. (Dahlgren Public Affairs Office)

The high energy Mk 45 Mod 4 Gun Mount, installed on the USS *Winston S. Churchill* (DDG-81) with an all-digital fiber-optic interface to the onboard Mk 160 Gun Computer System (GCS), fires an ERGM slug to test the impact of 18 megajoule shots on the warship's structure. (Dahlgren Public Affairs Office)

A Tactical TOMAHAWK Cruise Missile launches from the guided missile destroyer USS *Stethem* (DDG-63) during a live warhead test in February 2004. (Dahlgren Public Affairs Office)

USS *Ronald Reagan* (CVN-76) joined the fleet in 2005 as the most technologically advanced ship in the world. The NSWC Dahlgren CVN-76 Strike Group Support Team was part of the team of engineers that helped certify the *Reagan* as combat-ready and combat worthy. The *Reagan* Combat Direction Center includes an SSDS Mk 2 based combat system designed to respond with a rapid reaction, anti-air defense capability against high-speed, low-flying anti-ship missiles, Cooperative Engagement Capability (CEC), Common Data Link Management System (CDLMS), and Shipboard Gridlock System (SGS).

Researchers at the Naval Surface Warfare Center, Dahlgren, Human Performance Laboratory (HPL) developed a testbed for future command and control concepts, the Integrated Command Environment (ICE) facility, that also serves as a vehicle to solicit valuable feedback from members of the fleet.

8

Chapter

Dahlgren Forever, 1995-2003

By mid-1995, Dahlgren had survived four rounds of defense base realignments and closures (BRAC) by becoming the Navy Department's representative "Warfare Systems Engineer" and by providing technical solutions to America's warfighters during international crises. BRAC was not the only government attempt to realize a post-Cold War "peace dividend" that wreaked havoc on the station's workforce. Mandatory hiring and promotion freezes and personnel "reductions-in-force" (RIFs) drove morale lower, and key personnel departed, leaving a generational and technological chasm within the organization that would take years to close.

Although the Clinton administration's extension of deep military budget cuts and efforts to "reinvent government" only made things worse, Dahlgren's civilian and military leadership succeeded in stabilizing the organization and restructuring it to better reflect the Navy's post-Cold War RDT&E emphasis on joint, theater, littoral, and asymmetric warfare. This reconfiguration proved fortunate for the Navy since the world was becoming more chaotic and dangerous than it had been during the Cold War. Multiple regional threats replaced the single threat of Soviet domination, and from the Middle East a vicious new strain of Islamic militancy mushroomed into a global terrorist network that sought to strike at the heart of America itself. When the devastating attack came suddenly on 11 September 2001, Dahlgren was poised to help fight a new global war on terrorism.

RESTRUCTURING DAHLGREN

The year 1995 began with Executive Director Tom Clare and commanding officer Captain John C. Overton, who had succeeded Captain Norm Scott in August 1994, still struggling with Dahlgren's multiplicity of problems. A short-term restructuring team had met in late 1994 to confront a number of problem areas, including an inadequate and unevenly distributed workforce, a lack of interdepartmental marketing coordination, and an absence of project priorities. Despite intense discussions, the team made no real progress because of the political uncertainties of the Clinton administration.[1]

By March 1995, however, Clare and Overton were ready to unveil a more comprehensive long-term restructuring program designed to unify the division, heighten Dahlgren's systems engineering reputation, strengthen its science culture, and increase its flexibility amid internal and external changes. The program reflected Dahlgren's need to adapt to the Navy's newly conceived strategy for the twenty-first century. Called *From the Sea* by its chief authors, CNO Admiral Frank B. Kelso and Marine Commandant General Carl E. Mundy Jr., and first published in 1992, the new strategy shifted the Navy's focus away from the global, blue-water warfare envisioned in its *Maritime Strategy* of the 1980s to regional theater, littoral, and expeditionary operations. Drawing heavily upon the Navy's experiences in its Middle East operations of the 1980s and the first Persian Gulf War in 1991, Clare and Overton's long-term restructuring program also bowed to DOD's mandate for jointness by promoting the ongoing warfare analysis work in "J" Department. Further realignment and personnel reductions notwithstanding, Clare and Overton ultimately envisioned a division with two sites, Dahlgren and Panama City, and a total workforce of between 3,800 and 4,200 people. They set 1 October 1996 as their goal for implementing all of the elements of the long-term restructuring program, subject to other significant decisions made by the Navy before that date.[2]

By June 1995, however, only three months after he and Overton launched the long-term restructuring program, Clare realized that Dahlgren was "not in a healthy position to move into the future" and was too quickly losing the flexibility vital to performing new work and anticipating the Navy's future technology needs. He believed that if NSWCDD waited until October 1996, Dahlgren would suffer serious damage to its creativity and lose its ability to respond to long-range fleet warfare demands. That summer Clare called all of his department heads and program managers into a series of discussions at both Dahlgren and Panama City calculated to determine the division's current degree of flexibility to handle new work and to adjust its resources to meet shifting demands. His goal was to frame strategy options that could be presented at a Division Council Workshop that fall.[3]

Clare got his workshop, but it could not stop external forces from converging to hinder his efforts to preserve Dahlgren's flexibility and to accelerate the restructuring. The first problem was a financial shortfall that occurred after Clare and Overton underestimated NSWCDD's income projections for 1996. They miscalculated because DOD's funding levels changed faster than they had anticipated, a direct result of the Clinton administration's drive to balance the federal budget and also to make the government a more efficient and less costly enterprise for American taxpayers. Since Clinton and Vice President Albert Gore Jr., who spearheaded the Reinventing Government movement, viewed Americans as customers rather than as beneficiaries, his fiscal policies ensured that administrative overhead costs in particular were scrutinized and targeted for reduction throughout the government to eliminate waste.[4]

At Dahlgren, overhead costs had routinely been shifted within projects and departments as necessary over the past few years to balance the books, but under the Clinton mandate to redefine how the government does business, this method was no longer effective. Consequently, in March, Clare and Overton ordered all hands to change the NSWCDD's approach to reducing overhead costs from one of "adjustment" to one of "redefinition" and "elimination." They planned to do this using a systems perspective, since their experience showed that all overhead functions were interlinked with other overhead and direct program functions. Moreover, Clare and Overton expected NSWCDD to use its systems engineering heritage to help reallocate functions to new and different combinations of people, processes, and machines, and also to eliminate some functions entirely. Anticipating the measure's effect on already low morale, Clare and Overton appealed to duty, reminding employees that "The Navy needs us to fulfill our mission under the restraints of these new and more stringent financial times," and implored them "to be patient and to understand why we are doing what we are doing."[5]

The Navy Department's 2 February call for additional civilian RIFs might have made the belt-tightening process more clear-cut, but it intensified Dahlgren's already severe staffing crisis. Under the plan, to take effect in September, Dahlgren stood to lose 25 employees, 10 through cash payment Voluntary Early Retirement Authority (VERA)/Separation Incentive Programs (SIP) and 15 others "involuntarily separated" from the service. Panama City faired worse—120 employees were targeted there. Although the layoffs at both sites were comparatively mild, they could not have come at a worse time for Clare and Overton.[6]

A BIG "MESS"

If the funding shortfall and RIF were not troublesome enough, President Clinton's program for Federal Laboratory Reform and congressional demands for additional laboratory cuts further complicated matters for NSWCDD's leadership. In early May 1994, President Clinton directed the National Science and Technology Council (NSTC) to review the government's three largest laboratory systems at the Department of Energy, National Aeronautics and Space Administration (NASA), and DOD and suggest reforms. Exactly a year later, NSTC released a report entitled *Interagency Federal Laboratory Review* recommending that the agencies clarify and better focus their laboratories' mission assignments, streamline agency and lab administration, and reduce excessive agency oversight. Clinton implemented these recommendations as *Guidelines for Federal Laboratory Reform* in his Presidential Decision Directive (PDD)/NSTC-5 of 24 September 1995.[7]

The directive required all three agencies to coordinate and integrate laboratory resources and facilities on an interagency and interservice basis, eliminating unnecessary duplication and establishing joint management where appropriate. This dovetailed with post-BRAC congressional efforts to force cross-service sharing of labs and centers within the armed services. Congress, it turned out, had been dissatisfied with the low level of cross-servicing that came out of BRAC '95. Its frustration was fueled by the findings of DOD's Laboratory Joint Cross-Service Group, which charged that the separate services undermined the BRAC process by protecting their own labs, leaving DOD with 35 percent more laboratories than it needed because of duplicated work.[8]

In a 1996 report, the U.S. General Accounting Office supported this contention and further argued that laboratory infrastructure reductions had not been cut as deeply as funding, personnel, and combat forces. Congress responded with Section 277 of the Fiscal Year 1996 National Defense Authorization Act, signed into law on 9 February 1996, which required the Secretary of Defense to develop a five-year plan to consolidate and restructure military laboratories and centers. The law's stated goal was to obtain recommendations for consolidating all military laboratories and T&E centers into as few facilities as possible by 1 October 2005.[9]

On 30 April 1996 the Secretary of Defense submitted a plan, called *Vision 21*. Resting upon three integrating pillars of infrastructure reduction, organization and administration restructuring (intra-service and cross-service), and revitalization of aging critical laboratories, *Vision 21* aimed for 20 percent reductions within DOD's laboratory infrastructure beyond BRAC 1995, to be implemented between fiscal years 2001 and 2005. The plan was approved, and DOD contracted with the accounting firm of KPMG Peat Marwick to undertake a cost study for *Vision 21* and to develop a more ambitious data call process than that used during the BRACs. DOD set July 1998 as the deadline for a detailed downsizing and consolidation process plan, which would serve as the basis for President Clinton's fiscal year 2000 budget.[10]

Vision 21 appeared viable in theory. In practice, though, Dahlgren's corporate staff found it nightmarish. To begin with, budget cuts, RIFs, and demands for overhead reductions had left NSWCDD's administration thoroughly decimated. When Captain Overton proposed using the same team members that had worked on BRAC for the *Vision 21* data calls, he found that most were gone due to downsizing. One staffer lamented that "the support staffs are already overworked and this will put everyone over the top."[11]

Staffing aside, the process itself was also problematic. "K" Department line manager Rob Gates, one of the few remaining at Dahlgren with BRAC experience, had recently come over to administration to help plan and track Dahlgren's responses to government initiatives. He became Dahlgren's lead manager on the *Vision 21* effort. Gates quickly realized that the *Vision 21* task, an across-the-board comparison of all of DOD's laboratories, was extremely complicated and so required extremely complicated data calls. The data calls, as developed by KPMG, dictated an enormous taxonomy, in which everything about a lab (personnel, facilities, equipment, etc.) was classified according to the technical area for which he, she, or it was associated. The lab's finances would then be broken down according to a specific technical area, so that the costs of a given function could be computed.[12]

DOD hoped the procedure would enable it to compare the different service labs at a fundamental level and to identify and eliminate duplication wherever possible. Gates thought that it only created a big "mess," with rampant confusion running all up and down the line. When he and his divisional counterparts met with NSWC central staff to develop consistent response strategies for the data calls, they often admitted to not understanding what KPMG was asking for. NSWC either made spot judgments or attempted to get clarifications and then pass them on to *Vision 21* divisional representatives. As a result, even though everyone in NSWC and its laboratories might have had the same understanding, that understanding was usually not the same as KPMG's.[13]

Additionally, KPMG attempted to adopt a common accounting methodology for *Vision 21*. Again, this seemed like a reasonable way to compare costs at the various defense labs. Unfortunately, each service operated its own labs and centers differently and used different cost accounting methods. As Gates later explained, the common accounting scheme was nearly impossible to use since Navy labs are 100 percent industrially funded, some Army labs are industrially funded and some are not, and no Air Force labs are industrially funded. It was therefore easy to determine the cost of a given function in a Navy lab but impossible with others, which only added to the confusion.[14]

KPMG's excessive secrecy also mystified Gates. He later remembered that the firm's *Vision 21* office was as tightly guarded as any compartmentalized program, including SLBMs, that he had ever worked in. "You couldn't get into the room or you couldn't see any of the stuff" relating to laboratory data, he recalled. KPMG's reticence made interpretation of their general instructions and data calls in Dahlgren's particular context—and even getting answers when needed—extremely difficult, since "you couldn't actually talk to anybody" during the process.[15]

In the end, KPMG and DOD never got the final data call answers they were looking for. *Vision 21* stalled in 1998 when draft enabling legislation, which would have established a BRAC-like commission to implement the additional reductions, was rolled into Secretary of Defense William Cohen's Quadrennial Defense Review (QDR). QDR called for two new BRAC rounds, but Congress balked and left the issue hanging for more than three years, until George W. Bush became president in 2001. Under Bush, QDR gave way to the Efficient Facilities Initiative in July 2001. In December 2001

Congress authorized BRAC 2005. Both initiatives promised to carry out cross-service laboratory reductions similar to those planned in *Vision 21*.[16]

Vision 21 and the Federal Laboratory Reform directive tied up Dahlgren's scarce resources and manpower just at the time when Clare and Overton were trying to finish their long-term restructuring program. Nevertheless, they succeeded in meeting their initial 1 October 1996 goal of restructuring NSWCDD to serve the Navy better under its warfighting doctrine, now called *Forward . . . From the Sea* after a 1994 revision. On 13 September 1996, Clare and Overton informed Dahlgren's Board of Directors that all outstanding organizational issues had been resolved and that the long-term restructuring process was complete. The new Dahlgren Division organization would stand up on 1 October.[17]

THEATER WARFARE

Clare and Overton's internal shake-up was the most comprehensive since Barney Smith's 1968 reorganization, with a whole range of administrative and management functions shifted among Dahlgren's various technical departments. One of the most significant changes involved the creation of a new "T" Department for "Theater Warfare Systems," to bring systems engineering to the theater level of warfare for the Navy of the future. "T" Department was largely the brainchild of "F" Department's Thomas C. Pendergraft, who like Clare had been schooled at Dahlgren under Barney Smith and Jim Colvard. Pendergraft's career began in 1963 with a four-year enlistment in the Navy that included two cruises off Vietnam. Leaving the service in 1967, he earned his bachelor of science degree in electrical engineering from Christian Brothers College in Memphis, Tennessee, and came to Dahlgren in 1971, where he specialized in radar, electronic warfare, and sensor systems, garnering him a patent in May 1976 for his invention of a Radar Signature Generator. Pendergraft studied graduate level electronics and systems engineering at Virginia Tech and the University of Virginia, and rose through Dahlgren's management ranks to become a member of the government's Senior Executive Service.[18]

While head of "F" Department in the mid-1990s, Pendergraft realized that his department's future lay at a higher level than developing individual systems for the fleet. SPAWAR had failed to reshape the fleet into a complete warfare system as originally intended, largely because its reliance on "jointness" was anathema to Navy conservatives in the 1980s. Much had happened in the intervening decade, and by 1996 the Navy was beginning to accept the need to extend the total systems approach to theater warfare since ships were no longer operating only by themselves or with other ships, but with other services and satellites. Pendergraft and Clare both sensed the Navy's changing mood and believed the time was right to reintroduce higher-order systems engineering to the service.[19]

Blessed by the Board of Directors, Dahlgren's new "T" Department, headed by Pendergraft, merged the electronic warfare oriented "F" Department with elements of "K" Department (Strategic and Strike Systems) and "A" Department. This was not the old comptroller's "A" Department of the Barney Smith era, but a much more recent warfare analysis and modeling organization similar to "J" Department that had been formed around some of Dahlgren's key GS-15 level managers and headed by Chris Kalivretenos. The new department came together smoothly under "a true partnership sort of arrangement," as Kalivretenos's former deputy Joe Francis recalled, in which leading senior program managers teamed with individual division heads and supported one another. This innovative dual management scheme brought the new department's senior managers, who normally had to focus on administrative and financial issues, back into the world of nuts-and-bolts level scientists and engineers and ensured that important programs and projects would be completed as quickly and efficiently as possible, with a minimum of bureaucratic fuss.[20]

Flexibility was another early hallmark of "T" Department. All of Dahlgren's expertise could hardly be concentrated within a single department. Therefore, Clare and Pendergraft decided early on that "T" Department would work problems not only vertically within the organization but also horizontally across department lines to bring all of Dahlgren's assets to bear on the theater warfare systems problem. In enabling "T" Department to work horizontally, Clare and Pendergraft ensured that it would quickly move to the cutting edge of key programs. Francis and future "T" Department head Barry Dillon later lauded Clare and Pendergraft's foresight in structuring the department as they did, making its capabilities more available to the country at large, and in light of the growing regional and terrorist threats to the country.[21]

"T" Department's flexibility was vital for its success since the problem now transcended systems engineering for individual AEGIS ships and centered around interoperability and electronic netcentric coordination of battle groups, comprised of many ships, operating in regional waters. Dillon later described the problem more fully: "It is essential that these ships be interoperable . . . in a joint manner with coalition forces, and that they achieve all the dynamics, all the flexibility, all the interoperability that can be taken from that level of operation." This included integration of Army and Air Force elements as well, since they would likely contribute forces to future regional operations similar to DESERT STORM, or even the 1980 EAGLE CLAW hostage rescue attempt, in the Middle East and elsewhere.[22]

To help achieve interoperability, "T" Department managers and engineers had to look ahead and develop Concepts of Operations (CONOPs) based upon forecasts of future theaters in which the Navy and the other services might have to fight. Using the forecasts and CONOPs, "T" Department then developed prescribed scenarios that could be run simultaneously on existing hardware at up to fourteen different Navy, Army, and Air Force sites across the country. The simulations gave commanders clear pictures of the integrated battle groups' capabilities and revealed their limitations, all the while preparing them for a possible full interoperability engagement within their assigned theaters of operation. Every battle group scheduled for deployment, from senior commanding officers down to senior enlisted chiefs, was trained this way before leaving port so that they could quickly develop their own CONOPs as necessary, based upon a corresponding simulation.[23]

Interoperability was not the only issue that concerned "T" Department after its establishment. Fire control for precision strike, or land attack, became another key focus of the department's activities since the Navy was developing warships that could fire a variety of weapons, besides SLBMs, landward at ranges exceeding 1,200 miles. The ultimate goal, according to Dillon, was to "put the right weapons in the right place, precisely," using a wide array of sensors and satellite guidance technology. This type of mission required "T" Department engineers to understand not only tactical systems but also strategic systems, and to "overlay" those and to process, filter, integrate, and transmit the necessary data to the fleet. As a result, "T" Department found new netcentric ways of coupling and disseminating information within and among battle groups using satellite communication networks. This allowed battle groups to accurately discern between friendly and hostile aircraft, to launch or fire a vast array of weapons together through a variety of means in a cohesive, organized fashion.[24]

To give the Navy the capability to precisely hit targets at great distances without inflicting significant collateral damage, "T" Department physicists and mathematicians also maintained the accuracy of DOD's Global Positioning System (GPS) satellites, work rooted in Charles Cohen's pioneering geodesy studies in the 1950s and 1960s. In Dillon's estimation, "T" Department's work made the military much more agile and would change the way DOD used the Army and Marine Corps, since in the old days "we used to send everything over, put it on four wheels, and then move it around the country. . . . You don't have to do that today," he boasted, because "you can shoot it from the ship that got it there in the first place."[25]

Indeed, "T" Department's interoperability simulations and precision strike fire control work were so successful that the Navy and DOD tasked the department with doing all the theater warfare assessments and analyses, and even occasional thought pieces, for the Joint Chiefs of Staff, the Chief of Naval Operations, and the senior levels of the Office of the Secretary of Defense (OSD). One noteworthy cost-and-effectiveness analysis performed in "T" Department became the basis for the Navy's anticipated DD-21 *Zumwalt* class Land Attack Destroyer. Designed primarily as a mobile sea-based artillery platform to provide high-precision fire support for ground forces up to a range of one hundred miles, the DD-21 was to be armed with the Mk 45 Mod 4 5-inch, 62-caliber gun and the 155-mm advanced gun system (AGS) capable of firing rocket-assisted extended range guided munitions (ERGM). Although the Navy canceled DD-21 in 2001, the service transferred the destroyer's baseline technology to the program's replacement, DD(X), which DOD envisioned as a multi-mission family of warships rather than a single class with a single mission. "T" Department also began doing anti-submarine analyses for OSD after Russia and China moved to improve their submarine fleets and also proliferated the technology to regional troublemakers, particularly Iran and North Korea, who sought to project their naval power within their respective theaters.[26]

Sea-based theater and national missile defense became critical to "T" Department's mission after the government grew alarmed in the late 1990s by the prospect of nuclear blackmail or, worse, a surprise missile attack on America or one of its allies by rogue nations such as North Korea, Iran, and Libya. The first Persian Gulf War, in which the Iraqis attacked Saudi Arabia, Bahrain, and Israel with SCUD missiles, had already given the United States a glimpse of what it could expect in future conflicts. During the 1990s, the threat only grew as the proliferation of ballistic missile technology among former Soviet republics, client states, and Third World countries gave many of them a real strategic strike capability against regional American interests. As a result, the United States reoriented its ballistic missile defense thinking away from President Reagan's ambitious Cold War era "Star Wars" *Strategic Defense Initiative* toward Theater Missile Defense (TMD), also called Ballistic Missile Defense (BMD), of not only forward-deployed American forces but also the territories of regional allies.[27]

The Army had assumed early responsibility for the resulting TMD program because of its highly visible PATRIOT air defense system, which had engaged incoming SCUDs during the Gulf War. In the mid-1990s, the service began flight-testing a Theater High Altitude Area Defense (THAAD) system with a range of several hundred miles. THAAD was designed to be the upper tier of a multi-layered defense against short-, medium-, and long-range theater ballistic missiles. At the same time, the Navy began parallel programs for sea-based Navy Area Defense and Navy Theaterwide Defense systems based upon the AEGIS/STANDARD missile combination, which some analysts later argued were the more sensible options for theater missile defense.[28]

By the late 1990s, attention returned to the issue of national missile defense after a commission chaired by once and future Defense Secretary Donald H. Rumsfeld warned in 1998 that America had underestimated the long-range ballistic missile threat posed by hostile countries and was vulnerable to a surprise intercontinental attack. The Clinton administration had supported theater missile defense but not national missile defense because of potential costs and possible conflict with the 1972 Anti-Ballistic Missile (ABM) Treaty. The Rumsfeld Commission's report galvanized Congress, however, and Clinton agreed to support a "limited" national defense system, authorized in the National Missile Defense Act of 1999. Rapid advances in anti-missile technology soon blurred the distinctions between theater and national missile defenses, as defined by the ABM Treaty, and the new Bush administration subsequently eliminated what it called the "artificial distinctions" between the two technologies. Abandoning the ABM Treaty as an outmoded relic of the Cold War, and determined to make national ballistic missile defense a cornerstone of his defense policy, Bush proposed in August 2002 the deployment of an enormous system built around ground- and sea-based interceptors and upgraded PATRIOT (PAC-3) units that would be interfaced with land-, sea-, and space-based sensors. THAAD, Airborne Lasers, and the Navy Theaterwide Defense systems would be incorporated into the "missile shield," but not the Navy Area Defense system, which was canceled in 2001 because of poor performance and cost overruns. Initial deployments were scheduled to begin in 2004 and 2005.[29]

Because of Dahlgren's long experience in ballistic trajectory computation and its long-standing work in geoballistics, satellite geodesy, and AEGIS systems engineering, "T" Department had been the natural lead laboratory for the Navy's theater missile defense programs. When Theater Missile Defense evolved into National Missile Defense, the department assumed a greater responsibility for engineering the sea-based component. It thus worked closely with both the Missile Defense Agency (MDA) and the Army's PATRIOT program to develop new methods of early detection, situation awareness, and target queuing so that decisions could be made quickly in advance of emerging threats. This involved all-aspect trajectory calculations, as well as computation of all atmospherics and environmental conditions, and enormously complex simulations that had never been done before. Likewise, the department would handle all battle management communications, command, and control to detect, engage, and kill a target if necessary on a global basis. Dillon later commented, "There's probably nothing more complex. You think about Star Wars? It's nothing compared to what we're on the verge of doing. . . . And not only can we provide the missile defense," he continued, "but we can ensure that it will work with coalition [partners] and the other services, in whatever interoperability [configuration] or method they want." To Dillon, that represented a tremendous asset and a capability that was uniquely Dahlgren.[30]

VINDICATION

While "T" Department stood up and became a major player in naval *theater* operations after the 1996 reorganization, "J" Department came into its own after DESERT STORM seemingly vindicated the Goldwater-Nichols reforms and DOD began instituting jointness with even greater enthusiasm in the mid-1990s. Emblematic of this was the record of the Joint Warfare Center (JWC), established in 1986 in response to Goldwater-Nichols to develop computer simulations for joint exercises and training programs for the chairman of the Joint Chiefs of Staff (JCS) and all the unified commanders in chief. In 1993 Congress approved the creation of the Joint Warfighting Center (JWFC) by combining JWC with the Joint Doctrine Center to analyze "lessons learned," develop joint doctrine further, train theater commanders, and improve computer wargaming using modeling and simulations.[31]

While JWFC and JWAC were standing up, JCS developed the conceptual framework for improving interoperability and conducting joint warfare in the twenty-first century. JCS chairman General John M. Shalikashvili released the plan in July 1996. Entitled *Joint Vision 2010*, it sought to "achieve full spectrum dominance" of future adversaries through the transformation of the U.S. armed forces into a fully integrated, technologically advanced, joint force by the year 2010. Under the mantle of *Joint Vision 2010*, and in recognition of USACOM's successful joint training program, JCS transferred JWFC, along with the Joint Warfare Analysis Center (JWAC), the Joint Command and Control Warfare Center, the Joint C⁴ISR Battle Center, and the Joint Communications Support Element, from their direct control to USACOM in October 1998. After the transfer, JWFC's focus expanded to include Joint Task Force commander training, joint interoperability training, and North Atlantic Treaty Organization (NATO) and Partnership for Peace training. By September 1999, USACOM had evolved and grown so much that JCS gave it a new name, the U.S. Joint Forces Command (USJFCOM).[32]

Although DOD was heading full-throttle into the realm of *joint* warfare, the Navy Department moved a bit slower. Up until 1996, Gallaher had not been allowed to use the word "joint" in his department's name, but he continued to fight hard for a name change that unequivocally reflected "J" Department's joint R&D mission. *Joint Vision 2010* enabled Gallaher to argue effectively that elements at Dahlgren had been doing "joint transformation" since 1980. Seeing that Dahlgren was well ahead of the curve in *joint* warfare, Gallaher's organization was accordingly renamed the "Joint Warfare Applications Department" during the 1996 Dahlgren Division reorganization. Acceptance of this name change had its limits, though, and as Gallaher later recalled, there was no real recognition of the department's joint mission within NAVSEA, where the motto was "we are ships."[33]

Despite NAVSEA's continued reluctance to embrace "J" Department's contribution, Gallaher and his team continued the special work in asymmetric and non-lethal warfare analysis for DOD that they had started in the 1980s. "J" Department began doing more and more work for joint commands including JWAC. One DESERT STORM carryover program that "J" Department developed and JWAC funded was the Collateral Damage Estimation Tool (CDET), which won the 1998 Defense Modeling Simulation Award. CDET allowed commanders to estimate incidental civilian damage.[34]

In 2002, JWAC and "J" Department developed an even better collateral damage tool, called the Fast Assessment Strike Tool-Collateral Damage (FAST-CD) but ingloriously nicknamed BUG SPLAT, after the shapes of projected blast patterns on computer screens that resembled bugs "splatting" against car windshields. Before hostilities opened in Iraq in 2003, an engineering team was sent to teach key theater commanders and their subordinates how to use FAST-CD. The technology and training were timely, and the tool played an important part in reducing collateral damage across a large target spectrum.[35]

Over the years, "J" Department "reverse-engineered" collateral damage tools and subsequently used them to support "force protection." Indeed, "J" Department's collateral damage tools have given the U.S. military the ability to determine defensive measures such as the proper placement of concrete barriers against car and truck bombs to redirect explosive forces away from friendly forces or innocent bystanders.[36]

In the late 1990s, "J" Department also became heavily involved with a number of other joint organizations and programs associated with its nontraditional warfare mission areas. Among these were the Marine Corps' Joint Nonlethal Weapons Directorate (JNLWD) at Quantico, Virginia, which oversaw the development of nonlethal equipment for use in volatile, politically sensitive situations, and the Dahlgren-based Naval Operations Other Than War Technology Center (NOOTW-TC). Established by Vice Chief of Naval Operations Admiral Harold W. Gehman Jr. in July 1997, NOOTW-TC became especially important in the Navy's drive to improve its anti-terrorist and force protection (AT/FP) capabilities following the 12 October 2000 terrorist attack on the USS *Cole* (DDG-67), which killed seventeen sailors and nearly sank the destroyer in the port of Aden, Yemen.[37]

NOOTW-TC's approach to finding solutions to asymmetric threats of this nature was unusual. Engineer Teiji Epling told *National Defense Magazine* in January 2004 that part of his job involved searching Internet chat rooms, reading journals, and contacting professional organizations to find untapped or undeveloped, yet effective, technologies for the Navy's use in asymmetric environments. He also noted that the National Archives was a useful source for rediscovering old concepts that were technologically unfeasible in the past but possibly attainable today. Furthermore, with respect to "J" Department's emphasis on emergency, must-have-now projects during military operations, the center only pursued technologies that could be matured and deployed in only six months. Citing the case of the *Cole*, Epling explained that after the attack the center worked closely with U.S. intelligence agencies to ascertain the exact nature of the asymmetric threat to the Navy. Within the allotted six months, Epling was proud to report, NOOTW-TC developed effective detection systems and countermeasures that could be deployed aboard ships to guard against small explosives-laden boats such as the one that attacked the destroyer.[38]

Among the novel products arising from NOOTW-TC's R&D effort were Unambiguous Warning Devices (UAWDs), nonlethal, tactical blast and stun munitions developed in response to the *Cole* attack. The Dahlgren designers believed that this warning system, essentially a cross between a 50-caliber gun mount and an aircraft flare dispenser, could help watchstanders determine the intent of inbound vessels, but they had to be tested shipboard in a realistic setting. NOOTW-TC representatives therefore first demonstrated the UAWDs in September 2002 aboard the USS *Blue Ridge* (LCC-19) during exercises in which security boats supplied by commander, Fleet Activities (COMFLEACT) in Yokosuka, Japan, made simulated low- and high-speed attack runs against the ship from random approaches, starting at 600 meters out. During the scenarios, the *Blue Ridge* fired 211 nonlethal munitions against the target boats, which reported extremely impressive concussive and ultra-flash effects at ranges of 200 meters and greater. The *Blue Ridge's* commander concluded that the nonlethal munitions could be easily integrated into the ship's force protection plans and procedures by merely equipping its current security personnel with the devices. As a result, the Navy expected to fully qualify and certify UAWDs in fiscal year 2003 and distribute them throughout the fleet shortly thereafter.[39]

"J" Department's focus on infrastructure analysis, nonlethal warfare, and FP/AT inevitably expanded into the national arena, through such organizations as its Joint Program Office for Special Technology Countermeasures (JPO-STC). Chartered by OSD in 1990, JPO-STC played a particularly vital role in the security of the nation's critical infrastructures, beginning with its Infrastructure Assurance Program (IAP), which provided combatant commanders with information concerning their dependencies on commercial and military infrastructures, assessed any disruptions to DOD missions, and identified options for mitigating disruptions. Based on its developed IAP capability, JPO-STC became the overall technical agent for DOD's Critical Infrastructure Protection (CIP) program. CIP started in earnest in 1998 after President Clinton issued Presidential Decision Directive-63, which ordered the identification and assessment of DOD's vital internal, commercial, and cyber-based infrastructures and the development of remediation strategies in the event of loss.[40]

Integrated Vulnerability Assessments (IVAs) are critical to the CIP process. They look at various elements, including computer network defense, physical security and force protection, continuity of operations, and commercial dependencies, and incorporate them into a single comprehensive package for further analysis from a systems perspective. JPO-STC is key to the IVAs, identifying single points of service that could be vulnerable to loss through terrorist acts or natural or man-made natural disasters. JPO-STC and CIP were up and running just before the greatly exaggerated Year 2000 (Y2K) computer bug crisis, in which a programming glitch in a universally used computer operating system was supposed to cause a worldwide crash of private and military computer network systems. The crash never happened, but CIP and JPO-STC gained a great deal of experience in managing and protecting DOD's infrastructure in the event of a real breakdown in the future.[41]

Dahlgren's cyber-security experience led to work with the Missile Defense Agency (MDA) and the Office of Naval Research (ONR) to develop a new computer network intrusion detection system called Secondary Heuristic Analysis for Defensive Online Warfare, or SHADOW for short. SHADOW came about after DOD and the Navy recognized that twenty-first century warfare would not only be fought on battlefields but also in the digital information arena. By the late 1990s, computer hackers were lurking along the so-called "information superhighway," probing for vulnerabilities and launching cyber-attacks against government and private network systems for any number of sinister reasons. To deal with the threat, MDA asked Dahlgren to engineer a multisite, network-based intrusion detection system to detect computer network attacks efficiently, report them quickly, and analyze them to help prevent future intrusions. As a result, SHADOW stood up in May 2001, and because of its public ownership, the Navy made it available to everyone for free via the Internet, with available custom enhancements for other federal agencies or private corporations on a cost reimbursement basis.[42]

Elsewhere on the national scene, Dahlgren's Systems Research and Technology Department, or "B" Department, became a central player in DOD's Counterdrug Technology Development Program as part of President Clinton's National Drug Control Policy. After NSWCDD became the program's Executive Agent in 1996, "B" Department began working closely with military and civilian counterdrug operational forces to determine their technology needs in detecting, monitoring, and restricting the flow of illegal drugs into the country. Once those needs were identified, the department then coordinated with the private sector, academia, and other government laboratories to develop and supply new detection and enforcement technologies, such as UAVs and prototype interdiction equipment, to civilian law enforcement agencies and the U.S. Coast Guard, as well as intelligence and communications support, in the national battle against illegal drug trafficking. Some of "B" Department's solutions, including one developed in "J" Department, could also be considered decidedly low-tech. Dahlgren's skipper from March 2001 to April 2004, Captain Lyal Davidson, later described one simple but effective "J" Department proposal for interdicting suspicious-acting speedboats. According to Davidson, Gallaher's folks suggested throwing a specially made rope or net in front of the boats and entangling their propellers. The Coast Guard was impressed with the method's simplicity. Shortly thereafter, Davidson noted that helicopter-borne Coast Guardsmen chasing alleged drug runners near Fort Lauderdale were observed popping a "J" Department entangling device in front of the suspects' speedboat, and "Whoosh . . . end of the run!"[43]

The legitimization of jointness at Dahlgren was aided by the development of the 1998 Dahlgren Division strategic plan. Mary E. Lacey, at the time the head of "B" Department, wrote "Leveraging Naval Expertise to Meet National Needs" into the plan for 1998-1999 as one of Dahlgren's six strategic goals. The "National Needs" element highlighted all of the ongoing work with joint nonlethal weapons, CIP, JPO-STC, DOD's Counterdrug Technology Program, and Operations Other Than War, and openly suggested that a number of "B" and "J" Department's capabilities were not only appropriate, but also could be used jointly for the greater good of the Navy, DOD, and the nation at large. NAVSEA began to agree, and in 2000, as Gallaher later related, the National Needs element became a virtual product area within NAVSEA. Al Qaeda's dramatic and murderous terrorist attacks within the United States on 11 September 2001 brought the National Needs efforts onto center stage and led to NAVSEA's creation of a Homeland and Force Protection product area within its organization. The vision of "J" Department had proven true. Gallaher felt vindicated, commenting later that the Global War on Terrorism, and much of the work of this new product area began with Bob Hudson and himself over twenty years ago with the Measured Response Options (MRO) program. He further added that DOD's twenty-first century focus on transformation was what the MRO efforts were all about. Said Gallaher, "We have been transforming for over twenty years."[44]

A NEW FORUM

After 1996, Tom Clare spent much of his remaining tenure helping Dahlgren recover from the accrued effects of the BRACs, the hiring and promotion freezes, and the RIFs. The laboratory was fortunate in that leaders like Clare understood the long-term impact of the "peace dividend" and positioned Dahlgren very well to weather the storm. Rob Gates later described how good planning during the lean years ultimately reaped benefits later. According to Gates, Clare confronted the management gap caused by the earlier hiring and promotion freezes by planning a much more rigorous workforce development program, funding it each year with discretionary money for leadership and academic training. After the freeze ended, he accelerated the promotion of those lower level managers who normally would not have been ready in the natural progression. At the same time, Dahlgren's other senior managers began planning budgets and rates very conservatively. As intended, the laboratory began receiving more money than planned, and since the planned budgets for Dahlgren always included overhead to meet expenses, "everything that came on top of that was gravy." Moreover, said Gates, "we were in the right place with the right people with the right kind of background and facilities," and as a result, "everything fell into place."[45]

By 2000, with Dahlgren generating more money than management had planned, more discretionary funds became available for reinvestment in technical advancement, workforce development, and training, giving NSWCDD a decided advantage over the other, less provident NSWC divisions, several of which continued to struggle. Gates particularly remembered the reactions of the other divisional representatives at a 1999 meeting when Captain Vaughn E. Mahaffey, who had become Dahlgren's fortieth commanding officer in September 1997, briefed them on Dahlgren's plan to meet a NAVSEA-mandated 4 percent labor cost reduction. According to Gates, "their jaws dropped" when they saw exactly how much extra money Dahlgren was making and that it was reinvesting it primarily in people, which none of them were doing. As Gates recollected, "they were really impressed that we could actually manage to pull that off," but unfortunately, "nobody felt sorry for us after that when they saw how well we were doing."[46]

While Dahlgren recovered its financial footing, Clare began building new relationships among other Navy laboratories engaged in air systems, undersea systems, and command and control systems. He believed that the Navy's laboratories had never shared a common objective, the key part of a functioning system in his estimation, and that they needed to come together and focus on a common systems approach to important long-term technical issues outside of NSWC's administrative purview but still important for the future Navy. He and others also feared that the service had become fixated only on short-term responsiveness and had turned away from responsible science and engineering, as suggested by the recent changes in Navy RDT&E management.[47]

Jim Colvard, who was still active in Navy RDT&E as a manager, consultant, and teacher, explained in a 1995 white paper written at Clare's request that the Director of Naval Laboratories (DNL) had previously provided a forum to mediate between the Secretary of the Navy's R&D policy intent and OPNAV's command orientation. But the DNL had been disestablished in the Navy's big reorganization of 1992. A Navy Laboratory/Center Coordinating Group (NL/CCG), comprised of the Commanding Officers and Executive Directors of the Navy's four warfare centers, had been created to replace DNL, but it ultimately focused more on administration and always seemed to operate in "defensive mode," fending off external threats to the laboratories and their programs. Moreover, BRAC '91 had removed the warfare centers from SPAWAR's control and placed them firmly under the strict management of NAVSEA and its line officers. Altogether, this left Navy RDT&E on a very short leash, and without focus or a direct mediator with the civilian side of the Navy Department. Thus, like an Old Testament prophet of doom, Colvard had grimly warned Clare and his NSWC superiors that "The Navy is currently living off technical investments of the past, but is pushing a potential disaster into the future."[48]

The lack of an institutional forum to discuss system-wide technical issues and to develop common goals across the entire Navy RDT&E establishment perturbed Clare greatly. A quick phone poll showed that all of the other Navy laboratory executive directors agreed. So, Clare invited them to Dahlgren for the first meeting of the Naval Warfare Systems Forum, which he hoped would forge the executive directors into a single, tight-knit group that could develop common strategies for dealing with Navy-wide systems engineering issues. The COs were also invited to attend, but none did since they recognized that it was focused solely on the technical end of the business and not on management. That first forum, held in 1996, was a tremendous success, and successive meetings, designated like Super Bowls with Roman numerals, followed in the months and years ahead. The meetings got "people talking about things that were of a common interest" in Navy RDT&E, said Clare. He was particularly impressed that "people put it as a high priority on their calendars and that they showed up."[49]

The forums helped develop a unity among all of the Navy laboratories never seen before. After Clare's retirement, several of the labs, including Dahlgren, presented a joint proposal outlining a common approach in the land attack warfare area to a group of admirals. The admirals were used to the old competition that had characterized past laboratory interrelationships and, according to Clare, "were just flabbergasted" since they had never seen them come together like that before. "That was a good thing I did," he concluded.[50]

INTO THE TWENTY-FIRST CENTURY

The Naval Warfare Systems Forum hardly overcame all of the Navy's RDT&E policy shortcomings, but it did give the laboratories a unified voice with which to argue for greater systems integration at every level within the fleet. That they found receptive ears among key admirals and policy makers was reflected in the fact that their systems vision, originally conceived and shaped at Dahlgren, found its way into the Navy strategy for the twenty-first century, called *Sea Power 21*. Announced by CNO Admiral Vern Clark in the October 2002 *U.S. Naval Institute Proceedings* and fleshed out by Clark's subordinates in successive issues, *Sea Power 21* broadened the focus of *Forward . . . From the Sea* to include fully integrated U.S. naval forces operating jointly on a global basis. Three interwoven operational concepts lay at the heart of *Sea Power 21*. The first, Sea Strike, involved the projection of precision and persistent offensive firepower against regional and transnational threats whenever necessary. Next, Sea Shield extended naval defenses beyond the task force to protect the American homeland against ballistic and cruise missiles, control the battlespace off hostile coasts, and provide a defensive umbrella over coalition members and joint forces operating ashore in distant theaters. Finally, Sea Basing sought to reduce the vulnerability of U.S. joint forces and minimize their reliance on the shore establishment by placing them on secure, highly mobile, networked "sea bases" such as aircraft carriers, multimission destroyers, submarines, and pre-positioned transport ships.[51]

The three strands of *Sea Power 21* were bound together through ForceNet, created to realize the long-discussed concept of netcentric warfare. ForceNet tied warfighters, ships, weapons, sensors, satellites, facilities, and command, control, communications, and intelligence assets together into an enormous integrated combat force. The Navy anticipated that ForceNet would greatly accelerate accurate decision-making and provide warfighters with the information and tools needed to dominate any given battlespace. The commander of the Naval Network Warfare Command, Vice Admiral Richard W. Mayo, and Deputy Chief of Naval Operations for Warfare Requirements and Programs Vice Admiral John Nathman touted the initial success of ForceNet in a February 2003 *Proceedings* article. They noted that 80 percent of the targets destroyed by sea-based aircraft during the opening phases of Operation ENDURING FREEDOM in Afghanistan were identified and passed on to the pilots after they had left the carriers' decks and not during mission briefings as traditionally done in the past. None of this would have been possible without the pioneering systems engineering and management work that had been done at Dahlgren, or without the joint input of all the Navy laboratories through the medium of the Naval Warfare Systems Forum. The Navy recognized the forum's contribution to *Sea Power 21* by naming it as the single point of contact for the Navy Warfare Centers' laboratories and as the foundation for the ForceNet Development Center.[52]

Clare finally retired on 30 September 1998 after thirty years of government service. Tom Pendergraft from "T" Department succeeded him as NSWCDD's new executive director in February 1999. A close friend and colleague of Clare's, he inherited a corporate enterprise with an average annual budget of $1.4 billion and roughly 4,930 employees, some 3,800 of which were at Dahlgren Laboratory (making it the largest concentration of scientists and engineers in Virginia) and the rest at the Panama City Coastal Systems Station. Pendergraft was the obvious choice to succeed Clare. As the head of "T" Department, he had managed more than 560 employees and a budget of over $100 million. In 1997 he had won NSWCDD's John Adolphus Dahlgren Award for his management of the department after quickly bringing it on-line only a year earlier. Most importantly, he shared Clare's commitment to systems engineering and management and believed that the consistency of Dahlgren's institutions, paid for over many years by American taxpayers, added significant value not only to the Navy but also to the nation. Captain Lyal Davidson later characterized Pendergraft as "a very passionate speaker for the entire base and its operation." Not surprisingly, then, he would repeatedly call upon his Washington contacts and fight hard with NSWC, NAVSEA, and the Comptroller of the Navy for every penny to maintain those institutions in the years ahead.[53]

Under Pendergraft, NSWCDD continued its recovery and expanded for the first time since BRAC with the realignment of the Dam Neck, Virginia Combat Direction Systems Activity (CDSA) into its organization. Established in 1941 as an anti-aircraft range five miles south of Virginia Beach, Dam Neck had been commissioned in March 1963 as the Atlantic Fleet Anti-Air Warfare Training Center and tasked with planning, developing, testing, and delivering computer programs for shipboard combat direction systems. The Navy had upgraded the center in July 1971 and renamed it the Fleet Combat Direction Systems Support Activity (FCDSSA), Dam Neck. It kept this designation until the January 1992 reorganization, when the Navy aligned it as a detachment with the Naval Surface Warfare Center–Port Hueneme Division and placed it under NAVSEA's command. This administrative arrangement remained in place until NAVSEA disestablished the detachment, reconstituted it as NAVSEA Dam Neck, and realigned it under NSWCDD in December 2000. As an annex to the Oceana Naval Air Station, the 1,100-acre Dam Neck facility still specialized in all non-AEGIS combat direction systems and software. With more than 330 employees, Dam Neck also hosted the Fleet Combat Training Center and thirteen other military tenant commands, comprising over 5,600 instructors, students, and support personnel living or working there, making it a formidable addition to the Dahlgren organization.[54]

While NSWCDD assimilated Dam Neck, Dahlgren's six technical departments ("B" - Systems Research and Technology, "K"- Strategic and

Strike Systems, "G" - Weapons Systems, "N" - Combat Systems, "J" - Joint Warfare Applications, and "T" - Theater Warfare Systems) continued their primary missions of providing technical solutions to America's warfighters as the twentieth century ended and the twenty-first century began. "J" Department and "T" Department kept at the forefront of their respective warfare mission areas, while the others adapted to meet the changing needs of the fleet as U.S. naval strategy evolved from *From the Sea* to *Sea Power 21*. Among its numerous programs, "B" Department continued its R&D work in chemical and biological defenses for the Navy and the joint services, culminating in the opening of the Herbert H. Bateman Chemical and Biological Defense Center (named in honor of the late Virginia congressman who had supported Dahlgren throughout his nine terms in office and as a senior member of the House Armed Services Committee) in the fall of 2001. The $8.6 million, two-story, 35,000-square-foot center was built to develop new chemical and biological agent detectors, next-generation shipboard collective protection systems (CPS), and BW/CW attack computer simulations, and to plan new Navy responses against BW/CW warfare threats.[55]

As in the past, "N" Department concentrated primarily on its AEGIS and higher-ships systems engineering work, and "K" Department was busy performing the systems engineering, software development, and system-level testing on the Navy's new TACTICAL TOMAHAWK Weapon Control System. TACTICAL TOMAHAWK originated in 1990, before the first TOMAHAWK cruise missile was ever fired in combat, when a small group from Dahlgren approached the Office of Naval Research with a proposal to fund a new version of TOMAHAWK that could be flight programmed in under five minutes. Many Navy scientists and engineers outside of Dahlgren thought this an impossible task since the technology would have to be three orders of magnitude better than the current system. However, ONR accepted the challenge and funded the project, beginning in 1991. First known as the "Quick Strike TOMAHAWK" (QST), the wholly new system had to be transition friendly—compatible with existing ship and submarine hardware yet familiar to the current crews. Moreover, the system would be guided using DOD's Global Positioning System (GPS) for precision accuracy rather than the old terrain-following technology of the first generation TOMAHAWKs.[56]

By 1996, "K" Department had engineered an advanced new route planning guidance algorithm that far exceeded expectations. In one QST prototype, a fully automatic GPS mission could be programmed on an existing TOMAHAWK weapon control system computer in under thirty seconds, and it could update threats and no-fly zones in a matter of seconds. For the first time, QST showed the potential for shipboard mission planning. In February 1997 the Surface and Submarine Divisions of the Deputy Chief of Naval Operations (Resources, Warfare Requirements, and Assessments) jointly issued a memorandum officially announcing a requirement for rapid TOMAHAWK mission planning on surface warships.[57]

During the project's contract bidding phase, Lockheed Martin selected the QST prototype as its approach for the rapid planning requirement, and in May 1999 the Navy awarded a contract to the company for the development of the weapon's fire control system, which would incorporate the ONR-funded, Dahlgren-engineered rapid planning algorithm. Subsequent improvements in QST's guidance algorithm made it possible for the missile to loiter in an area and then be redirected to a new target in mid-flight.[58]

One potentially serious guidance problem did emerge during the system's design effort, however. The missile redirection method called for a mid-flight "aimpoint update" with only a new target location and a flight altitude. With an aimpoint update, the missile's embedded guidance program, rather than that supplied during pre-launch mission planning, controls its route trajectory and would simply have the missile make a single turn toward the new target and then fly a great circle trajectory to it. Simulations at Dahlgren showed that the original aimpoint update logic would prevent the missile from reaching its target if it were relatively close to the missile. "K" Department program manager Wayne Harman described the problem as being similar to "trying to run over an object with a car when the object is a few feet to the left of the driver's door. Making a sharp turn to the left will not work because the car turn radius is much larger than the distance to the object. The car would make a complete circle, come back to its starting point, and never be on a path to hit the object." In short, the redirected missile would run in circles. To solve the problem, "K" Department developed a new algorithm that ensured that the redirected missile could reach any target regardless of distance and orientation, proven during simulations and tests at Dahlgren and confirmed by the missile's prime contractor, Raytheon Missile Systems.[59]

Development went smoothly, and in December 2002, DOD approved the Navy's plan to buy 1,353 TACTICAL TOMAHAWK cruise missiles for nearly $2 billion over five years. The complete system's first test occurred on 5 April 2003, when engineers successfully programmed and launched a Raytheon TACTICAL TOMAHAWK missile from the *Arleigh Burke*-class destroyer USS *Stethem* (DDG-63) into the China Lake test range using Lockheed Martin's TACTICAL TOMAHAWK Weapon Control System and the ship's Vertical Launching System. Preliminary analysis indicated that the missile maintained its course as programmed aboard the *Stethem* and landed within its planned Circular Error Probable (CEP) impact zone. The engineers reported that the missile maintained two-way strike communications with the controller and was able to relay its system health and status and battle damage indicators, and successfully received and responded to flight modification commands during the mission. With this successful test, the system entered full production in Fiscal Year 2004.[60]

STORED KILLS

Given that work at Dahlgren had originally been gun and projectile based, it was perhaps a bit ironic that, long after missiles had become the predominant means of delivering ordnance to a target, "G" Department found a twenty-first-century role for shipboard guns. The Navy retired the last of its four *Iowa* class battleships with their mighty 16-inch, 50-caliber guns after the first Persian Gulf War ended in 1991, leaving the Marine Corps and future expeditionary forces with no long-range naval fire support besides that provided by air-dropped bombs, missiles, and the lightweight, short-ranged (thirteen nautical miles) 5-inch, 54-caliber guns carried by most surface warships. Although the battleships had hammered Iraqi defenses on the Kuwaiti coast into rubble, the Navy believed them too inaccurate, inefficient, and costly for the precision warfare needs of the twenty-first century. Under the *From the Sea* strategy, the Marines expected to launch future amphibious assaults from at least twenty-five miles from shore and ideally would be protected by naval fire support from between forty-one and sixty-three nautical miles out. However, the Corps was willing to accept the risks associated with the Navy's decision, but the Senate Armed Services Committee balked. During its Fiscal Year 1991 and 1992 hearings, the committee criticized the Navy for decommissioning the battleships and encouraged the service to adopt a sea-based version of the Army's over-the-horizon MULTIPLE LAUNCH ROCKET SYSTEM (MLRS) and its TACTICAL MISSILE SYSTEM (TACMS), which had been employed with deadly effect in DESERT STORM.[61]

In 1994 the Navy launched a two-phase Naval Surface Fire Support (NSFS) modernization program to remedy its shortfall in fire support. As part of the first phase, which was geared toward the short term, the Deputy Chief of Naval Operations (Resources, Requirements, and Assessments) ordered a round of tests of the so-called NAVAL TACTICAL MISSILE SYSTEM (NATACMS), based upon the Army's TACMS. The tests, using

GPS guidance, were conducted in early 1995 at the White Sands Missile Range in New Mexico and aboard the USS *Mount Vernon* (LSD-39). Post-flight analyses by "G" Department's Missile Systems Division showed that NATACMS could indeed be a "valuable weapon" in the naval inventory for NSFS and pre-invasion strikes. However, in May 1998, CNO decided to modify the existing STANDARD missile for a surface-to-surface land attack role with a planned range of approximately 150 miles rather than developing NATACMS. The Navy planned to install this modified missile on twenty-seven new *Arleigh Burke* class destroyers between 2001 and 2009 and twenty-two *Ticonderoga* class cruisers that had been selected for modernization between 2004 and 2009. However, the installation was canceled due to funding constraints.[62]

The fates of NATACMS and the modified land attack STANDARD missile underscored the fact that rockets and missiles are very expensive to produce: the least expensive is $100,000 per shot, while TOMAHAWKs cost more than $1 million each. Furthermore, they are bulky and warships cannot carry very many of them. Since missile-only ships become impotent after all of their weapons are fired, and because they cannot be reloaded at sea, their efficiency is limited in a combat situation. These drawbacks were clear to the Navy at the start of its NSFS program, so in light of fire support requirements of the littoral and theater focused *From the Sea* doctrine, the service turned to the special 5-inch, 62-caliber gun, called Mk 45 Mod 4, and special extended range, guided munitions (ERGM, pronounced "ur-gum") to meet its short-term NSFS requirements.[63]

Dahlgren's "G" Department became the Technical Direction Agent for the development of both the new gun and ammunition, with United Defense Industries (UDI) and Raytheon as prime contractors, respectively. The UDI Mk 45 Mod 4 gun was a lengthened 5-inch, 54-caliber gun that was strengthened to handle the high-energy propellant needed to shoot an ERGM round over the horizon at distances of up to sixty-three nautical miles. Essentially a rocket-assisted projectile (RAP), ERGM was derived from existing technology, specifically a combination of the 8-inch semi-active laser (SAL) guided projectile developed in "G" Department during the Vietnam War and, ironically enough, the long-range guided projectile concept promoted by Dr. Gerald Bull at the same time. Missiles had eclipsed guided projectiles in the 1970s and 1980s, though, and the technology had come to naught. However, the Navy's NSFS program and *From the Sea* put guns and guided projectiles back in business in the mid-1990s, and "G" Department picked up the work that it had started nearly thirty years earlier.[64]

Times had changed. Back in the good old days, Dahlgren scientists used to have complete control over weapon designs from start to finish, producing the drawings and doing the calculations themselves, and then handing everything over to a manufacturer for production. By the 1990s, because of increasing privatization pressures, the process had evolved to the point that design and control were almost completely contracted out, with Dahlgren providing direction only through critical design reviews. The situation was discouraging from the perspective of Dahlgren personnel, since a contractor's primary motive was often profit, whereas it was the laboratory's business to make sure the Navy received a good product. Moreover, contractors had increasingly taken a disparaging view of government laboratories over the years. With the shift of work to the private sector, some at the labs believed they had been marginalized, often at the peril of specific projects.[65]

ERGM was an example of the tensions that can arise between a government laboratory and defense contractors. As "G" Department division head Tommy Tschirn later recalled, Dahlgren's relationship with Raytheon was difficult during the project's first few years when the overly confident missile giant ignored requested design changes and Dahlgren's advice, even though the laboratory was ERGM's Technical Design Agent. Raytheon ultimately realized that a gun-fired missile, which is what ERGM really was, presented a whole different set of challenges than conventional missiles. After two years of mounting failures and escalating costs, Raytheon finally began accepting "G" Department's input. Drawing upon its thirty-year corporate memory, "G" Department soon put the program back on track and ERGM sprang to life. In the end, "G" Department designed most of ERGM's critical components: the guidance and control system, the optical tracker, the tail-fin assembly, the positive-stop device, and its warhead. Tschirn later insisted that "If we hadn't been there, it would not have worked. They could not make it happen without us."[66]

ERGM subsystem testing started at the White Sands Missile Range in New Mexico, followed by a series of "Control Test Vehicle" (CTV) flight tests in late 2001. On 25 June 2002 the project team reached an important milestone by successfully firing from the 5-inch, 62-caliber gun a GPS-guided projectile 38.5 nautical miles to its target, a world record for a guided gun-launched munition of this type. During the all-up round test, designated Guided Gunfire-1, ERGM achieved the tactical gun-launched acceleration of 10,100 Gs, and engineers observed that all of ERGM's Dahlgren-designed flight control systems worked perfectly. Further, they believed that the projectile's flight could have been stretched beyond 50 nautical miles if the gun had been positioned differently. The projectile's "terminal performance" was especially significant since its onboard GPS system accurately guided it to the target, fulfilling all expectations and signaling that ERGM was ready for performance, safety, and environmental qualifications, which began in Fiscal Year 2003, before entering its initial operational phase in 2006.[67]

According to Tschirn, ERGM was far superior to a conventional missile even though its warhead is not as big. A DD(X) destroyer could carry 220 ERGM "stored kills" in its magazine and the cruiser would be able to store 600 of them. Moreover, they can be reloaded at sea, and because of their exceptional GPS-aided accuracy, they can also kill the vast majority of targets in a combat zone without the need for larger warheads or repeated shots. Since individual warships can carry greater numbers of ERGMs, each of which can destroy a target with only one shot, fewer ships are needed, and at only $50,000 a shot, said Tschirn, "That gun becomes exactly the right device to get the job done."[68]

A RAIL GUN REVIVAL

"G" Department also participated in the development of a GPS-guided, 155-mm Long Range Land Attack Projectile (LRLAP) for the DD(X) warship's Advanced Gun System (AGS), with a range of up to 100 miles, as part of the Navy's second phase of its NSFS program. However, the U.S. General Accounting Office criticized the program's long-term goals as still insufficient for meeting the Marines' requirements for range, lethality, and volume of fire. Furthermore, both the House National Security and Senate Armed Services Committees raised concerns about the extent to which the Navy had considered different gun alternatives. As a result, the Navy looked beyond ERGM and LRLAP for its twenty-first century gunnery needs and began assessing various ONR demonstration projects that explored both maturing and emerging technologies that could be developed to fulfill the Marines' long-range NSFS needs. Breakthroughs in gun and electrical power technology soon provided a potentially powerful solution with the electric, or "rail," gun.[69]

Rail guns operate by generating tremendous electromagnetic forces along two parallel conductors, or "rails," that are bridged by an electrically conductive sliding armature. When a very large current pulse of millions of amps is applied to the rails, a powerful magnetic field is induced that interacts with the armature current, thereby hurtling the armature forward and accelerating a projectile down the gun's barrel to hypersonic speeds of over Mach 7 (2,500 meters per second). Rail guns are conceivably capable of obliterating sea- or land-based targets with GPS-guided projectiles at ranges of over 200 miles and with very short flight times. On top of their lethality, rail guns offer a number of logistical and safety advantages to the Navy. First, the projectiles are compact, measuring thirty inches long and weighing fifty pounds at most. A ship magazine that could only accommodate several hundred bulkier conventional or rocket-assisted rounds could hold thousands of rail gun rounds. Further, automatic loading could be infinitely simplified because of the projectiles' low weight, and the projectiles would be much safer and more convenient for the crew to handle, since in the absence of propellant, there would be no danger of accidental detonations from electromagnetic radiation or Electrostatic Discharge (ESD). Finally, rail gun magazines can be replenished at sea, allowing equipped ships to stay on station indefinitely.[70]

The rail gun concept was nothing new. In 1917 Frenchman Andre Fauchon-Villeplee built a model "electric cannon" in which a magnetic field propelled a 50-gram wing-fitted "flechette" projectile down a conductor-wound gun barrel at a velocity of 200 meters per second. Power technology was far too primitive at the time, though, for the device to be further developed, but during World War II the German Luftwaffe expressed an interest in electromagnetic devices as possible high-velocity anti-aircraft guns. According to one account, the Germans first experimented with a 40-mm electromagnetic gun, sporting a 10-meter-long barrel wrapped in conductive coils and mounted on the undercarriage of a 125-mm anti-aircraft gun. The Germans intended the gun to propel 6.5 kilogram projectiles high into the air at velocities approaching 1,980 meters per second with a rate of fire of 6,000 rounds per minute, but the energy requirements for such an achievement, 1,590,000 amperes at 1,345 volts, doomed the chances for success. Intensive tests conducted late in the war with a smaller 20-mm gun were more successful. Fired at the slopes of Wetterstein Mountain in the foothills of the Alps, it achieved muzzle velocities of over 2,000 meters per second as specified, but still required a considerable amount of energy for its operation. The German engineers developed a new type of condenser that would hopefully improve the gun's efficiency, but the war ended before further tests were done. The U.S. Army captured the experimental gun but lost interest when its scientists quickly found that each gun required a complete power station to operate. Dahlgren's old Experimental Department likewise surveyed the German "magnetic gun" technology in November 1945 but could not overcome the power problem either, and so the concept languished.[71]

In the early 1970s, engineers at the Dahlgren Naval Weapons Laboratory briefly revived the idea and actually drew up plans for a prospective electromagnetic gun. Power requirements were still prohibitive, however, and the project went nowhere. During Reagan's Strategic Defense Initiative of the 1980s, DOD took a close look at rail guns as a possible defense against Soviet ICBMs, and in 1985 the Army initiated research to develop mobile, ground-based rail gun systems capable of defeating future armored vehicles. Technology was still lagging, and as late as 1997 Tschirn and "G" Department advised CNO not to waste money on electric guns because the technology simply was not there yet. Three enabling technologies appeared by 2000, though, that made rail guns feasible and changed everyone's mind. The first was SECNAV's decision to build the new DD(X) warships using Integrated Power Systems (IPS) and fully electric drives. Generating an expected eighty megawatts of electrical power for the DD(X) ships, IPS would allow electrical propulsion motors, sensors, and electric weapons to share power, which could easily be reallocated as needed depending upon changing tactical situations. Using IPS, more than enough power would be available to fire fifteen-to-thirty megawatt rail guns at sustained rates of six to twelve rounds per minute without any loss of ship performance.[72]

The second technological enabler was the advance in precision GPS-guided projectile technology, as seen in ERGMs and "barrage" rounds, which significantly lowered the kinetic energy requirements by reducing warhead mass and eliminating rocket motors. The third enabler came from the Army, which had struggled with a troublesome barrel-wear problem that had given its rail guns a barrel life of only one round. Army-sponsored research at the Institute for Advanced Technology solved the gun barrel problem just as the other two technology enablers materialized. As a result, ONR and NSWCDD sponsored a number of studies that concluded the technologies had sufficiently matured to allow a full-scale proof of concept (POC) demonstration to validate key performance characteristics of both the electromagnetic launcher and the hypersonic guided projectile. CNO directed NAVSEA to incorporate a new Electromagnetic Weapons Division into the Navy Electric Weapons Office with which to manage the development of a full-scale POC rail gun. Additionally, Admiral Robert Natter, Commander, Fleet Forces Command, approved a 1/8-scale rail gun demonstration.[73]

In 2002, "G" Department engineers traveled to the United Kingdom's Electromagnetic Laboratory at Kirkcudbright, Scotland, to help prepare the 1/8-scale demonstration and to design a projectile that could withstand the high acceleration forces expected during hypersonic launches. The rail gun team fired a series of initial proof shots in February and March 2003, and on 24 April executed the official demonstration, attended by Admiral Natter and

Chief of Naval Research Rear Admiral Jay Cohen, with terrific results. Two projectiles were fired through "witness screens" one kilometer downrange using 7.3 megajoules of electrical energy. Each projectile recorded a muzzle velocity of over 2,000 meters (or 6,000 feet) per second and exhibited stable flight characteristics with a .3 mil dispersion.[74]

Since the Navy's envisioned operational rail gun will require 64 megajoules of energy, almost an order of magnitude more than the 1/8-scale gun in Scotland, a great deal of engineering work lies ahead before the full-sized weapon is developed and enters service with a later class of DD(X) warships. As of 2003, Tschirn estimated that the naval rail gun was about fifteen years away from becoming operational, complete with GPS-guided hypervelocity projectiles that are expected to be able to hit targets with devastating accuracy at ranges up to 250 miles. Despite the program's youth, many in the Navy have already grasped its implications. During a visit to Dahlgren in June 2003, commander of the Sixth Fleet Vice Admiral Scott A. Fry exclaimed to Tschirn, "Why do I want to fly an airplane in if I can go in 200 miles with a projectile? I've got 4,000 of them on one ship! What I could do with that would just change the way we fight wars!" Fry's suggestion that rail guns represented progress and could change the Navy not only reflected how far gun technology had evolved over the years and where it was going, but also showed that the service was coming full circle by returning to the gun as its preferred attack weapon for the future.[75]

THE INTERCEPTOR CHALLENGE

While "G" Department nurtured the renaissance of naval gun technology, it was also instrumental in the development of endoatmospheric and exoatmospheric interceptors for Theater Missile Defense (TMD) and National Missile Defense (NMD). This work was accomplished for the STANDARD Missile Program Office and for MDA.[76]

A successful defense against ballistic missiles required the development of an anti-air weapon system with an ability to neutralize the complete range of payloads, including high-explosive, nuclear, biological, and chemical warheads. By the early twenty-first century, ballistic missiles had entered the strategic forces of many of America's potential enemies and had in fact already been deployed. The 1991 DESERT STORM attacks by Iraqi ballistic missiles particularly demonstrated the great need to develop TMD interceptors and assess their lethality and effectiveness. "G" Department, under the direction of the STANDARD Missile Program Office, worked with MDA to establish a lethality program that included not only a means to measure and assess damage at the intercept point, but also a way to assess the post-impact effects of the ballistic missile payloads while defending ground assets. MDA's Corporate Lethality Program led the post-impact effects effort. Assessing the effectiveness of a weapon is critical for two reasons. During the design phase, the amount of damage required at the intercept point is the input that drives the system accuracy requirements and ordnance system design. Additionally, the weapon system's total effectiveness, or the ability to defend against the incoming threat, is the overall reason to deploy a defensive missile system. The Navy and DOD, therefore, must be able to assess weapon system effectiveness to determine its viability.

Along with this extensive assessment program, "G" Department scientists and engineers also participated in the Navy's shipboard validation tests of endo- and exo-atmospheric interceptors. These very difficult tests have proven the potential of these advanced interceptors to significantly contribute to the future national defense, as potentially hostile regional powers are expected to increase the ranges of their ballistic missiles far enough to directly threaten the American mainland in the twenty-first century.

The Navy's role in TMD has had a major impact on the tools and testing techniques used for lethality and endgame effectiveness. The high intercept altitudes and expected high closing velocities have resulted in formidable challenges for simulation and ground testing of typical engagements. Accordingly, Dahlgren test engineers devised unique methods of testing blast-fragment and direct hit TMD warheads at simulated high endo- and exoatmospheric altitudes. Further, Dahlgren was also instrumental in developing new techniques for testing and damage propagation measurements at the Light-Gas Gun Test Facility at Arnold Engineering Development Center in Tullahoma, Tennessee, and at the Sled Track Test Facility at Holloman Air Force Base at Alamagordo, New Mexico.

Along with advanced test techniques, the Navy also had to develop advanced computational tools with which to analyze data for its missile interceptors. Commonly known as hydrocodes, these tools are more correctly described as shock physics analysis packages. These codes solve the conservation equations of mass, momentum, and energy in an energy regime where large deformations, high strain rates, and/or strong shocks occur. Hydrocodes have been extensively validated for penetration, perforation, high explosive detonation, and fragmentation phenomena. These tools have been used extensively to simulate final impact effects for both the warhead-equipped STANDARD Missile-2 Block IVA and the direct-hit STANDARD Missile-3. Dahlgren engineers use the data from these analyses to develop more meaningful tests and assess overall weapon effectiveness in these highly dynamic encounters between incoming ballistic missiles and defending interceptors.

Additionally, building on these established computational tools, new methods of developing and linking Computational Fluid Dynamics Codes with Finite Element Analyses Models have provided new insight into the design of highly complex missile components. Previously, designs of this complexity were only realized through time-consuming and costly trial and error methods. State-of-the-art tools, like those described above, have given the Navy's engineers valuable insights and feedback methods for future complex missile development and assessment.

DAHLGREN STRIKES BACK

On the clear blue morning of 11 September 2001, two hijacked U.S. airliners slammed into the twin towers of the World Trade Center in New York City, while a third plowed into the Pentagon in Washington, D.C. A fourth, believed to be heading for the Capitol building in Washington, nose-dived into the western Pennsylvania countryside after its passengers rallied and attempted to overpower their hijackers. The towers toppled and the Pentagon was severely damaged. All told, nearly 3,000 people were killed. "9/11," as that day became known to the public, shocked America and most of the civilized world by the suddenness and ferocity of the attacks, which were reminiscent of the Japanese *kamikaze* tactics of World War II. The U.S. government soon learned that nineteen Islamic terrorists acting under the orders of Al Qaeda mastermind Osama bin Laden were responsible.[77]

The thunderclap of 9/11 jarred the nation wide awake to the terrorist threat overseas and within its borders. Even before the dust settled over the ruins of the World Trade Center and before the fires inside the Pentagon were fully extinguished, President George W. Bush and Secretary of Defense Donald H. Rumsfeld decided that the attacks constituted outright war, and that America should wage a ruthless campaign of annihilation against Al Qaeda and its affiliates. Shortly thereafter, Bush announced a "Global War on Terrorism" aimed at the complete destruction of the Al Qaeda terrorist network and punishment of those governments that harbored them, beginning with the Islamic fundamentalist Taliban regime in Afghanistan, which sheltered bin Laden and his organization.[78]

Immediately after the attacks, the Navy put to sea with all hands at battle stations. When orders soon arrived to take the fight to the terrorists, the service mobilized its resources for a long fight, fully recognizing that it would have to quickly adapt to the new asymmetric threat posed by Al Qaeda. NAVSEA commander Vice Admiral G. P. Nanos Jr. declared that 9/11 "made obsolete all previous standards for accelerating change," and that its impact "truly transformed the world, our global view, and the importance of ensuring that the Fleet is always ready."[79]

At Dahlgren, Pendergraft and new commanding officer Captain Lyal Davidson swung NSWCDD into a war footing and brought its considerable technical resources to bear in support of the American armada that steamed toward south central Asia in the early fall of 2001. All of Dahlgren's technical departments contributed to the effort, called Operation ENDURING FREEDOM, particularly Barry Dillon's "T" Department, which began coordinating the combat operations for two full Navy battle groups in the Indian Ocean using the systems capability that Pendergraft and his engineering team had built in the late 1990s. Gene Gallaher and his team in "J" Department were, of course, ready for this type of asymmetric conflict. During the ensuing campaign, "J" Department supported combatant commanders by developing emergency asymmetric solutions, most of which remain classified, for American forces fighting on Afghanistan's asymmetric battlefields, which required a true joint effort, with a heavy emphasis on multiservice and multinational Special Forces working in conjunction with American air power and friendly Afghan warlords.[80]

On the home front, reports suggested that bin Laden and his senior associates had likely planted "sleeper cells" throughout the country, as well as in Europe and Asia, and had planned massive follow-on attacks against the United States. According to Captain Davidson, Dahlgren recorded several thousand ideas from inside and outside the government for confronting the asymmetric threat within America. To deal with this overwhelming number of concepts and to bring all of the ongoing smaller projects into much better focus and prioritize them in light of the Global War on Terrorism, Gallaher, with Pendergraft and Davidson's strong backing, organized for DOD the National Innovative Technology Mission Assurance Center (NITMAC) at Dahlgren. NITMAC's mission was to establish a single technology clearinghouse in the fight against terrorism and, using Dahlgren's tried-and-true systems methodology, integrate into a single headquarters all of its affiliated joint and anti-terrorist technology resources such as JPO-STC, NOOTW-TC, JNLWD, SHADOW, "B" Department's Chemical-Biological Warfare Defense Systems, the Counterdrug Technology Program Office, and Electromagnetic Environmental Effects (E^3) to ensure that America is not out-played technologically and asymmetrically.[81]

NITMAC's creation had been approved in 1998 and ground broken on its $11.3 million, 70,000-square-foot building six months before 9/11. After the attacks, however, work accelerated as not only Dahlgren's military customers but also officials from the new civilian Department of Homeland Security began calling Dahlgren to inquire about "J" Department's potent anti-terrorist capabilities. Construction was completed in April 2003, and several NITMAC projects settled into their new offices well before the 25 August ribbon-cutting ceremony. Said Congresswoman Jo Ann Davis of the 1ˢᵗ Virginia District, "NITMAC will be an indispensable asset to stop those intent on doing us and our children harm."[82]

Following the Taliban's fall in early 2002, and in the midst of the ongoing Global War on Terrorism, President Bush ordered General Thomas R. "Tommy" Franks of the U.S. Central Command (CENTCOM) to prepare plans for the invasion of Iraq as part of the administration's new doctrine of preemption, announced in the wake of 9/11. Under Bush's preemption doctrine, the United States would not wait for an attack against American interests to materialize before taking military action against possible threats, terrorist or otherwise. Accordingly, on 19 March 2003, American and coalition forces stormed into Iraq after President Bush authorized CENTCOM to launch Operation IRAQI FREEDOM.

Unlike Operation ENDURING FREEDOM, primarily an asymmetric conflict that one veteran characterized as a "Special Forces Olympics," Operation IRAQI FREEDOM was a more conventional campaign with regular infantry, heavy armored vehicles and tanks, and air and naval forces attacking modern, relatively well-equipped forces defending themselves from behind fixed strong points and bunkers. However, in contrast to DESERT STORM, IRAQI FREEDOM was a far more joint operation from the American perspective, as coalition forces enjoyed a high level of operational integration among air, ground, and naval units that had never been seen before. This much-improved "single-team" aspect of the invasion resulted in lightning advances up the Tigris and Euphrates river valleys by Anglo-American forces, punctuated by precision air strikes against key components of the Iraqi infrastructure, culminating in the fall of Baghdad on 9 April.

Although brief, the military campaign did have an asymmetric component, as Anglo-American forces repeatedly encountered thousands of fanatical Hussein loyalist militiamen called "Sadaam Fedayeen" who were not part of the regular Iraqi Army but fought as civilian guerillas in Baathist stronghold cities. The coalition defeated or bypassed these irregular forces during their drive to Baghdad using joint warfare techniques, including Psychological Operations (PsyOps) and precision infrastructure strikes against their leadership, most notably in Basra, when two American F-15E attack fighters, acting on instant intelligence, dropped laser-guided bombs on a Baathist headquarters in the heart of the city, killing some two hundred Fedayeen Sadaam guerillas and their commanders while sparing nearby structures.[83]

As in DESERT STORM and ENDURING FREEDOM, Dahlgren's technical departments supported coalition naval and land operations during IRAQI FREEDOM and, not surprisingly, with a joint flavor. "N" Department, in coordination with the AEGIS Training and Readiness Center, once again supported AEGIS operations within the five battle groups, while "B" Department supported the fleet with its Chemical, Biological & Radiological Defense systems, which fortunately were not needed despite widespread concern about a purported WMD stockpile. "K" Department specifically provided software and analysis support for the first forty TOMAHAWK cruise missiles launched into Iraq from U.S. naval forces operating in the Persian Gulf and Red Sea on 19 March 2003, and for eight hundred more fired over the next three weeks. To facilitate the TOMAHAWK strikes, "T" Department completely integrated the operations of five large battle groups, something never accomplished before. To accomplish this exceedingly difficult task, "T" Department organized, aligned, and repositioned the battle groups as needed based upon the threats and the fire control geometry required for successful strikes, and also advised the groups on their effective capability limitations within their operating areas. These advisories were critical for the collective battle group commanders because in some instances, their ships and equipment were twenty to thirty years old and not as good as the more recent systems. Said Dillon, "It's essential that they know what those limitations are so that they not get hurt or killed" in combat.[84]

By the end of April 2003, all of Iraq was under American and coalition control, and on 1 May, President Bush landed aboard the aircraft carrier USS *Lincoln* (CVN-72) to announce the end of major combat operations. Although the conventional war had ended, a difficult and lengthy occupation remained. Continuing unrest in post-Hussein Iraq aside, Dahlgren performed its mission extremely well. Pendergraft and Davidson extended their congratulations to their workforce in the spring issue of *The Dahlgren Leading Edge*, noting that "The many scientists and engineers at NSWCDD who have for years made significant contributions to the military's ability to fight, win, and come home safely can stand a little prouder, and a little taller these days, after the world witnessed U.S. warfighter readiness and operational superiority during Operation IRAQI FREEDOM."[85]

Dahlgren likewise received a hearty "job well done" from NAVSEA commander Vice Admiral Phillip M. Balisle when he addressed the laboratory's military and civilian staff during an all-hands meeting on 6 June 2003. "I have been your customer for thirty-three years," he said. "I have sailed the ships you build and you develop and you maintain," and "I have benefitted from the sailors that you have influenced. . . . We're in the Global War on Terrorism," Balisle continued, and although "the enemy is very different than in the Cold War," he was equally adept and maintained a passion in his belief that was so strong that he was willing to send his children to suicidal death to achieve his goals. The war would be a long one, he predicted, lasting decades perhaps, but ultimately, "our children" will win. Unspoken, but understood, was that it was incumbent upon Dahlgren to be even better in the difficult campaign ahead.[86]

ECHOES OF THE PAST

To win the Global War on Terrorism, Balisle announced yet another, even more radical realignment of NAVSEA's shipyards and warfare centers—including Dahlgren—in only a hundred days. The realignment would be executed according to the principles of *Sea Power 21* as well as *Sea Enterprise*, CNO's new program to further reform the Navy's business practices by breaking down production boundaries, eliminating R&D "stovepipes," increasing efficiency, and recapitalizing the new ships and systems that will be needed to continue the transformation of the service in the twenty-first century. Said Balisle, "It is a tough time in our history. We're facing a lot of tough choices in the process." The Vice Admiral concluded with an unsettling forecast, "When the dust settles, we will look different. We may be smaller. . . . We will certainly be different."[87]

Balisle's warning was occasioned by the Bush administration's relentless drive to further shed DOD's surplus infrastructure capacity, which equaled in intensity its fierce determination to fight terrorists. Specifically, Defense Secretary Rumsfeld hoped to close another 20-25 percent of America's military bases and installations for an estimated savings of $3 billion per year. As a result, in mid-December 2001 Congress authorized a new round of BRAC, slated for 2005, and the old familiar cycle of data calls began all over again when DOD publicly announced on 6 January 2004 the first of several such requests from commanders for quantified information concerning their respective bases and facilities. Under the new legislation, DOD would submit its recommendations to the BRAC Commission no later than 16 May 2005.[88]

Rumors of closure began to proliferate in early 2003 when the Navy transferred part of Dahlgren's contract to service AEGIS's software to Lockheed Martin's facility in New Jersey. With the contractor preparing to deliver the final AEGIS systems to the fleet, it was difficult to justify supporting two AEGIS computer centers, so the Navy hinted that it might close the one at Dahlgren. Continued defense industry consolidation also detracted from Dahlgren's *raison d'être*, as contractors merged and began co-opting the systems engineering concept. This reduced both private competition and the need for government laboratories to oversee complex systems design and management, and for a Navy Department Warfare Systems Engineer, as Tom Clare had originally envisioned it in 1992. Additionally, residential growth pressured Dahlgren, much as it had White Oak in the early 1990s. Ever since its establishment in 1918, the station had been isolated and its community something of a "frontier town." However, by the mid-1990s, suburban sprawl had crept southward from LaPlata, Maryland, and westward from Fredericksburg, Virginia, toward Dahlgren. Not surprisingly, there were complaints from new residents who did not work at the laboratory and took exception to the gun testing and ammunition proofing that is still done at Dahlgren. At one point, in response to the concerns, and to remind the community of the laboratory's vital national defense work, a sign appeared outside Dahlgren's front gate that read, "Don't Mind Our Noise: It's the Sound of Freedom!"[89]

Pendergraft and Davidson added to this crisp expression in calming local fears. Pendergraft told the Fredericksburg *Free Lance-Star* that he expected Dahlgren "will always be a place for Navy and other military officials to go when they need a technical solution to their war-fighting problems." Citing brand-new Dahlgren facilities, such as NITMAC, the Distributed Engineering Plant, which is an integrated network linking all of the Navy's shore-based hardware and software combat systems test laboratories into a virtual land-based battle group to insure interoperability before ships go to sea, and the Open Architecture Test Facility, which allows let the Navy build a computer system capable of running several different weapon systems using different software, Pendergraft noted that "We've been involved in this stuff for a long time, but it's all been designed to protect the military. After 9/11, he added, "The technical capability developed at Dahlgren took on a whole new meaning."[90]

Indeed, from its establishment in 1918 through to the present, Dahlgren has always demonstrated a remarkable resiliency and ability to adapt to changing times in order to survive. Its institutional flexibility has allowed it to progress from an isolated gun testing facility, to the Navy's primary computing center and one of only three places in the country where ballistic trajectories could be computed, and finally to the Navy's premier RDT&E laboratory and systems engineering complex. During another difficult period in Dahlgren's history, Jim Colvard mused about the inherent tensions between the fleeting and the enduring. "Someone once said, 'there will always be an England,'" he remarked, asserting with equal certainty that there would always be a Dahlgren. Dahlgren's legacy to the Navy and the nation included not only guns, projectiles, and missiles, but also complete integrated systems with which to deploy and target them with precision and accuracy. With its contributions to the transformation of the military, Dahlgren took the technical skills and systems understanding developed in the world of ordnance to much broader problems of war fighting, strategy, and communication. As an institution, Dahlgren survived and its mission expanded because it adapted to the changing face of the Navy, to the evolving threats to the nation's security, and to the pace of constantly evolving technology. Whatever the future holds, the Dahlgren idea will endure whenever naval surface warfare systems are tested, targeted, deployed, and fired. The sound of freedom begins at Dahlgren—how far it will carry is only a matter of potential and possibility.[91]

Postscript

The Way Ahead, 2004-2006

The year 2004 began much the way 2003 had ended, with dramatic downward pressure on NAVSEA's laboratories. On 3 June 2003, the newly installed commander of NAVSEA, Vice Admiral Phillip M. Balisle, had announced a large-scale, one-hundred-day realignment of the Navy's shore establishment in accordance with CNO Admiral Vern Clark's *Sea Power 21* and *Sea Enterprise* strategies.

The Navy, under tremendous pressure to respond to DOD's requirement of increased short-term readiness and efficiency, looked internally to find the resources to transform and recapitalize the fleet. Its transformation effort included moving management of its shore installations to a newly created command, Commander Naval Installations (CNI). In October 2003, the base at Dahlgren transferred to Naval District Washington West Area, and NSWCDD became a tenant command of the base it had run and owned since its inception.[1]

LEADERSHIP TURNOVER

In April 2004 Captain Joseph L. McGettigan succeeded Captain Lyal Davidson, who retired from active duty and accepted a position in the defense contracting community. McGettigan, a native of Pennsauken, New Jersey, and a 1980 graduate of the U.S. Naval Academy and the Naval Postgraduate School, holds a bachelor of science degree in naval architecture, a master of science degree in undersea warfare technology, and a master of arts degree in national security and strategic studies. During his career, he had served in a number of important technical billets both at sea and ashore. These included, among others, command of the Surface Combat Systems Center at Wallops Island, Virginia, from January 1999 until October 2001 and then Aircraft Carrier and Large Deck Combat System Project Manager in the Program Executive Office for Integrated Warfare Systems before coming to Dahlgren.

Later that same year, Thomas Pendergraft retired from the Senior Executive Service and joined the ranks of the defense contracting community. Following Pendergraft's retirement, the Navy named Stuart Koch as NSWCDD's first Technical Operations Manager in September. Prior to his selection, Koch served in several senior leadership positions for the Force Warfare Systems Department, first as the Deputy Department Head and then as the Acting Department Head. He was also appointed and served as the acting NAVSEA Technical Warrant Holder for Radar, Infrared, Radio Frequency, and Electro-Optic Sensors (except submarine systems) at that time.

Vice Admiral Balisle's realignment was only the first step in NAVSEA's greater transformation program, and happened during a period of significant change in Navy leadership, both up and down the chain of command. In July 2004, Rear Admiral Archer M. Macy, Jr. became the new NSWC commander when his predecessor, Rear Admiral Alan B. Hicks, was reassigned as the Deputy Director for Combat Systems and Weapons (N76F) in the Surface Warfare Directorate of the Office of the Chief of Naval Operations. Vice Admiral Balisle, the initial driver of the NAVSEA transformation, retired in June 2005 and was succeeded by Vice Admiral Paul Sullivan. Sullivan moved quickly to ensure that no momentum was lost during the command turnover. Within the first six months, he had issued a number of tasking memos and Commander's Guidance papers to the Warfare Centers. In October, on his first visit to the Dahlgren laboratory, Sullivan told the managers in an all-hands meeting, "We must continue on our path of change. I need you to stay light on your feet."[2]

The summer of 2005 also saw the retirement of the Chief of Naval Operations, Admiral Vern Clark. His successor, Admiral Mike Mullen, took the Navy's helm in July. In his initial guidance to the Navy, Mullen had made it clear that he planned to continue pursuing Sea Power 21 and the Navy's transformation and recapitalization process. The admiral was succinct in his posture statement before the House Armed Services Committee, telling the congressmen that "With our partners in industry, the acquisition community, OSD, and the interagency, and with the continuing support of the Congress, the Navy will build a force that is properly sized, balanced—and priced for tomorrow."[3]

FUNDAMENTALS OF SUCCESS

Shortly after he assumed his post, McGettigan tapped senior civilian leaders on base to help him find the laboratory's "way ahead." Deep Navy budget cuts and continuing efforts to realign the service made long-term planning a difficult enterprise, but McGettigan was determined to stabilize the organization and restructure it to better suit the CNO's plan for the future. The resulting *Strategic Plan for the Naval Surface Warfare Center, Dahlgren Laboratory* took more than a year to complete, but was crafted with the combined vision and experience of the center's technical leaders and the captain's desire to prepare the center for the uncertainties ahead. The plan incorporated the Strategic direction from OPNAV, NAVSEA, NSWC and the various Product Areas and was an attempt to show each employee how their work supported the larger Navy.

The plan clearly spelled out McGettigan's vision for Dahlgren Laboratory to become "the Department of the Navy's leading warfare system architect and system engineer, recognized as the technical leader in delivering innovative, affordable, and effective solutions for the Navy, Joint Forces, and the Nation." McGettigan also determined that one of the guiding principles that would shape the laboratory's future rested on Dahlgren's investment in its workforce. Simply stated, it declared that "Our people and their competence are fundamental to our success." Firmly ensconced in McGettigan's leadership style and credo, this principle was in line with Mullen's stated guidance to the Navy upon becoming CNO: "Our success in the defense of this nation depends upon the men and women of the United States Navy—active, reserve, and civilian—and their families. Personal and family readiness is vital to combat readiness. Our strength and our future also rely on our diversity."[4]

Indeed, early in his tenure, Mullen had written to his subordinates and commanders that they must consider diversity an every day issue, and that the Navy must begin thinking of diversity in new ways. McGettigan needed no prompting, and he pursued an aggressive policy aimed at fostering it in Dahlgren's workforce. In June 2005, he challenged Dahlgren's managers and supervisors to ensure equal opportunity for all employees. He stated that "as an organization comprised of scientists and engineers, we value the by-product of diversity; the application of varied life experiences to the complex challenge of finding innovative solutions to the nation's technical warfighting needs."[5]

He was so successful that in 2005 Admiral Mullen presented him with the Nathaniel Stinson Equal Employment Opportunity Award. During the ceremony at the Navy Memorial in Washington, D.C., Mullen recognized McGettigan for creating a work place at Dahlgren "which is acknowledged as being a model for creating equal opportunity, and valuing diversity." "From his first day of command," Mullen continued, "he has integrated the strategic management of diversity into Dahlgren's total force strategy with a number of successful initiatives, including a diversity policy, recruitment program, and observance programs for Hispanic Americans, Martin Luther King, Jr., African-Americans, and Asian Pacific Americans as well as many others."[6]

Along with his diversity efforts, McGettigan also implemented the Dahlgren Academic Incentive Program by December 2005—essentially allowing employees, even non-technical ones, to pursue academic degrees and professional certifications partially on the clock and receive a monetary bonus upon successful completion of the program.[7]

BRAC 2005

The 2005 BRAC process led to many sleepless nights and innumerable frayed nerves at Dahlgren. As in previous naval cutback periods and BRAC proceedings, rumors circulated early on that Dahlgren would be closed, based on the old enduring myth that all the station did was test guns and ammunition that the Navy no longer needed. However, after two years of capacity, military value, and scenario data calls, DOD identified Dahlgren in May 2005 as a "specialty site" for Naval Surface Warfare that was unique to the services and a "centroid" for Navy surface ship developments. DOD thus recommended keeping Dahlgren open but wanted to move its Guns and Ammunition programs to Picatinny Arsenal in New Jersey, its Weapons and Armaments (missiles and missile components) program to the Naval Air Weapons Station at China Lake, California, and its C4ISR programs to Point

Loma, California. DOD also recommended moving Dahlgren's Chemical and Biological Research and Development work to the Edgewood Chemical Biological Center at Aberdeen Proving Ground in Maryland, even though a $7.8 million, 19,000-square-foot chemical and biological defense facility had recently been completed at Dahlgren—the Herbert H. Bateman Research Center for Chemical and Biological Defense. Dahlgren in turn would gain the Sensors work from both, Point Loma, California, and Charleston, South Carolina, and the Combat Systems Testing from San Diego. If the BRAC commission accepted this comprehensive recommendation, then Dahlgren would stand to lose 351 civilian jobs and much of its character as a surface warfare RDT&E center.[8]

After the DOD realignment recommendations were issued on 15 May, the Fredericksburg Regional Chamber of Commerce took the lead in the subsequent campaign to overturn the "losing" recommendations. The chamber, in fact, had been preparing for BRAC since 2002 and spent more than $600,000 over the summer to convince the BRAC commission to spare Fredericksburg area military programs, especially those at Dahlgren.

The King George County Board of Supervisors, the local congressional delegation, and the Chamber of Commerce met twice with the commission's staff before decision day. The local officials argued strongly against moving Dahlgren's Guns and Ammunition programs to Picatinny, noting that "DOD's proposal ignored the goals of operational efficiency, enhanced synergy, and reduced excess capacity through consolidation of technical facilities while retaining at least two geographically separated sites." Moreover, they pointed out that its transplanted personnel would have to frequently return there from Picatinny (which has neither big guns nor a test range) to conduct necessary testing, and also that Navy guns are integrated parts of a warship and differ from relatively stand-alone Army guns. In short, the officials argued that a forced move to Picatinny would hurt the Navy's ability to engineer and integrate shipboard combat systems. Finally, they warned the commission of a potential employee brain drain and a "loss of intellectual capital," predicting that no more than 20-25 percent of Dahlgren's professional and technical staff would be willing to relocate to New Jersey.[9]

Concerning DOD's recommendation to move Dahlgren's missile and missile component work to China Lake, Dahlgren's defenders told the BRAC staff that it conflicted with DOD's other recommendation that designated Dahlgren as a "specialty site" for Naval Surface Warfare and that it would impair the Navy's warfighting capability if implemented. And as far as the Chemical and Biological Research and Development work went, they argued against separating its experts from a "fleet-focused environment," especially since Dahlgren had a certified laboratory and offered a unique shipboard testing environment. They also pointed out that a close working relationship already existed between Dahlgren and Edgewood Arsenal, and that relocation to Maryland was unnecessary.[10]

In the end, the BRAC commissioners agreed with most of the arguments in favor of keeping the targeted programs and jobs at Dahlgren. On 25 August they met in a televised session and voted to overturn three of DOD's four recommendations for transferring the programs in question out of Dahlgren. Dahlgren's Chemical and Biological Defense and C4ISR programs remained in place, as well as its Guns and Ammunition RDT&E program. Indeed, during deliberations, the Potomac River Range operations particularly impressed the commissioners, whose chairman at one point interjected that "I don't think that can be replicated at Picatinny." In their final report to the President, the BRAC commissioners concluded that "NSWC Dahlgren has a unique capability to test large over-water guns, and that it possesses most of the expertise in Research, Development, and Acquisition and Testing and Evaluation of these large guns and of the weapons systems integration." "It made more sense," the commissioners wrote, "to retain the life-cycle management of these guns at a single location."[11]

Dahlgren lost its missile and missile component programs to China Lake but gained the Sensors and Combat Systems Testing work as recommended by DOD, for a net gain of jobs. Ultimately, both President Bush and Congress accepted the BRAC commission's recommendations, and Dahlgren emerged in good shape when the process was finally over. Chamber of Commerce President Linda Worrell told the Fredericksburg *Free Lance-Star* that "This is a home run! It's unbelievable." King George County Supervisor Joe Grzeika added that "cooperation between local businesses, and elected and base officials also paid off," and that "the whole community's effort is what made the difference."[12]

A "CROWN JEWEL"

While the Navy Department, DOD, and Dahlgren's leadership labored with BRAC, the station's workforce continued its mission of delivering technical solutions to the Navy's warfighters. This was not an easy task. The overarching demands of GWOT and the Iraq War compelled the Navy to continually revise its RDT&E priorities as Congress and the Bush administration shifted more money away from the sea service to fund Army and Marine combat operations abroad.

Amid the changes and uncertainties, there have been some moments of special recognition for Dahlgren and its capabilities. On 9 August 2004, the chairman of the Senate Armed Services Committee, Senator John Warner of Virginia, toured the Northern Virginia area's three military installations, including Dahlgren. He met with Captain McGettigan and Dahlgren's senior managers and then reviewed some of its most important programs and capabilities. These included the Electric Rail Gun, Special Operations, Homeland and Force Protection, Integrated Command Environment (ICE), TACTICAL TOMAHAWK, NITMAC, and the Distributed Engineering Plant (DEP).[13]

After finishing his tour, Warner issued a fitting valedictory for the station. He commented that "I have seen some absolutely fascinating technology in the minds of the core of civilians and military that operate here, producing things that are saving lives of men and women in our armed forces all over the world." He added, "I think Dahlgren is on a good course and speed to continue to provide not only the Navy, but across the board in a joint way–the armed forces of the United States–with the finest and the best thinking and imagination. . . . I don't think there's anything that duplicates Dahlgren that can be found anywhere in the entire military structure of our country." "I describe it as one of the crown jewels of American defense," Warner concluded, a sentiment shared no doubt by the thousands of servicemen and servicewomen who rely on Dahlgren's technologies to defend them and their country against foreign and domestic threats every day.[14]

Appendices

Appendix I

Roll of Dahlgren's Military Commanders

The **United States Naval Proving Ground** at Dahlgren, Virginia, is established in 1918.

Title: **Inspector of Ordnance**
Commander Henry E. Lackey, January 1917-April 1920
Captain John W. Greenslade, April 1920-May 1923
Captain Claude C. Bloch, June 1923-September 1923
Commander Andrew C. Pickens, September 1923-November 1925
Captain Harold R. Stark, November 1925-September 1928
Captain Herbert F. Leary, October 1928-May 1931
Commander Garrett L. Schuyler, May 1931-July 1934
Captain William R. Furlong, July 1934-May 1936
Captain C. R. Robinson, June 1936-December 1938
Captain J. S. Dowell, December 1938-April 1941
Captain David I. Hedrick, April 1941-April 1943

Title Changed to **Commanding Officer** (April 1943)
Captain David I. Hedrick, April 1943-June 1946
Rear Admiral Charles T. Joy, June 1946-November 1948

Title Changed to **Commander** (November 1948)
Rear Admiral Charles T. Joy, November 1948-August 1949
Rear Admiral Willard A. Kitts, III, September 1949-June 1951
Rear Admiral Irving T. Duke, July 1951-June 1952
Captain James F. Byrne, June 1952-June 1956
Captain R. D. Risser, July 1956-September 1956
Captain G. H. Wales, September 1956-August 1957
Captain R. D. Risser, August 1957-October 1957
Captain M. H. Simmons, Jr., October 1957-August 1959

Dahlgren Naval Proving Ground becomes **Dahlgren Naval Weapons Laboratory** (NWL), 15 August 1959

Bureau of Ordnance becomes **Bureau of Naval Weapons** (BuWeps), 1 December 1959

Appendix I

Roll of Dahlgren's Military Commanders (Continued)

NWL assigned from BuWeps to the **Chief of Naval Materials**, 1 April 1966
  Captain A. R. Faust, September 1959-March 1960
  Captain Thomas H. Morton, March 1960-August 1961
  Captain Robert F. Sellars, September 1961-June 1964
  Captain George G. Ball, July 1964-September 1964
  Captain William A. Hasler, Jr., September 1964-July 1968
  Rear Admiral John D. Chase, August 1968-July 1969
  Captain Steven N. Anastasion, July 1969-January 1972
  Captain John H. Burton, January 1972-August 1972
  Captain Robert F. Schniedwind, August 1972-July 1973
  Captain Robert B. Meeks, Jr., July 1973-September 1974

Dahlgren Naval Weapons Laboratory becomes **Naval Surface Weapons Center,** September 1974
  Captain Robert Williamson, II, September 1974-March 1975
  Captain Conrad J. Rorie, March 1975-September 1977
  Captain Paul L. Anderson, September 1977-August 1981
  Captain James E. Fernandes, August 1981-June 1983
  Captain J. R. Williams, June 1983-August 1986
  Captain Carl A. Anderson, August 1986-June 1988

Naval Surface Weapons Center becomes **Naval Surface Warfare Center,** 1 August 1987
  Captain Robert P. Fuscaldo, June 1988-June 1991
  Captain Norman S. Scott, June 1991-August 1994

Realigned to **Naval Surface Warfare Center, Dahlgren Division,** 2 January 1992
  Captain John C. Overton, August 1994-September 1997
  Captain Vaughn E. Mahaffey, September 1997-March 2001
  Captain Lyal B. Davidson, March 2001-April 2004
  Captain Joseph L. McGettigan, April 2004-Present

Appendix II

Roll of Dahlgren's Civilian Directors

Chief Physicist, Naval Proving Ground
Dr. Louis T. E. Thompson, October 1923-June 1942

Title Changed to **Official in Charge of Laboratories**
Dr. Ralph Sawyer, December 1944-August 1945

Title Changed to **Director of Research**
Dr. Charles C. Bramble, June 1951-January 1954
Mr. Nils A. M. Riffolt, January 1954-August 1956

Title Changed to **Technical Director**
Dr. Russell H. Lyddane, September 1956-August 1964
Mr. Bernard Smith, August 1964-June 1973
Dr. James E. Colvard, July 1973-April 1980
Mr. Ronald S. Vaughn, August 1980-January 1984
Dr. Lemmuel L. Hill, January 1984-February 1989
Dr. Thomas A. Clare, February 1989-July 1992

Title Changed to **Executive Director**
Dr. Thomas A. Clare, July 1992-September 1998
Mr. Thomas C. Pendergraft, February 1999-September 2003

Appendix III

Chronology of Milestones in the History of Dahlgren

13 August 1842	Lobbied by Secretary of the Navy Abel P. Upshur, Congress authorizes the reorganization and modernization of the U.S. Navy, including the creation of a Bureau of Ordnance and Hydrography charged with developing and constructing shipboard weapons and projectiles, as well as surveying and charting the sea floor and coastlines for navigation.
12 February 1844	A 12-inch gun, called the "Peacemaker," explodes aboard the new steam frigate USS *Princeton*, killing Secretary of State Abel P. Upshur and Secretary of the Navy Thomas W. Gilmer, as well as three other officials, two sailors, and President John Tyler's valet.
1847	The Navy assigns Lt. John A. Dahlgren to the Washington Navy Yard. Lt. Dahlgren begins implementing a more scientific and methodical approach to naval gunnery and ordnance testing at his new Experimental Battery along the Anacostia River.
5 July 1862	Congress transfers the hydrographic functions of the Bureau of Ordnance and Hydrography to a new Bureau of Navigation, leaving the newly styled Bureau of Ordnance focused solely on naval guns and ordnance.
1872	The Bureau of Ordnance formally establishes an experimental battery and proving ground at Greenberry Point on the Severn River near Annapolis.
1891	The Navy establishes the U.S. Proving Ground at Indian Head, Maryland, and transfers all ordnance testing from the Annapolis facility.
1902	The Chief of the Bureau of Ordnance, Rear Admiral Charles O'Neil, warns for the first time that because of the greater power and longer ranges of new guns

Appendix III

Chronology of Milestones in the History of Dahlgren *(Continued)*

	Indian Head is quickly becoming obsolete, and that a more isolated location will eventually be necessary.
1910	Because of Indian Head's unfitness as an "experimental station" and its inability to host a new, congressionally funded, experimental program in high-powered gunnery, the Bureau of Ordnance deviates from standard ordnance practice by employing the monitor *Tallahassee* as an experimental ship and the condemned ram *Katahdin* as a floating target platform.
1 August 1914	Germany declares war on Russia, escalating a regional conflict between Austria-Hungary and Serbia into World War I.
8 October 1915	Chief of the Bureau of Ordnance Rear Admiral Joseph Strauss warns that Indian Head has become completely unsuitable for modern testing and experimental work and urgently requests the establishment of a new proving ground.
6 October 1916	The new chief of the Bureau of Ordnance, Rear Admiral Ralph Earle, warns that World War I has changed the character of proof work very greatly, and that the point has been reached in which the government must secure an additional proving ground or the Navy's efficiency will be crippled.
6 April 1917	The United States declares war on Germany and enters World War I.
26 April 1918	Public Law No. 140, 65th Congress, authorizes the President to condemn land for a new proving ground. For the site, Rear Admiral Earle has already chosen land adjacent to Machodoc Creek, a small tributary of the Potomac River near Lower Cedar Point Light.

Appendix III

Chronology of Milestones in the History of Dahlgren (Continued)

28 May 1918	The Navy begins constructing "a great and complete proving grounds" to replace the inadequate facilities at Indian Head with a "complete battery of guns of all sizes, firing down a clear water range of nearly 40,000 yards."
10 June 1918	By Presidential Proclamation #1458, 994.3 acres of the designated site are formally appropriated for the new proving ground.
16 October 1918	Marines supervised by Lt. Commander H. K. Lewis successfully test fire an Army 7-inch, 45-caliber tractor-mounted gun at the new "Lower Station," the first shot ever fired at Dahlgren.
4 November 1918	Presidential Proclamation #1494 attaches the adjacent 372-acre Arnold Farm to the first tract.
11 November 1918	Germany and the Allied powers sign an armistice ending World War I.
December 1918	Naval officers recommend the additional purchase of the 70-acre Blackistone Island, situated some 30,000 yards downriver from the new "Lower Station," for use as a target for heavy-caliber projectiles as well as an observation station.
15 January 1919	The Secretary of the Navy submits "Dahlgren" as the proving ground's identification to the Postmaster General, who soon directs that the post office at the Lower Station be called by that name.
4 March 1919	By Presidential Proclamation #1514, the Navy assumes the title to Blackistone Island.

Appendix III

Chronology of Milestones in the History of Dahlgren (Continued)

May 1919	Under Navy contract, Carl Norden moves his "flying bomb" experiments, using pilotless, explosives-laden aircraft, to Dahlgren from Amity, New York.
18 June 1919	The Navy takes formal possession of Blackistone Island, as well as a spotting range location at Piney Point, Maryland, and five other spotting range locations at strategic positions along the Potomac River shore.
5 August 1919	The Mk II, 14-inch, 50-caliber railway gun is successfully tested at the Lower Station before an audience of Army and Navy officials and prominent engineers.
1920	The Range Section, consisting of a section head, a clerk, and two instrument men, is established. Further, the Bureau of Ordnance asks former Sperry engineer Carl L. Norden to improve the Mk III bombsight.
1921	All gun testing is shifted from Indian Head to Dahlgren.
10 March 1921	Dahlgren submits its first powder test report to the Bureau of Ordnance.
25 July 1921	Construction of the Plate Battery is completed, and ordnance officers conduct the first armor plate firing test, using a 9-inch Class A plate for the USS *Indiana* (BB-50).
April 1923	Civilian physicist and ballistician Dr. L. T. E. Thompson accepts a job as Dahlgren's new Chief Physicist and starts a vigorous experimental program at the station.
1924	Carl Norden and his partner Theodore H. Barth deliver three prototype Mk XI bombsights to Dahlgren for testing.

Appendix III

Chronology of Milestones in the History of Dahlgren (Continued)

15 September 1924	Lt. Ballantine conducts the first successful take-off to landing flight, lasting twelve minutes, of a remotely controlled N9 seaplane.
1926	Small-scale expansion of the Main Battery begins.
1927	Thompson first recommends that a laboratory be built for development tests and experimental work on armor, projectile, and assorted systems at small scale.
1931	Carl Norden, working under Navy contract, begins research, development, and flight testing of the Mark XV Norden Bombsight.
1932	The Bureau of Ordnance formally separates Dahlgren from Indian Head and makes it a separate command.
1936	Five range stations in Virginia are added to the river range and an Experimental Laboratory is established.
1939	Because of increasing danger from bombing tests near State Route 301, Congress appropriates $100,000 to purchase a 6,000-foot "safety zone" around the Naval Proving Ground's simulated aircraft carrier "deck" target.
1940	L. T. E. Thompson and Experimental Officer Lt. Commander William S. Parsons propose the creation of new laboratory dedicated to fundamental research of the metallurgical properties of armor and projectiles.
November 1941	The Armor and Projectile Laboratory is completed and, under newly arrived physicist Dr. Ralph Sawyer, begins conducting reduced-scale tests of armor and projectiles.

Appendix III

Chronology of Milestones in the History of Dahlgren *(Continued)*

7 December 1941   Japanese naval warplanes attack Pearl Harbor, pulling the United States into World War II.

January 1942   L. T. E. Thompson leaves Dahlgren to become scientific director for the Lukas-Harold Corporation, a subsidiary of Carl L. Norden, Inc., which mass-produces bomb and gun sights for the Navy.

February 1942   Naval reservist and physics instructor Dr. Ralph Sawyer becomes senior scientist at Dahlgren.

April 1942   Dahlgren Commanding Officer Captain David I. Hedrick urgently requests that the Bureau of Ordnance procure for Dahlgren a Bush differential analyzer to improve existing ballistic data and to launch a program of ballistic refinement to improve naval fire control.

September 1942   With no differential analyzer available, the Bureau of Ordnance permits Hedrick to form a new exterior ballistics group under Naval Academy mathematician and reservist Dr. Charles C. Bramble to "polish" data supplied by the Massachusetts Institute of Technology and to generate new range tables based on the data.

1943   The Bureau of Ordnance establishes the Aviation Experimental Laboratory within Dahlgren's Aviation Ordnance Department to develop and test new, more exotic types of bombs.

1 January 1943   The Machine Gun Battery enters commission.

4 January 1943   The VT radio proximity fuze, developed and tested at Dahlgren by Parsons and scientists from the National Defense Research Committee's Section "T," is used in combat for the first time.

Appendix III

Chronology of Milestones in the History of Dahlgren *(Continued)*

July 1943 As part of the Manhattan Project, Parsons and Dr. Norman Ramsey first tests the ballistic qualities of the "Thin Man" gun assembly atomic bomb design at Dahlgren using scale, sewer pipe shaped models dropped from 20,000 feet.

October 1943 Hedrick asks the Bureau of Ordnance to explore the possibility of developing a more advanced differential analyzer specifically for the Naval Proving Ground's use.

January 1944 The Navy establishes the Gunner's Mates Training School at Dahlgren.

March 1944 In its final land acquisition, the Naval Proving Ground annexes the Pumpkin Neck Test Area at the mouth of Machodoc Creek.

September 1944 A Rocket Laboratory is built at Dahlgren to test rocket motors.

11 September 1944 At a conference at Dahlgren, Hedrick and Bramble meet military, academic, and private sector experts, including Naval Reserve and Harvard electrical engineering professor Commander Howard H. Aiken, to discuss the preliminary designs of new computing equipment for the proving ground.

October 1944 The Bureau of Ordnance authorizes Hedrick to contract with Harvard University for the design and construction of a controlled sequence calculator, to be designed and built by Aiken for the proving ground.

February 1945 Aiken begins building a new electro-mechanical, sequence-controlled calculator, called the Aiken Relay Calculator Mk II, under contract to Dahlgren.

Appendix III

Chronology of Milestones in the History of Dahlgren (Continued)

6 August 1945	Former Dahlgren experimental officer Captain William S. Parsons serves as "weaponeer" aboard the B-29 bomber *Enola Gay* and arms the "Little Boy" gun assembly atomic bomb in-flight on the way to Hiroshima.
2 September 1945	Japan surrenders to the Allies, ending World War II.
1946	Civilian physicist Dr. Russell H. Lyddane becomes both head of the Armor & Projectile Laboratory and the station's senior scientist after Sawyer leaves Dahlgren to become Technical Director for Operation CROSSROADS.
1947	Bramble reorganizes his exterior ballistics group into the new Computation and Ballistics Department to manage Aiken's Mark II Relay Calculator, which is still being "de-bugged" at Harvard.
March 1948	The Mark II Relay Calculator arrives at Dahlgren, along with a permanent technical team from Harvard, including Ralph A. Niemann, to help Bramble's staff operate and service the machine.
October 1948	The Bureau of Ordnance chooses the Dahlgren Naval Proving Ground as the primary ballistics test and evaluation facility for the light case (LC, or ELSIE) gun assembly, ground-penetrating atomic bomb.
March 1950	Aiken's hybrid Electronic Calculator Mk III arrives at Dahlgren from Harvard.
1950	Using Aiken's relay calculator, Dr. Charles J. Cohen develops the world's first operational six-degree-of-freedom trajectory simulation for unguided rockets, making the development of guided ballistic missiles possible.

Appendix III

Chronology of Milestones in the History of Dahlgren (Continued)

June 1950	The North Korean army invades South Korea, igniting the Korean War.
1951	Dr. Charles Bramble becomes Dahlgren's first "Director of Research."
27 July 1953	An armistice between United Nations forces and the North Koreans and Chinese effectively halts the Korean War.
1954	Charles Bramble retires as Director of Research and is succeeded by Nils Riffolt, while Russell Lyddane becomes Assistant Director of Research.
1955	At a cost of some $2.5 million, IBM builds the Naval Ordnance Research Calculator (NORC) and installs it at Dahlgren, where it begins generating long-range trajectory computations for the first U.S. ballistic missile system—the Army's JUPITER.
9 September 1955	Chief of the Bureau of Ordnance Rear Admiral Fredric S. Withington designates the Dahlgren Naval Proving Ground as the Bureau's prime agency for the respective scientific fields of computation, exterior/rigid body/terminal ballistics, and warhead characteristics. He also authorizes the creation of a new Computation and Exterior Ballistics Laboratory ("K" Laboratory), a Warhead and Terminal Ballistics Laboratory ("T" Laboratory), and a Weapons Development and Evaluation Laboratory ("W" Laboratory).
November 1955	The Navy's Special Projects Office (SPO) is established to oversee the high priority development of fleet ballistic missiles (FBMs).

Appendix III

Chronology of Milestones in the History of Dahlgren (Continued)

1956	Impressed by Dahlgren's eagerness to diversify, Withington assigns the Hazards of Electromagnetic Radiation to Ordnance (HERO) program to "W" Laboratory. This is the beginning of Dahlgren's lengthy involvement with electromagnetic environmental effects and safety for the Navy.
1957	After developing the first rigorous mathematical descriptions of the Earth's gravitational field, which DOD had adopted for all original long-range missile trajectories, "K" Department's Charles J. Cohen and David R. Brown, Jr., capture a trajectory computational role for Dahlgren in SPO's FBM project.
October 1957	The Soviets launch Sputnik I into orbit, sparking a panic within the United States.
1958	Impressed by "K" Department's early work in FBM trajectory computation, the Department of Defense (DOD) assigns Dahlgren full responsibility for preparing and supplying all geoballistic computations and operational aiming data for POLARIS submarine-launched ballistic missiles (SLBMs).
24 May 1959	Since Dahlgren possesses the only Navy computer capable of processing satellite orbital data, the Naval Space Surveillance Operations Center is established in the Computation and Analysis Laboratory to monitor foreign satellites passing over the United States.
15 August 1959	At the request of Technical Director Russell Lyddane and his staff, the Bureau of Ordnance authorizes the Dahlgren Naval Proving Ground to change its name to the U.S. Naval Weapons Laboratory.

Appendix III

Chronology of Milestones in the History of Dahlgren (Continued)

18 August 1959	Congress merges the Bureau of Ordnance with the Bureau of Aeronautics to create the new Bureau of Naval Weapons.
1960	Dr. Charles Cohen and Richard Anderle verify the Earth's pear-shaped gravity field.
22 July 1960	The ballistic missile submarine USS *George Washington* successfully conducts the first underwater launch of a POLARIS missile off Cape Canaveral.
15 November 1960	The USS *George Washington* departs Charleston, South Carolina, for its first operational patrol carrying sixteen POLARIS A1 missiles and some 300,000 targeting cards prepared at Dahlgren.
3 February 1961	The Naval Space Surveillance Operations Center at Dahlgren is redesignated as the Naval Space Surveillance System (NAVSPASUR).
1962	IBM's latest computer, the 7030 STRETCH, arrives at the Naval Weapons Laboratory to help relieve the computing strain on NORC caused by increasing satellite geodesy analyses and missile trajectory work.
	Dahlgren scientists pioneer the development of the General Geodetic Solution, which leads to DOD's World Geodetic System (WGS-62).
1963	NAVSPASUR is designated as a backup computational facility for NORAD's Space Defense Operations Center Computational Center in Cheyenne Mountain, Colorado.
December 1963	Russell Lyddane retires from government service, and the Bureau of Naval Weapons encounters difficulty in finding a replacement at Dahlgren.

Appendix III

Chronology of Milestones in the History of Dahlgren (Continued)

1964	The new Computation and Analysis Building is completed for "K" Laboratory, which helps to give the Naval Weapons Laboratory a more scientific appearance.
	Dahlgren's "K" Laboratory contributes the critically needed guidance method that makes POSEIDON feasible and targetable.
August 1964	While North Vietnamese naval forces attack the USS *Maddox* in the Gulf of Tonkin and Congress gives President Lyndon B. Johnson broad authority to escalate the Vietnam War, former chief engineer of the Bureau of Naval Weapons Bernard Smith arrives at Dahlgren to succeed Russell Lyddane as the new technical director.
1965	DOD assigns Dahlgren responsibility for development and operational responsibilities for POSEIDON as accrued for POLARIS.
20 December 1965	The Navy establishes a Director of Navy Laboratories to represent its laboratories' interests outside the military chain of command.
1966	As part of Operation CONSHOT, Dahlgren engineers construct a 2,600-foot-long conical shock tube to simulate and analyze 20-kiloton nuclear blasts without generating hazardous radiation.
March 1966	Secretary of the Navy Paul Nitze and Defense Secretary Robert McNamara reorganize the Navy's material management system, abolishing the old bureaus and replacing them with Systems Commands. Nitze places the Navy laboratories, including Dahlgren, within the Naval Material Command under a Director of Navy Laboratories to protect laboratory interests against neglect and misuse.

Appendix III

Chronology of Milestones in the History of Dahlgren *(Continued)*

21 October 1967   Soviet-built *Komar*-class missile boats of the Egyptian Navy sink the Israeli destroyer *Eilat* off Port Said with three SS-N-2 STYX missiles, leading the U.S. Navy to reevaluate its fleet anti-cruise missile defenses and to subsequently develop its own offensive anti-ship cruise missile systems, such as HARPOON and TOMAHAWK.

1968   Dahlgren Technical Director Barney Smith oversees a major reorganization of the Naval Weapons Laboratory, abolishing the three-laboratory system in favor of five technical departments and a number of command, administrative, and support departments.

1969   Dahlgren hosts the first annual Naval Gunnery Conclave to discuss the field of gunnery and to determine if it has a future in modern warfare.

December 1969   In response to the *Eilat* sinking, the Navy's Defense Capability Plan (DCP) 16 launches the AEGIS combat systems program.

1970   The Navy designates Dahlgren as its lead laboratory for biological and chemical warfare defense and countermeasures.

15 April 1970   The commander of Naval Ordnance Systems Command designates the Dahlgren Naval Weapons Laboratory as the "Lead Lab for Surface Weapons, with Total Responsibility for the Development of Surface Gunnery Systems."

April 1971   Dahlgren engineers install a 105-mm howitzer aboard an AC-130 Spectre gunship, and it becomes the largest gun ever successfully fired from an American aircraft.

Appendix III

Chronology of Milestones in the History of Dahlgren (Continued)

1972	The Navy formally assigns the surface warfare mission to Dahlgren and gives the station development and operational responsibilities for TRIDENT as accrued for POLARIS and POSEIDON. Computing requirements increase further, and the Naval Weapons Laboratory acquires a Control Data Corporation (CDC) 6700 mainframe computer, designed by CDC chief engineer Seymour Cray, to replace NORC.
January 1972	Secretary of Defense Melvin Laird orders the development of the Strategic Cruise Missile, which is later called the Submarine-Launched Cruise Missile, and ultimately TOMAHAWK.
June 1973	As the Vietnam War ends, Barney Smith retires and is succeeded by James E. Colvard.
1 July 1974	The Naval Ship Systems Command merges with the Naval Ordnance Systems Command to create the new Naval Sea Systems Command (NAVSEA).
1 September 1974	Spurred by ASN (R&D) Dr. David Potter, the Navy consolidates the Dahlgren Naval Weapons Laboratory and the White Oak Naval Ordnance Laboratory to create the Naval Surface Weapons Center (NSWC), headquartered at Dahlgren.
1975	General Dynamics' Pomona Division works closely with Dahlgren's "G" Department to produce a prototype of the PHALANX close-in ship defense system.
1976	The Secretary of the Navy designates NSWC as the lead laboratory for the proposed new AEGIS Combat System, sending Dahlgren into the new field of systems engineering.

Appendix III

Chronology of Milestones in the History of Dahlgren *(Continued)*

January 1977	Flight testing for the TRIDENT (C-4) Submarine Launched Ballistic Missile (SLBM) begins, using Dahlgren-developed fire control and guidance systems.
1 January 1978	Captain Paul Anderson and Dahlgren's Board of Directors "complete the merger" with White Oak by eliminating the separate management structure for the Dahlgren and White Oak sites.
1979	The Navy assigns all TOMAHAWK targeting software development to Dahlgren's "K" Department.
19 March 1980	During tests, the destroyer USS *Merrill* becomes the first surface vessel to successfully launch a TOMAHAWK sea-launched cruise missile.
1980	The first operational PHALANX system is deployed aboard the aircraft carrier USS *Coral Sea*.
April 1980	Colvard leaves Dahlgren to become deputy Chief of Naval Material and is succeeded by Ronald S. Vaughn.
24 April 1980	The multi-service mission to rescue fifty-two American hostages in Teheran, Iran, Operation EAGLE CLAW, ends in disaster because of exceptionally poor planning, training, coordination, communications, and command. The debacle provides a catalyst for change within DOD and leads to a new drive within the U.S. military for more joint operations.
1982	The Navy established a high-tech AEGIS Computer Center at Dahlgren for the development of AEGIS software and to provide facilities and computer engineering services for vessels equipped with the system.

Appendix III

Chronology of Milestones in the History of Dahlgren (Continued)

23 January 1983	Designed and built using a new systems engineering approach pioneered at Dahlgren, AEGIS officially enters service when the Navy commissions the first guided-missile cruiser of its class, the USS *Ticonderoga*.
1 October 1983	Secretary of the Navy John Lehman commissions the Naval Space Command, which assumes control over the Naval Space Surveillance Center.
1983	Dahlgren engineers install the first modern Collective Protection System (CPS) aboard USS *Belleau Wood* (LHA-3). Almost every major warship today uses the Dahlgren CPS as the cornerstone for shipboard chemical, biological, and radiological warfare defense.
1984	Ongoing work in the TOMAHAWK program leads to the establishment of the Cruise Missile Weapon Systems Division at NSWC.
January 1984	Lemmuel Hill succeeds Ron Vaughn as NSWC's technical director.
9 November 1984	The Navy establishes the AEGIS Training Center (ATC) at Dahlgren to prepare and instruct select officers and sailors from the fleet and from Allied navies in the operation and maintenance of AEGIS.
April 1985	Navy Secretary Lehman reorganizes the Navy's laboratory establishment. To eliminate a reporting layer and to decentralize Navy acquisition management, he disestablishes the Naval Material Command and transfers the Director of Naval Laboratories and the warfare centers, first to the Chief of Naval Research, and then to the new Naval and Space Warfare Systems Command (SPAWAR).

Appendix III

Chronology of Milestones in the History of Dahlgren (Continued)

September 1985	The U.S. Space Command (USSPACECOM) is established and assumes space surveillance and space defense missions from NORAD, but NAVSPASUR continues serving as the alternate to USSPACECOM's Space Surveillance Center.
30 September 1986	President Ronald W. Reagan signs the Nichols-Goldwater Defense Reorganization Act into law, strengthening the authority of the chairman of the Joint Chiefs of Staff (JCS) and mandating an enormous reorganization of the U.S. defense establishment to better institute and manage "joint" operations in the future. The Joint Warfare Center (JWC) is established shortly afterward to develop computer simulations for joint exercises and training programs for the chairman of the JCS and all the unified commanders in chief.
21 November 1986	The U.S. Space Command assigns the Navy responsibility for establishing and maintaining an Alternate Space Defense Operations Center (ASPADOC) to support USSPACECOM's primary operations center in case of natural disaster, equipment outage, or hostile attack, a task which fell to NAVSPASUR.
December 1986	NAVSEA designates NSWC as the lead laboratory for STANDARD missile R&D.
March 1987	After two Iraqi EXOCET missiles strike and damage the American frigate USS *Stark* (FFG-31) in the Persian Gulf, Dahlgren engineers participate in the incident's investigation.
17 August 1987	The Chief of Naval Operations approves Dahlgren skipper Captain Carl A. Anderson's request to change the Naval Surface Weapons Center's name to the Naval Surface Warfare Center.

Appendix III

Chronology of Milestones in the History of Dahlgren (Continued)

3 July 1988	After the AEGIS cruiser USS *Vincennes* accidentally shoots down Iranian Airlines Flight 655 with a STANDARD missile, the ship's AEGIS tapes are returned to Dahlgren and studied at the Wallops Island facility.
February 1989	Soviet forces withdraw from Afghanistan following a disastrous eight-year guerrilla war with U.S.-backed Muslim mujahadeen fighters.
27 February 1989	Dr. Thomas A. Clare succeeds Lemmuel Hill as NSWC's technical director.
19 April 1989	An explosion rips through turret two of the battleship USS *Iowa*, killing forty-seven sailors. During the subsequent inquiry, Dahlgren investigators demonstrate that over-ramming propellants into a gun breech could create enough pressure to cause premature detonations.
9 November 1989	The Berlin Wall falls, marking the beginning of the end of the Cold War.
1990	DOD charters the Joint Program Office for Special Technology Countermeasures at NSWCDD to assess the military's dependence on commercial and defense-related infrastructures within the country and to determine how disruptions could affect its ability to perform its missions.
13 February 1990	Commander-in-Chief of the Atlantic Fleet Admiral Paul D. Miller asks Director of Naval Laboratories Gerald Schiefer to establish a technical department and a program office at Dahlgren to foster research in joint and asymmetric warfare and operations other than war. A new Warfare Systems Department, or "J" Department, under Gene Gallaher, "stands up" shortly thereafter.

Appendix III

Chronology of Milestones in the History of Dahlgren (Continued)

2 August 1990	Iraqi dictator Sadaam Hussein's army seizes Kuwait and threatens Saudi Arabia, sparking a regional crisis and an enormous American and allied mobilization in the Middle East under Operation DESERT SHIELD. NSWC's technical departments support early naval operations in a myriad of ways, including, among other things, the upgrade of AN/SLQ-32V electronic warfare threat libraries for Saudi ships, safety certification of ammunition and load-out for the battleships USS *Wisconsin* and USS *Missouri*, and the development of chemical/biological/radiation detection and defensive systems for both fleet and shore-based forces.
16 January 1991	American and coalition forces in the Middle East launch Operation DESERT STORM to evict the Iraqis from Kuwait. Air-to-ground friendly fire among coalition forces becomes a military and political issue, so "J" Department develops a special Identification-Friend-or-Foe (IFF) device for ground vehicles that helps reduce the number of incidents during the campaign's final weeks.
27 February 1991	Following a six-week air campaign and a four-day ground war, American and coalition forces defeat the Iraqi army and liberate Kuwait.
6 March 1991	In the aftermath of DESERT STORM and in view of the dwindling Soviet threat, President George H. W. Bush boasts that a "New World Order" has emerged and signals that a difficult period of realignment and downsizing for the U.S. military is about to begin.
26 December 1991	The Supreme Soviet officially dissolves the Soviet Union, bringing the Cold War to a close.

Appendix III

Chronology of Milestones in the History of Dahlgren *(Continued)*

2 January 1992   As recommended by the Defense Base Closure and Realignment Commission (DBCRC), the Navy's RDT&E establishment is realigned and streamlined into four megacenters, with Dahlgren, White Oak, and the Panama City Coastal Systems Station aligned into the new Dahlgren Division of the Naval Surface Warfare Center. The former proving ground and weapons laboratory becomes known as Dahlgren Laboratory. As part of the larger Navy RDT&E reorganization, the office of the Director of Naval Laboratories is abolished and SPAWAR relinquishes control of the new NSWCDD to NAVSEA.

1993   NAVSPASUR is consolidated with the Naval Space Command, which assumes direct responsibility for operating the naval space surveillance network. Additionally, Congress approves the creation of the Joint Warfighting Center (JWFC) by combining JWC with the Joint Doctrine Center to analyze lessons learned, further develop joint doctrine, train theater commanders, and improve computer wargaming using modeling and simulations. Senior Navy officials at DOD decide that a portion of "J" Department is of such high value to joint warfighting that it should become a new operational command at Dahlgren. Consequently, a portion of "J" Department becomes a separate command and tenant as the new Naval Warfare Analysis Center (NAVWAC).

October 1993   Defense Secretary Les Aspin designates the U.S. Atlantic Command (USACOM), based in Norfolk, Virginia, as the Joint Force Command (JFC) to integrate, train, and oversee most conventional forces based in the continental United States (CONUS).

1994   NAVWAC becomes a joint command and begins reporting to the Joint Chiefs of Staff. As a result, its

Appendix III

Chronology of Milestones in the History of Dahlgren *(Continued)*

	name is changed to the Joint Warfare Analysis Center (JWAC).
28 February 1995	During the third round of Base Realignment and Closure (BRAC), Defense Secretary William Perry announces DOD's recommendation to close the White Oak laboratory.
1996	DOD designates NSWCDD as the executive agent for its Counterdrug Technology Development Program Office, while Tom Clare organizes and hosts the first Naval Warfare Systems Forum at Dahlgren to discuss system-wide technical issues and to develop common goals across the entire Navy RDT&E establishment. Also, "K" Department engineers develop an advanced new route planning guidance algorithm for the Navy's latest cruise missile, TACTICAL TOMAHAWK, capable of quick shipboard targeting and guided by DOD's satellite global positioning system.
30 April 1996	DOD submits a plan called Vision 21 that aims for 20 percent reductions within DOD's laboratory infrastructure beyond BRAC.
July 1996	JCS Chairman General John M. Shalikashvili releases a plan entitled *Joint Vision 2010,* which aims to achieve full spectrum dominance of future adversaries through the transformation of the U.S. armed forces into a fully integrated, technologically advanced, joint force by the year 2010.
1 October 1996	Executive Director Tom Clare and Commanding Officer Captain John Overton complete their post-BRAC internal long-term restructuring of NSWCDD, creating a new "T" Department to focus on Theater Warfare Systems and changing "J" Department's name to the Joint Warfare Applications Department.

Appendix III

Chronology of Milestones in the History of Dahlgren *(Continued)*

1997   Vice Chief of Naval Operations Admiral Jay L. Johnson establishes the Naval Operations Other Than War Technology Center (NOOTW-TC) at Dahlgren to study measured response options for the Navy and to improve its anti-terrorist and force protection (AT/FP) capabilities.

31 July 1997   The White Oak laboratory formally closes as required by BRAC '95.

1998   NSWCDD's strategic plan includes "Leveraging Naval Expertise to Meet National Needs," highlighting, among other things, "J" Department's ongoing work in non-lethal, infrastructure protection, and asymmetric warfare programs that can be used jointly for the greater good of the Navy, DOD, and the nation at large. Also, DOD assigns JWAC to the U.S. Atlantic Command.

30 September 1998   Tom Clare retires as NSWCDD's executive director. Thomas C. Pendergraft from "T" Department succeeds him.

September 1999   The Joint Chiefs of Staff rename the U.S. Atlantic Command as the U.S. Joint Forces Command (USJFCOM).

2000   NSWCDD's "National Needs" thrust becomes a virtual product area within NAVSEA.

12 October 2000   Al Qaeda terrorists attack the USS *Cole* (DDG-67) in the port of Aden, Yemen, killing seventeen sailors and nearly sinking the destroyer. The Navy responds by stepping up R&D at NSWCDD into anti-terrorist and force protection technologies.

Appendix III

Chronology of Milestones in the History of Dahlgren (Continued)

December 2000 The Navy's non-AEGIS computer programming center at Dam Neck, near Virginia Beach, Virginia, is realigned from the Naval Surface Warfare Center, Port Hueneme Division, to Dahlgren Division.

11 September 2001 Al Qaeda terrorists hijack four U.S. commercial airliners and crash two of them into the World Trade Center, one into the Pentagon, and the fourth into the Pennsylvania countryside, killing nearly 3,000 people. President George W. Bush soon launches a "War Against Terrorism" to eradicate the militant Islamic threat to America.

October 2001 As Operation ENDURING FREEDOM gets under way, NSWCDD begins supporting naval and special joint operations against the Taliban regime in Afghanistan.

December 2001 Congress authorizes a new round of BRAC, slated for 2005.

25 June 2002 With NSWCDD acting as technical design agent, "G" Department and Raytheon engineers successfully fire an extended-range, guided munition (ERGM) from the 5-inch/62-caliber gun 38.5 nautical miles to its target, a world record for a guided gun-launched munition of this type.

12 July 2002 Naval Space Command and Naval Network Operations Command merge to form the Naval Network and Space Operations Command.

22 August 2002 Navy officials dedicate the Herbert H. Bateman Chemical Biological Defense Center at Dahlgren and charge it with leading the Navy's RDT&E efforts against chemical and biological warfare threats.

Appendix III

Chronology of Milestones in the History of Dahlgren *(Continued)*

19 March 2003

Under his doctrine of preemption, President Bush authorizes DOD's Central Command to launch Operation IRAQI FREEDOM. During the war, "T" Department integrates, organizes, aligns, and repositions five separate battle groups to achieve maximum strike effectiveness, something never achieved before in combat.

9 April 2003

Baghdad falls to advancing American and coalition forces, signaling the end of Sadaam Hussein's regime in Iraq.

24 April 2003

"G" Department engineers demonstrate a 1/8-scale electromagnetic rail gun at the United Kingdom's Electromagnetic Laboratory at Kirkcudbright, Scotland.

1 May 2003

From the aircraft carrier USS *Lincoln*, President Bush announces the end of major combat operations in Iraq.

6 June 2003

In an all-hands meeting at Dahlgren, NAVSEA Commander Vice Admiral Phillip M. Balisle announces a radical realignment of NAVSEA's shipyards and warfare centers in only a hundred days.

25 August 2003

At Dahlgren, the ribbon is cut on "J" Department's newest facility, the National Innovative Technology Mission Assurance Center (NITMAC). DOD tasks NITMAC to work with nearly all other federal government departments, agencies, and laboratories, as well as private industry and academia, to focus on operational planning, mission assurance, crisis response, critical infrastructure and force protection, weapons of mass destruction defense, and counterdrug technologies, using the integrated systems engineering approach.

Appendix III

Chronology of Milestones in the History of Dahlgren (Continued)

26 August 2003   New National Innovative Technology and Mission Assurance Center (NITMAC) building is dedicated. The facility enables Dahlgren to co-locate many of its homeland defense and force protection programs, and better accomplish its chartered leadership responsibilities in critical infrastructure protection and counterdrug technology for the Department of Defense, chemical and biological warfare defense systems for joint and naval applications, and as the clearinghouse for technologies used in naval operations other than war.

1 October 2003   Chief of Naval Installations is established, resulting in NSWC Dahlgren becoming a tenant command. Management of the Dahlgren base is transferred from NSWC Dahlgren to Naval District Washington (NDW). The base is renamed Naval District Washington West Area.

October 2003   A new structure for the NAVSEA Warfare Center Enterprise is implemented. It includes a combined NSWC/NUWC Board of Directors, 12 Product Area Directors assigned with national responsibilities, and technical Operations Managers at each site.

23 April 2004   Captain Joseph McGettigan takes over for Captain Davidson as NSWCDD Commanding Officer.

27 May 2004   The new TOMAHAWK Block-4 weapon system is approved for Initial Operational Capability (IOC) using Dahlgren's weapon control and mission planning software.

9 August 2004   Senator John Warner (R–Virginia), the chairman of the Senate Armed Services Committee, visits Dahlgren and declares NSWC Dahlgren Laboratory one of the "crown jewels of American defense."

Appendix III

Chronology of Milestones in the History of Dahlgren (Continued)

November 2005   The *Strategic Plan for NSWC Dahlgren Laboratory, 2005 – 2010*, is issued, defining three new focus areas: *Global Strike, Integrated Air and Missile Defense,* and *Full Spectrum Operations*.   Two cross-cutting Leadership Areas are also defined: *System Engineering and Warfare Analysis* and *Science and Technology*.

October 2006   Dahlgren realigns to a new strategic plan, reducing the number of technical departments from six to five.

2 October 2006   NSWC Dahlgren's successful rail-gun test is the first time a rail gun is fired by the Navy in the United States.   This was an extraordinary effort by several departments.

Appendix IV

PUBLIC LAW NO. 140, 65ᵀᴴ CONGRESS, 2ᴺᴰ SESSION
26 APRIL 1918

An Act to authorize the Secretary of the Navy to increase the facilities for the proof and test of ordnance material, and for other purposes.

Be it enacted by the Senate and House of Representatives of the United States of America in Congress assembled, That the Secretary of the Navy is hereby authorized to expend the sum of $1,000,000, or any part thereof, in his discretion, for the purpose of increasing the facilities for the proof and test of ordnance material, including necessary buildings, construction, equipment, railroad, and water facilities, land, and damages and losses to persons, firms, and corporations resulting from the procurement of the land for this purpose, and also all necessary expenses incident to the procurement of said land: *Provided*, That if such lands and appurtenances and improvements attached thereto, can not be procured by purchase within one month after the passage of this Act the President is hereby authorized and empowered to take over for the United States the immediate possession and title of such lands and improvements, including all easements, rights of way, riparian, and other rights appurtenant thereto, or any land selected by him to be used for the carrying out of the purposes of this Act. That if said land and appurtenances and improvements shall be taken over as aforesaid, the United States shall make just compensation therefor, to be determined by the President, and if the amount thereof so determined by the President is unsatisfactory to the person entitled to receive the same, such person shall be paid seventy-five per centum of the amount so determined by the President and shall be entitled to sue the United States to recover such further sum, as, added to the said seventy-five per centum, will make up such amount as will be just compensation therefor, in the manner provided for by section twenty-four, paragraph twenty, and section one hundred and forty five of the Judicial Code. Upon the taking over of said property by the President as aforesaid, the title to all such property so taken over shall immediately vest in the United States. For the purposes of this Act there is hereby appropriated out of any money in the Treasury of the United States not otherwise appropriated the sum of $1,000,000, or so much thereof as may be necessary: *Provided*, That no railroad shall be built in the District of Columbia under this Act, until Congress has approved the point from which such road may start and also the route to be followed in the District of Columbia.

Approved, April 26, 1918.

Appendix V – Letters

Wallbrook Circle
Scarsdale, N.Y.,
March 18th, 1936.

Capt. W.R. Furlong, U.S.N.,
Inspector of Ordnance in Charge,
U.S. Naval Proving Ground,
Dahlgren, Virginia.

Dear Furlong,

Your recent letter, requesting information as to how the Dahlgren Proving Ground got its name, reached me in due course, and I am pleased to be able to recount the following circumstances:

In September of 1918, upon reporting for duty in the Bureau of Ordnance, I was assigned to Desk H, the Armor and Projectile Section. Among its responsibilities was the handling of the Bureau's correspondence in connection with the construction of the new proving ground.

The new proving ground was then referred to as the "Lower Station", and all correspondence, materials, and ordnance went to and through Indian Head; but as its work and permanent population expanded, the necessity for a separate identity became obvious. And it was growing at full speed, as Indian Head was swamped with proof work.

There was a long established post office on the property designated as "Dido", but our plans contemplated its removal; and our correspondence to that end with the Post Office Department had included the designation of the new post office as "Machodoc Creek". But when we were informed that there already existed a post office in Virginia with that designation, the choice of another name became necessary.

It was at that time, late in 1918, that I submitted a memorandum to Admiral Earle, the Chief of the Bureau, in which I recommended that we abandon the procedure that had been followed in the establishment of Indian Head, of employing an existing geographical name, and name the new station after some Naval officer who had been eminent in the development of Naval ordnance, suggesting Stockton, Dahlgren, Dashiell, Alger, and one or two others. Admiral Earle added Sampson and Converse to the list, and in the discussions that followed I have no doubt that other names were considered.

The name Dahlgren was finally selected by Admiral Earle, and upon his recommendation the Secretary of the Navy, on January 15th, 1919, proposed it to the Postmaster General, and on January 24th, 1919, the Postmaster General accepted and directed that designation.

Cordially yours,
<signed>
Logan Cresap
Commander, U.S.N. (Ret.)

Appendix V – Letters (Continued)

March 23, 1936

Commander Logan Cresap, U.S. Navy (Ret),
71 Broadway, Room 411,
New York, New York.

My dear Logan:

I have your letter of the eighteenth of March with a copy of your letter to our good shipmate, Captain Furlong, who is now in command of the prize naval station. I note that he is to give it up in June, and I feel that he must regret that greatly.

You are quite correct in your letter relative to how Dahlgren received its name. It took me rather a long time to discover how to make a new post office. However, the way was simple, once learned. I chose Rear Admiral Dahlgren because I considered him the father of modern ordnance, for it was he who really pulled the Service out of a rut in ordnance in which the Service had been since the War of 1812, and built and advocated heavy ordnance.

He did a great deal of his work at the Naval Gun Factory, was the first Chief of the Bureau of Ordnance, so that I considered it eminently fitting that an Ordnance and Gunnery building at the Naval Academy and our great Naval Proving Ground on the Potomac should carry the name of Dahlgren.

It was nice to hear from you, and I hope that all goes well with you and yours. Someday, I hope to attend one of our Naval Academy graduates' luncheons and dinners at New York, and then perhaps I may have an opportunity to see you. Or possibly, you may be in this vicinity in the summer, and if so, I hope we may be able to meet.[1]

With best wishes, I remain, as always,

Very sincerely yours,

<signed>

Ralph Earle
President

[Worcester Polytechnic Institute,
Worcester, Massachusetts]

---

[1] Both letters are maintained by Ms. Patricia Albert in the Museum Historical Collection, Folder 1936, NSWCDD, Dahlgren, Virginia.

Appendix VI

Oral History Interviews Conducted

Interviewee	*Interviewer*	*Date*
Captain Paul L. Anderson	Dr. Rodney P. Carlisle	26 September 2003
Dr. Thomas A. Clare	Dr. Rodney P. Carlisle	1 May 2003
Captain Lyal Davidson	Dr. Rodney P. Carlisle	5 February 2004
Dr. Armido DiDonato	James P. Rife	13 August 2003
Barry Dillon and Joe Francis	James P. Rife	4 June 2003
C. Eugene Gallaher	Dr. Rodney P. Carlisle	2 May 2003
C. Eugene Gallaher	James P. Rife	14 May 2004
Dr. Robert V. Gates	James P. Rife	4 June 2003
Dr. Lemmuel Hill	Dr. Rodney P. Carlisle	9 May 2003
Charles Roble	James P. Rife	29 July 2003
Captain Norman S. Scott	Dr. Rodney P. Carlisle	22 May 2003
Thomas Tschirn	James P. Rife	25 June 2003
Sheila Young	James P. Rife	5 June 2003

Tapes and unedited transcripts are maintained at NSWCDD, Dahlgren, Virginia.

Appendix VII

Selected Acronyms

A&P	Armor and Projectile
ABL	Armored Box Launcher
ABM	Anti-Ballistic Missile
ADEC	Aiken Dahlgren Electronic Computer
AEC	Atomic Energy Commission
AGS	Advanced Gun System
ALCM	Air-Launched Cruise Missile
AMCGS	Advanced Minor-Caliber Gun System
APL	Applied Physics Laboratory
ARC	Aiken Relay Calculator
ARPA	Advanced Research Projects Agency
ASD/HD	Assistant Secretary of Defense for Homeland Defense
ASN	Assistant Secretary of the Navy
ASN (R&D)	Assistant Secretary of the Navy for Research and Development
ASN (RD&A)	Assistant Secretary of the Navy for Research, Development, and Acquisition
ASPADOC	Alternate Space Defense Operations Center
ASROC	Anti-Submarine Rocket
ASSC	Alternate Space Surveillance Center
ATC	AEGIS Training Center
AT/FP	Anti-Terrorist and Force Protection
ATRC	AEGIS Training and Readiness Center
BRAC	Base Realignment and Closure
BSC	Base Structures Committee
BUAER	Bureau of Aeronautics
BUENG	Bureau of Engineering
BUORD	Bureau of Ordnance
BUSHIPS	Bureau of Ships
BUWEPS	Bureau of Naval Weapons
BW/CW	Biological Warfare/Chemical Warfare
CAD	Cartridge Actuated Device
CBR	Chemical/Biological/Radiological
CDC	Control Data Corporation

Appendix VII

Selected Acronyms *(Continued)*

CDET	Collateral Damage Estimation Tool
CDTDO	Counterdrug Technology Development Office
CENTCOM	United States Central Command
CEP	Circular Error Probable
CHAG	Chemical Hazard Assessment Guide
C^3I	Command, Control, Communications, and Intelligence
C^4ISR	Command, Control, Communications, Computers, and Intelligence/Surveillance Reconnaissance
CIA	Central Intelligence Agency
CIC	Combat Information Center
CIP	Critical Infrastructure Protection
CIWS	Close-In Weapon System
CMP	Cruise Missiles Project
CNM	Chief of Naval Material
CNO	Chief of Naval Operations
CO	Commanding Officer
COMFLEACT	Commander, Fleet Activities
CONOPS	Concept of Operations
CONUS	Continental United States
CPS	Collective Protection Systems
CPU	Central Processing Unit
CSS	Coastal Systems Station
CTV	Control Test Vehicle
DASA	Defense Atomic Support Agency
DBCRC	Defense Base Closure and Realignment Commission
DCP	Defense Capability Plan
DDR&D	Director of Defense Research and Development
DIA	Defense Intelligence Agency
DNA	Defense Nuclear Agency
DNL	Director of Naval Laboratories
DOD	Department of Defense

Appendix VII

Selected Acronyms (*Continued*)

DPO/MA	Defense Program Office for Mission Assurance
EASY	Emulation Aid System
ECM	Electronic Countermeasures
ED	Executive Director
EDO	Engineering Duty Only
E^3	Electromagnetic Environmental Effects
EI	Electronic Intelligence
EMP	Electromagnetic Pulse
EMPASS	Electromagnetic Performance of Aircraft and Ship Systems
EMR	Electromagnetic Radiation
EMV	Electromagnetic Vulnerability
ENIAC	Electronic Numerical Integrator and Computer
ERGM	Extended Range Guided Munition
EPA	Environmental Protection Agency
ESD	Electrostatic Discharge
FBM	Fleet Ballistic Missile
FLIR	Forward Looking Infrared Radar
GPS	Global Positioning System
HERO	Hazards of Electromagnetic Radiation to Ordnance
HULTEC	Hull-to-Emitter Correlation
IAP	Improved Accuracy Program
IAP	Infrastructure Assurance Program
IBM	International Business Machines
ICAD	Integrated Cover and Deception
ICBM	Intercontinental Ballistic Missile
IED	Independent Exploratory Development
IFF	Identification of Friend or Foe
IPS	Integrated Power System
JATO	Jet-Assisted Take-Off
JCMPO	Joint Cruise Missile Program Office
JCS	Joint Chiefs of Staff
JLCTCG	Joint Logistics Commanders Technical Coordinating Group

Appendix VII

Selected Acronyms (*Continued*)

JNLWD	Joint Nonlethal Weapons Directorate
JPO-STC	Joint Program Office for Special Technology Countermeasures
JWA	Joint Warfare Applications
JWAC	Joint Warfare Analysis Center
JWC	Joint Warfare Center
JWFC	Joint Warfighting Center
LC	Light Case
LRLAP	Long-Range Land Attack Projectile
MAF	Marine Amphibious Force
MDA	Missile Defense Agency
MDL	Mine Defense Laboratory
MIRV	Multiple Independently-targetable Reentry Vehicles
MIT	Massachusetts Institute of Technology
MLRS	Multiple Launch Rocket System
MRV	Multiple Reentry Vehicles
NACA	National Advisory Committee for Aeronautics
NASA	National Aeronautics and Space Administration
NATACMS	Naval Tactical Missile System
NATC	Naval Aviation Test Center
NATO	North Atlantic Treaty Organization
NAVAIR	Naval Air Systems Command
NAVELEX	Naval Electronic Systems Command
NAVMAT	Naval Material
NAVORD	Naval Ordnance Systems Command
NAVOSH	Naval Occupational Safety and Health
NAVSEA	Naval Sea Systems Command
NAVSHIPS	Naval Ships Systems
NAVSPACOM	Naval Space Command
NAVSPASUR	Naval Space Surveillance
NAVSPASURFAC	Naval Space Surveillance Facility
NAWC	Naval Air Warfare Center
NCCOSC	Naval Command, Control, and Ocean Surveillance Center

Appendix VII

Selected Acronyms (Continued)

NDRC	National Defense Research Committee
NIH	Not Invented Here
NIRA	National Industrial Recovery Act
NITMAC	National Innovative Technology Mission Assurance Center
NIVA	Naval Integrated Vulnerability Assessments
NLAAG-V	Navy Laboratory Analysis Augmentation Group Vietnam
NL/CCG	Navy Laboratory/Center Coordinating Group
NMD	National Missile Defense
NOL	Naval Ordnance Laboratory
NOOTW-TC	Naval Operations Other Than War Technology Center
NORAD	North American Air Defense
NORC	Naval Ordnance Research Calculator
NOTS	Naval Ordnance Test Station
NPG	Naval Proving Ground
NPL	National Priorities List
NRAC	Naval Research Advisory Council
NRDU-V	Navy Research and Development Unit Vietnam
NRL	Naval Research Laboratory
NSAP	Navy Science Assistance Program
NSFS	Naval Surface Fire Support
NSTC	National Science and Technology Council
NSWC	Naval Surface Weapons Center (1974), later Naval Surface Warfare Center (1987)
NSWCDD	Naval Surface Warfare Center, Dahlgren Division (1992)
NTC	Naval Test Center
NUWC	Naval Undersea Warfare Center
NWAC	Naval Warfare Analysis Center
NWL	Naval Weapons Laboratory
ONR	Office of Naval Research
OPNAV	Office of the Chief of Naval Operations
OSD	Office of the Secretary of Defense

Appendix VII

Selected Acronyms *(Continued)*

OSRD	Office of Scientific Research and Development
PDD	Presidential Decision Directive
PE	Photo Electric
PEO	Program Executive Officer
POC	Proof of Concept
PTCCS	POLARIS Target Card Computer System
QDR	Quadrennial Defense Review
QST	Quick Strike TOMAHAWK
RAP	Rocket Assisted Projectile
RCA	Radio Corporation of America
RDT&E	Research, Development, Test, and Evaluation
RF	Radio Frequency
RFP	Request for Proposal
RIF	Reduction in Force
RV	Reentry Vehicle
SAL	Semi-active Laser
SEAL	Sea, Air, and Land
SECDEF	Secretary of Defense
SECNAV	Secretary of the Navy
SES	Senior Executive Service
SHADOW	Secondary Heuristic Analysis for Defensive Online Warfare
SIP	Separation Incentive Program
SLBM	Submarine Launched Ballistic Missile
SLCM	Sea-Launched Cruise Missile
SPASUR	Space Surveillance
SPAWAR	Space and Naval Warfare Systems
SPO	Special Projects Office (1955)
SRC	Space Research Corporation
SSBN	Sub-Surface, Ballistic, Nuclear
SSN	Space Surveillance Network
SSP	Strategic Systems Programs (1987)
SSPO	Strategic Systems Project Office (1968), later Strategic Systems Program Office (1984)
STAG	Special Task Air Group

Appendix VII

Selected Acronyms (Continued)

SUBROC	Submarine Rocket
SWPS	Stabilized Weapons Platform System
SYSCOM	Systems Command
TACMS	Tactical Missile System
TD	Technical Director
TDR	Target Drone, Radio-controlled
TERCOM	Terrain Comparison
THAAD	Theater High Altitude Area Defense
TMD	Theater Missile Defense
TQL	Total Quality Leadership
UAV	Unmanned Aerial Vehicle
UAWD	Unambiguous Warning Device
UDI	United Defense Industries
UDT	Underwater Demolition Team
ULMS	Undersea Long Range Missile System
UN	United Nations
USACOM	U.S. Atlantic Command
USJFCOM	U.S. Joint Forces Command
USSOCOM	U.S. Special Operations Command
USSPACECOM	U.S. Space Command
VDEQ	Virginia Department of Environmental Quality
VDWM	Virginia Department of Waste Management
VERA	Voluntary Early Retirement Authority
VLAP	Vietnam Laboratory Assistance Program
VLS	Vertical Launch System
VT	Variable Timed
WAVES	Women Accepted for Voluntary Emergency Service
WGS	World Geodetic System
WMD	Weapon of Mass Destruction

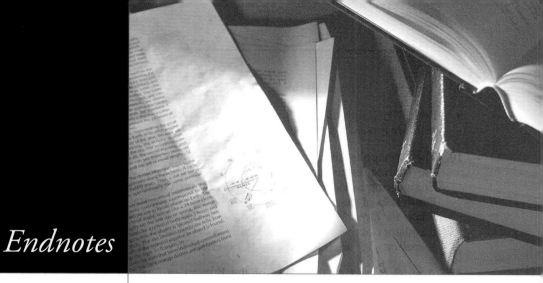

Endnotes

Chapter 1: *Introduction: Proving Ground to Warfare Center*

No Endnotes

Chapter 2: *Finding the Range, 1841-1932*

1. Paolo E. Coletta, "Abel Parker Upshur," in *American Secretaries of the Navy, Volume I, 1775-1913*, ed. Paolo E. Coletta (Annapolis, Md.: Naval Institute Press, 1980), 177-94.

2. House Committee on Naval Affairs, *Accident on Steam-Ship "Princeton,"* 28th Cong., 1st sess. (15 May 1844), Rep. 479, 4-8, 12-14.

3. Ibid., 7, 11.

4. Ibid., 5.

5. The Oregon never fired another shot after its certification tests and was subsequently removed from the *Princeton* after the accident and placed on display at the U.S. Naval Academy in Annapolis, Maryland, where it remains today.

6. House Committee on Naval Affairs, *Accident on Steam-Ship "Princeton,"* 28th Cong., 1st sess. (15 May 1844), Rep. 479, 1-3, 14.

7. Robert J. Schneller Jr., *A Quest for Glory: A Biography of Rear Admiral John A. Dahlgren* (Annapolis, Md.: Naval Institute Press, 1996), 68-70, 75-78.

8. Ibid., 80.

9. Ibid., 81; Ralph Earle, "John Adolphus Dahlgren (1809-1870)," *U.S. Naval Institute Proceedings* 51, no. 5 (March 1925): 424-36.

10. The Annapolis site later became the Engineering Experiment Station (later the Marine Engineering Laboratory), established by the Bureau of Steam Engineering in 1908; the Indian Head Proving Ground was established in 1890. For details of the establishment of the engineering station, see Rodney Carlisle, *Where the Fleet Begins* (Washington, D.C.: Naval Historical Center, 1998), 33-42; the history of the Indian Head facility is detailed in Rodney Carlisle, *Powder and Propellants*, 2nd ed. (Denton, Tex.: University of North Texas Press, 2002).

11. Annual Report of the Chief of the Bureau of Ordnance to the Secretary of the Navy (hereafter BuOrd Annual Report), 1891, 225-26. (Note: BuOrd and SecNav Annual Reports were published by the Government Printing Office in Washington, D.C.)

12. Carlisle, *Powder and Propellants*, 36, n. 51; BuOrd Annual Report, 1902, 33.

13. Carlisle, *Powder and Propellants*, 37.

14. Ibid., 46, nn. 14-17.

15. BuOrd Annual Report, 1910, 5-6; BuOrd Annual Report, 1912, 9; House Committee on Naval Affairs, *Hearings before the Committee on Naval Affairs of the House of Representatives on Sundry Legislation Affecting the Naval Establishment, 1922-1923*, 67th Cong., 2nd, 3rd, and 4th sessions, No. 59, "Statement of Captain H. E. Lackey, United States Navy," 27 February 1922, 455.

16. Carlisle, *Powder and Propellants*, 43-44, notes 1-11.

17. Ralph Earle, ed., *Navy Ordnance Activities, World War, 1917-1918* (Washington, D.C.: Government Printing Office, 1920), 251-52.

18. Ibid., 250.

19. The Board of Investigation determined that $549.66 would be a just amount necessary to repair the damage to the house and property. During the negotiations, Mr. Swann seemed to take the accident in stride, but his wife was not so mollified. She apparently talked to a lawyer and sought greater compensation for the damage (Naval Proving Ground to Bureau of Ordnance, 28 August 1916, National Archives and Records Administration, Washington, D.C. (hereafter NARA I), Record Group (RG) 80, 26893-237). For the full investigation report with accompanying photographs, see "Record of Proceedings of a Board of Investigation Convened at the Naval Proving Grounds, Indian Head, MD., By order of the Inspector of Ordnance in Charge. To Inquire into and Determine Circumstances Surrounding Damage Caused by Firing Sixteen (16) Inch Gun," 7 September 1916, NARA I, RG 80, 26893-237.

Ranging activities were also problematic. After 16-inch guns were installed and ranged later at Dahlgren, ordnance officers discovered that the guns' range tables, which had been calculated at Indian Head from a firing elevation of 8°, were off by 3,000 yards. Such an error, had it not been found and corrected, would have had disastrous consequences for the Navy if its capital ships were to ever engage a foreign battle fleet in a Jutland-scale action. See Admiral McVay's testimony in Senate, *Hearings before the Subcommittee of the Committee on Appropriations of the Senate on H.R. 11228, Navy Department Appropriation Bill*, 67th Cong., 2nd sess., 1923, 221 (hereafter "Senate Hearings on 11228").

20. Annual Report of the Secretary of the Navy to the President (hereafter SecNav Annual Report), 1919, 66-67.

21. Navy Department Library, Special Collections, Senior Commander Biographies, "Rear Admiral Ralph Earle"; for BUORD's official post-World War I account of the Navy's railway batteries, see Earle, *Navy Ordnance Activities*, 179-201; see also Lieutenant Commander Edward Breck's *The United States Naval Railway Batteries in France* (Washington, D.C.: Government Printing Office, 1922), published under the direction of Navy Secretary Edwin Denby; SecNav Annual Report, 1918, 46-47; Commander Ralph Earle to Secretary of the Navy on "Armament of New Battleships," 6 November 1916, NARA I, RG 80, 5039/239.

22. Breck, *The United States Naval Railway Batteries in France*, 7; Mr. Swepson Earle to the Editor of the *Baltimore Sun*, 1 May 1922, newspaper clipping found at NARA I, RG 74, General Correspondence, 1926-1943, Box 2984, Folder 37630.

23. House Committee on Naval Affairs, *Hearings before the Committee on Naval Affairs of the House of Representatives on Estimates Submitted by the Secretary of the Navy 1918*, 65th Cong., No. 4, "Statement of Rear Admiral Ralph Earle, United States Navy, Chief of the Bureau of Ordnance, Navy Department,"18 January 1918, 62-63, 73-77; House Committee on Naval Affairs, *Hearings before the Committee on Naval Affairs of the House of Representatives on Sundry Legislation Affecting the Naval Establishment*, 1922-1923, 67th Cong., 2nd, 3rd, and 4th sessions, No. 59, "Statement of Captain H. E. Lackey, United States Navy," 27 February 1922, 454-45.

24. Mr. Swepson Earle to the Editor of the *Baltimore Sun*, 1 May 1922, newspaper clipping found at NARA I, RG 74, General Correspondence, 1926-1943, Box 2984, Folder 37630; House Committee on Naval Affairs, *Hearings before the Committee on Naval Affairs of the House of Representatives on Sundry Legislation Affecting the Naval Establishment, 1922-1923*, 67th Cong., 2nd, 3rd, and 4th sessions, No. 59, "Statement of Captain H. E. Lackey, United States Navy," 27 February 1922, 454-55. Located on the north mouth of the River Thames close to Southend-on-Sea, Essex, the Shoeburyness range was BUORD's model for building its new proving ground at Machodoc Creek. Dahlgren's history, in fact, closely parallels that of Shoeburyness, which had become the British Army and Navy's primary proving ground in the 1840s after the artillery ranges at Plumstead Marshes, near Woolwich, became increasingly dangerous for gun testing because of the greater distances needed and their proximity to the heavily traveled Thames

shipping route. Today, the parallels continue as the Defence Evaluation and Research Agency (DERA) Shoeburyness does much of the same work for the British Ministry of Defence as the Naval Surface Warfare Center-Dahlgren Division does for the U.S. Navy and the Department of Defense.

25. House Committee on Naval Affairs, *Hearings before the Committee on Naval Affairs of the House of Representatives on Estimates Submitted by the Secretary of the Navy 1918*, 65th Cong., No. 4, "Statement of Rear Admiral Ralph Earle, United States Navy, Chief of the Bureau of Ordnance, Navy Department,"18 January 1918, 74-75; Public Law 140, 65th Cong., 2nd sess. (26 April 1918), see Appendix IV.

26. The Battle of Jutland was fought 31 May-1 June 1916 and represented the last clash between two "main battle fleets" comprising "Dreadnought" class battleships and a host of lesser vessels in a gunnery duel in the twentieth century. (The Battle of Tsushima Strait in the Russo-Japanese War on 27 May 1905 was the first in the twentieth century.) As such, Jutland was studied extensively at the time and since. It was clear that Earle and others examined the immediate reports from Britain and, after the war, from Germany in order to learn what improvements to ordnance might be made on the basis of the battle experience. In particular, the post-battle report of the Commander in Chief of the German High Seas Fleet, Vice Admiral Reinhard Scheer, was obtained and translated by the Office of Naval Intelligence and then sent to BUORD for review by U.S. ordnance officers. The report was subsequently filed away at BUORD and later became a part of its archival record. See "Report of the Commander in Chief of the German High Seas Forces Regarding the Naval Battle of Skagerrack," Office of Naval Intelligence Translations, 9 February 1920, NARA II, RG 74, Entry 1001, Box 3, Folder Scheer's Report on Jutland, 32156. The extensive literature on the battle has been supplemented in recent years by V. E. Tarrant's *Jutland, The German Perspective* (London: Cassell, 1995) and George Bonney's *The Battle of Jutland*, 1916 (London: Sutton Publishing, 2002). For ordnance experts, the immediate concerns were several: protection of powder magazines and elevators, improved armor, improved projectiles for penetrating armor, improved illuminating shells for night fighting, and, not incidentally, recognition of the role of submarines, aircraft, and radio as new elements of the naval battle scene. The outcome of the battle was much debated, as Britain lost more tonnage of ships and was generally out-gunned and out-officered by Germany, but the German navy decided not to risk another major engagement and remained largely in port through the rest of World War I. The failure of British fire control is discussed in Peter Padfield, Guns at Sea (London: Evelyn, 1973), 278 ff. For specific reference to Jutland before Congress, see House Committee on Naval Affairs, *Hearings before the Committee on Naval Affairs of the House of Representatives on Estimates Submitted by the Secretary of the Navy 1918*, 65th Cong., No. 4, "Statement of Rear Admiral Ralph Earle, United States Navy, Chief of the Bureau of Ordnance, Navy Department,"18 January 1918, Exhibit A, 106; SecNav Annual Report, 1918, 58.

27. Earle, *Navy Ordnance Activities*, 252-53; *Public Laws of the Sixty-Fifth Congress of the United States*, chap. 64, Public Law 140, 65th Cong., 2nd sess., 537-38 (see Appendix IV); Presidential Proclamation 1458, 10 June 1918, Library of Congress, Madison Law Library, CIS: Presidential Executive Orders and Proclamations (Microfiche), 1918-PR-1420 to 1918-PR-1458; Presidential Proclamation 1494, 4 November 1918, Library of Congress, Madison Law Library, CIS: Presidential Executive Orders and Proclamations (Microfiche), 1918-PR-1486 to 1918-PR-1502; Presidential Proclamation 1514, 4 March 1919, Library of Congress, Madison Law Library, CIS: Presidential Executive Orders and Proclamations (Microfiche), 1919-PR-1506A to 1919-PR-1515; Acting Secretary of the Navy Theodore Roosevelt to the Attorney General, 7 April 1921, NARA I, RG 80, 26266, 786. For the impact on local property owners, see James N. Payne, "Early Days of Dahlgren and Before," undated memoir, but ca. 1978, 3-4, NSWCDD Museum Historical Collection; Inspector of Ordnance in Charge to Chief BUORD, Subject: Development of Ordnance Shore Establishments-Naval Proving Ground," 15 January 1921, 8, National Archives and Records Administration– Philadelphia Branch (hereafter PNAB), RG 181, Records of the 5th Naval District and Shore Establishments: Indian Head, MD, Naval Powder Factory, 1907-1925, Box 15, File 2125-1-1, Folder 1.

28. Indian Head Annual Report for 1918, 20 September 1918, PNAB, RG 181, Records of the 5th Naval District and Shore Establishments: Indian Head, MD, Naval Powder Factory, 1907-1925, Box 12, File 2103-1, Folder 2; Mr. Swepson Earle to the Editor of the *Baltimore Sun*, 1 May 1922, newspaper clipping found at NARA I, RG 74, General Correspondence, 1926-1943, Box 2984, Folder 37630. For Swepson Earle's biography, see Albert Nelson Marquis, ed., *Who's Who in America: A Biographical Dictionary of Notable Living Men and Women of the United States*, vol. 20, 1938-1939, 805.

29. Indian Head Annual Report for 1918, 20 September 1918, 1, 18, PNAB, RG 181, Records of the 5th Naval District and Shore Establishments: Indian Head, MD, Naval Powder Factory, 1907-1925, Box 12, File 2103-1, Folder 2; BuOrd Annual Report, 1918, 492-93; Acting Secretary of the Navy Theodore Roosevelt to the Attorney General, 7 April 1921, Paragraph 5, NARA I, RG 80, 26266-786; Naval Proving Ground, Indian Head, MD, to BUORD, Subject: "Report of Operations of Construction Work at Lower Station of the Naval Proving Grounds, Machodoc Creek, Va.," 1 November 1918, PNAB, RG 181, Records of the 5th Naval District and Shore Establishments: Indian Head, MD, Naval Powder Factory, 1907-1925, Box 12, File 2103-1, Folder 2.

30. For the detailed history of the Navy's 7-inch, 45-caliber tractor batteries, see Earle, *Navy Ordnance Activities*, 203-11; "If You're an Old Timer, Then You'll Remember . . .," U.S. Naval Proving Ground *News Sheet*, 13 March 1952, Dahlgren, Va.; Mr. Swepson Earle to the Editor of the *Baltimore Sun*, 1 May 1922, newspaper clipping found at NARA I, RG 74, General Correspondence, 1926-1943, Box 2984, Folder 37630. After this first test at the Lower Station, the U.S. Army asked for thirty-six of the 7-inch tractor guns for service in France. However, the Armistice, signed on 11 November, intervened and the Navy only

delivered eighteen to the Army. Neither the 10th Marine Artillery Regiment nor the Corps' twenty guns ever left American soil, missing World War I entirely.

31. Indian Head Annual Reports, 20 September 1918, 1, 18, and 31 July 1919, 18-19, PNAB, RG 181, Records of the 5th Naval District and Shore Establishments: Indian Head, MD, Naval Powder Factory, 1907-1925, Box 12, File 2103-1, Folder 2; BuOrd Annual Report, 1919, 524-25.

32. BuOrd Annual Report, 1919, 522-25.

33. Earle, *Navy Ordnance Activities*, 198-201; Inspector of Ordnance in Charge to Chief BUORD, Subject: Development of Ordnance Shore Establishments-Naval Proving Ground," 12, PNAB, RG 181, Records of the 5th Naval District and Shore Establishments: Indian Head, MD, Naval Powder Factory, 1907-1925, Box 15, File 2125-1-1, Folder 1.

34. Inspector of Ordnance in Charge to Chief BUORD, Subject: Development of Ordnance Shore Establishments-Naval Proving Ground," 12, PNAB, RG 181, Records of the 5th Naval District and Shore Establishments: Indian Head, MD, Naval Powder Factory, 1907-1925, Box 15, File 2125-1-1, Folder 1; Mr. Swepson Earle to the Editor of the *Baltimore Sun*, 1 May 1922, newspaper clipping found at NARA I, RG 74, General Correspondence, 1926-1943, Box 2984, Folder 37630; Earle, *Navy Ordnance Activities*, 198-201; SecNav Annual Report, 1920, 78. Swepson Earle said that the Mk II railway gun test occurred on 5 August, but Inspector of Ordnance Greenslade reported to the BUORD Chief on 15 January 1921 that the test happened on 16 August.

35. BuOrd Annual Report, 1920, 636-37; SecNav Annual Report, 1920, 75.

36. Commander Logan Cresap, U.S.N. (Ret.) to Captain W. R. Furlong, U.S.N., Inspector of Ordnance in Charge, U.S. Naval Proving Ground, Dahlgren, Virginia, 18 March 1936, NSWCDD, Museum Historical Collection, Folder 1936 Base Name Ltr, 1 of 2; Ralph Earle to Commander Logan Cresap, 23 March 1936, NSWCDD, Museum Historical Collection, Folder 1936 Base Name Ltr, 2 of 2. Both of these letters have been transcribed in Appendix V.

37. Ibid. Earle thought so much of John Dahlgren that he later researched his hero's life and published a biographical article in the *Naval Institute Proceedings*. In the article, Earle remarked, rather poetically, that Dahlgren "found the Navy asleep in all but seamanship; he aroused it, none too soon, and entirely changed guns and gunnery as well as the construction of ships. He was a scientist and inventor as well as a sea-going officer of the first quality." See Earle, "John Adolphus Dahlgren (1809-1870)," 424-36.

38. Cresap to Furlong, 18 March 1936; Earle to Cresap, 23 March 1936; Secretary Daniels to Postmaster General, 15 January 1919, all in NARA I, RG 74, General Correspondence, Entry 25, Folder 34428; Charles Stewart, Supt. Naval Records and Library, to Admiral Ralph Earle, 5 February 1919, NARA I, RG 74, Entry 25, Folder 34428.

39. House Document 197, "Memorial Addresses Delivered in the House of Representatives of the United States in Memory of Ambrose E. B. Stephens," 47, 11 March 1928, 70th Cong., 1st sess., 1927-1928 (hereafter cited as "Stephens Eulogies"); Kenneth G. McCollum, ed., *Dahlgren* (Dahlgren: Naval Surface Weapons Center, June 1977), 3-5.

40. House Committee on Naval Affairs, *Hearings before the Committee on Naval Affairs of the House of Representatives on Estimates Submitted by the Secretary of the Navy 1919*, 66th Cong., pt. 2, no. 21 (hereafter 1919 Estimates Hearings); House Committee on Naval Affairs, *Hearings before the Committee on Naval Affairs of the House of Representatives on Sundry Legislation Affecting the Naval Establishment, 1921*, 67th Cong., 1st sess. (hereafter 1921 Sundry Legislation Hearings), 685-95.

41. 1921 Sundry Legislation Hearings, 814, 823.

42. House Committee on Naval Affairs, *Hearings before the Committee on Naval Affairs of the House of Representatives on Estimates Submitted by the Secretary of the Navy, 1920*, 66th Cong., No. 12, Naval Appropriation Bill, Testimony of Rear Admiral Ralph Earle and Captain Claude C. Bloch, 27 January 1920, 694; 1919 Estimates Hearings, 823.

43. Chief BUORD Charles B. McVay Jr. Circular Letter C-37, D38 to Commandants of Navy Yards and Stations and Inspectors of Ordnance in Charge, Subject: Amendment to Naval Appropriation Bill, 27 June 1921, PNAB, RG 181, Records of the 5th Naval District and Shore Establishments: Indian Head, MD, Naval Powder Factory, 1907-1925, Box 16, File 2125-1-4, Folder 1; "Rear Admiral Ralph Earle," Navy Department Library, Special Collections, Senior Commander Biographies; Senate Hearings on 11228, 385.

44. 1921 Sundry Legislation Hearings, No. 137, "Investigation of Appropriations and Expenditures at the Naval Proving Ground, Indianhead, MD., and Dahlgren, VA., by a Special Committee of the Committee on Naval Affairs of the House of Representatives," 25-26 July 1921, 669-735; also, No. 168, "Regulating the Expenditure of Naval Appropriations at the Naval Reservation at Dahlgren, VA.," House Committee on Naval Affairs Hearing on House Joint Resolution 198, 19 October 1921, 909-57.

45. Ibid.; *Congressional Record 62, pt. 6, Proceedings and Debates of the 2nd Session of the 67th Congress of the United States of America*, 13 April to 8 May 1922 (18 April 1922), 5673-76.

46. 1921 Sundry Legislation Hearings, 671-72, 675-80, 699, 915-16, 921-30.

47. *Congressional Record 62, pt. 6, Proceedings and Debates of the 2nd Session of the 67th Congress of the United States of America*, 13 April to 8 May 1922 (18 April 1922), 5676; 1921 Sundry Legislation Hearings, 675-80, 703-25.

48. *Congressional Record 61, pt. 6, Proceedings and Debates of the 1st Session of the 67th Congress of the United States of America*, 22 August to 20 October 1921 (24 August

1921), 5709; Navy Department Solicitor to BUORD, Subject: H.J. Res.-198, "Regulating the Expenditure of Naval Appropriations at the Naval Reservation at Dahlgren, Virginia," 29 August 1921; Congressman Thomas S. Butler to Secretary of the Navy Edwin Denby, 25 August 1921; House Joint Resolution 198, 24 August 1924; Secretary of the Navy Edwin Denby to Congressman Thomas S. Butler, 26 September 1921, all at NARA I, RG 80, Entry 25, Folder 8369, Document 234.

49. 1921 Sundry Legislation Hearings, 911, 915.

50. Ibid., 911-18.

51. Ibid., 919.

52. Ibid., 922-40.

53. Telephone Message from Mr. Carey, Private Secretary to the Secretary of the Navy and Telephone Message from Chief BUORD Charles B. McVay Jr., to Commander Green, 7 December 1921; Congressman Thomas S. Butler to Secretary of the Navy Edwin Denby, 8 December 1921; Secretary of the Navy Edwin Denby to Congressman Thomas S. Butler, 12 December 1921, all at NARA I, RG 80, Entry 25, Folder 8369, Document 234.

54. Ibid.

55. House Committee on Naval Affairs, *Hearings before the Committee on Naval Affairs of the House of Representatives on Sundry Legislation Affecting the Naval Establishment, 1922-1923*, 67th Cong., 2nd, 3rd, and 4th sessions, No. 59, "Statement of Captain H. E. Lackey, United States Navy," 27 February 1922, 453-61.

56. *Congressional Record* 62, pt. 6, 18 April 1922, 5671.

57. Ibid., 5676-77.

58. 1921 Sundry Legislation Hearings, 909-10; Senate Hearings on 11228, 224-227; "Arm Fight in Sight on Naval Proving Ground: Indian Head-Dahlgren Controversy Will Involve House and Senate and Two Delegations," *Baltimore Sun*, 4 May 1922, newspaper clipping found at NARA I, RG 74, Entry 25, General Correspondence, 1926-1943, Box 2984, Folder 37630.

59. Senate Hearings on 11228, 224-25, 385-95.

60. Ibid.

61. *Congressional Record* 62, pt. 9, 14 June to 29 June 1922 (16 June 1922), 8860-67; Swepson Earle to Ralph Earle, 7 May 1922; Navy Department Appropriation Bill, 31 May 1922, 224-25, 385-95; "Stephens Eulogies," 9. Dahlgren's congressional supporters were helped by Swepson Earle, who became a popular conservationist in Maryland after leaving the Navy. Although a Marylander, Earle was interested in doing the right thing for the country and

decided to use his local prestige to help Dahlgren by getting the facts out to the people. At the height of the congressional fight over Dahlgren, he crafted a thoughtful and objective letter to the *Baltimore Sun* outlining Dahlgren's history and contrasting the past problems at Indian Head with Dahlgren's successes. He then debunked the pro-Indian Head arguments that Dahlgren seriously interfered with navigation and the oyster and fishery industries on the Potomac by noting that he had laid out the installation precisely to avoid those very things from happening. Further, he argued that abandoning Dahlgren and returning to Indian Head would be a "decidedly false economy" and a "serious mistake." While "the permanent adoption of Dahlgren as proving ground and the abandonment of Indian Head would necessarily cause a loss to Maryland of those now employed at Indian Head," he believed that "Maryland [would] sacrifice her interests" when the cause justified it and "look to the interest of the whole country." Before submitting his letter, Earle met with the *Sun's* editor, who arranged to follow up with a favorable editorial, which appeared in the newspaper two days later. Shortly afterward, Earle wrote to former BUORD Chief Ralph Earle informing him of what he had done. The conservationist and former range officer predicted that "this has done some good and Maryland's public sentiment will help spike Mudd's progress." See Mr. Swepson Earle to the Editor of the *Baltimore Sun*, 1 May 1922, and Swepson Earle to Captain Ralph Earle, 7 May 1922, both at NARA I, RG 74, Entry 25, General Correspondence, 1926-1943, Box 2984, Folder 37630.

62. George W. Baer, *One Hundred Years of Sea Power: The U.S. Navy, 1890-1990* (Stanford, Calif.: Stanford University Press, 1994), 93-103.

63. Ibid. Regarding the Washington Conference, a 28 December 1921 *Chicago Tribune* editorial (reprinted in *U.S. Naval Institute Proceedings*, 48, no. 3 (March 1922): 471) lamented that "the attitude of Congress towards the navy is disquieting. The Administration agrees with Japan and Great Britain to junk part of it, but Congress wants to shoot the works and junk all of it, to decrease the personnel, make it impossible to man the ships and keep them in commission, and consequently make effective fighting of them impossible."

64. Indian Head Annual Report for 1921, 28 July 1921, PNAB, RG 181, Records of the 5th Naval District and Shore Establishments: Indian Head, MD, Naval Powder Factory, 1907-1925, Box 12, File 2103-1, Folder 3.

65. Indian Head Annual Report for 1922, 10 August 1922, 12, PNAB, RG 181, Records of the 5th Naval District and Shore Establishments: Indian Head, MD, Naval Powder Factory, 1907-1925, Box 12, File 2103-1, Folder 3.

66. Indian Head Annual Report for 1921, 28 July 1921, 16-17, PNAB, RG 181, Records of the 5th Naval District and Shore Establishments: Indian Head, MD, Naval Powder Factory, 1907-1925, Box 12, File 2103-1, Folder 3.

67. Inspector of Ordnance in Charge Captain John W. Greenslade to Chief of BUORD RADM Charles B. McVay Jr., 17 March 1922, PNAB, RG 181, Records of the 5th

Naval District and Shore Establishments: Indian Head, MD, Naval Powder Factory, 1907-1925, Box 6, File 2100-0. Greenslade made oblique reference to the Washington Naval Arms Limitation treaty, negotiated in November 1921 at a highly publicized international conference in that city. The treaty imposed a ten-year "holiday" on new ship construction, required the scrapping of some ships, and established a ratio of ship tonnage between the five major naval powers: Britain, the United States, Japan, France, and Italy. For a detailed and reliable account of the conference and its outcome, see Harold and Margaret Sprout, *Toward a New Order of Sea Power: American Naval Policy and the World Scene, 1918-1922* (Princeton, 1940; Greenwood Press, 1969).

68. Inspector of Ordnance in Charge Captain John W. Greenslade to Chief of BUORD RADM Charles B. McVay Jr., 17 March 1922, PNAB, RG 181, Records of the 5th Naval District and Shore Establishments: Indian Head, MD, Naval Powder Factory, 1907-1925, Box 6, File 2100-0, 2-3.

69. Ibid. Curtis later became head of the Computer Department of the Bureau of Standards and helped found the Office of Naval Research in 1946.

70. Ibid., 3.

71. Ibid., heading: Tests of Major Caliber Fuzes, 1-2.

72. Indian Head Annual Report for 1923, 27 July 1923, PNAB, RG 181, Records of the 5th Naval District and Shore Establishments: Indian Head, MD, Naval Powder Factory, 1907-1925, Box 12, File 2103-1, Folder 3.

73. Ibid.

74. "Proving Ground Regulations," NARA I, RG 74, Entry 25, Box 959, Folder NP9/580-A2-5 (hereafter 1928 Indian Head Regulations).

75. 1928 Indian Head Regulations.

76. Admiral George Hussey, interview by A. B. Christman, 1966, at Navy Operational Archives, NL/CCG, Record Collection 8, "Oral Histories," Box 1017, Folder Interview of Vice Admiral George F. Hussey and Dr. L.T.E. Thompson (hereafter Christman/Hussey interview), 27-29.

77. The fact that Patterson was senior to Thompson but Thompson held a doctorate while Patterson did not, and the fact that the two scientists served on the same technical liaison team may have been a factor contributing to Earle's decision to sponsor Patterson for an honorary doctorate at Worcester Polytechnic in 1932.

78. Dr. L. T. E. Thompson, biographical sketch, NSWCDD Museum Historical Collection; Christman/Hussey interview, 23, 55-56; Clark University, *Department of Physics and Ballistic Institute*, in Navy Operational Archives, Navy Laboratory/Center Coordinating Group (NL/CCG) Records, Record Collection (RC) 12 (Personal Papers), Thompson, Dr. LTE, Box 1015, Folder Dept. of

Physics Pamphlet, 1918-1919, 12-15; Dr. L. T .E. Thompson, interview by Albert Christman, November 1967, Re: "Mixed discussion covering early military rocket programs and a review of the planned treatment for the first section of the Early NOTS History," in Navy Operational Archives, NL/CCG Record Collection 8, "Oral Histories," Box 883, Folder LTE Thompson, November 1967 (hereafter Christman/Thompson interview, Nov. 1967), 42.

79. Christman/Hussey interview, 4.

80. Ibid.; Christman/Thompson interview, Nov. 1967, 39-41.

81. Christman/Hussey interview, 4-5. As to the assertion of "full-fledged," Christman in his interview drew Hussey into a discussion of the role of the chemists, including Patterson at Indian Head, and Hussey grudgingly admitted that they had scientific training, although none had earned the Ph.D.

82. Christman/Thompson interview, Nov. 1967, 42-43.

83. Albert B. Christman, *Sailors, Scientists, and Rockets: Origins of the Navy Rocket Program and of the Naval Ordnance Test Station, Inyokern*, vol. 1 of the *History of the Naval Weapons Center, China Lake, California* (Washington, D.C.: Naval History Division, 1971), 51-66; Christman/Thompson interview, Nov. 1967, 12; Louis Thompson, "Long-Range Gunnery and High-Altitude Bombing: A Comparison of Ballistic Merit," in *Journal of the Franklin Institute* 215, no. 2 (February 1933): 119-32.

84. Technical Memorandum 12-1, Erosion of Naval Guns and Effect of Chromium Plating, 15 March 1950, Naval Ordnance Test Station (China Lake), 1-4, in Navy Operational Archives, NL/CCG Records, Record Collection 12 (Personal Papers), Folder Thompson, Dr. L.T.E.

85. Ibid., 5, 6.

86. L. Thompson, "Ballistic Engineering Problems, Empirical Summaries," *U.S. Naval Institute Proceedings* 56, no. 327 (May 1930): 411-18.

87. Ibid., 412.

88. Christman/Hussey interview, 83.

89. Inspector of Ordnance in Charge Captain Claude C. Bloch to Chief BUORD RADM Charles B. McVay Jr., "Quarterly Preparedness Report for the Quarter Ending 30 September 1923," 16 October 1923, PNAB, RG 181, Records of the 5th Naval District and Shore Establishments: Indian Head, MD, Naval Powder Factory, 1907-1925, Box 15, File 2125-1-1, Folder 1; Inspector of Ordnance in Charge Captain Claude C. Bloch to Chief BUORD RADM Charles B. McVay Jr., "Quarterly Preparedness Report for the Quarter Ending 31 December 1923," 3 January 1924 (first page erroneously dated 3 January 1923); Indian Head Annual Report for 1924, 15 July 1924, PNAB 181, Records of the 5th Naval

District and Shore Establishments: Indian Head, MD, Naval Powder Factory, 1907-1925, Box 6, File 2103-1, Folder 3.

90.   Earle, *Navy Ordnance Activities*, 143-50.

91.   Ibid., 149-50.

92.   Memorandum for Captain Craven, Subject: "Flying Fields for Training Fleet Battleship Plane Pilots," 14 May 1920, NARA I, RG 72, Entry 25, General Correspondence, 1926-1943, File 3150, Document 19; Bureau of Supplies and Accounts to Commanding Officer, Balloon Detachment, Marine Barracks, Quantico, Virginia, Subject: Shipment of Three JN-4 Airplanes to Commanding Officer, Lower Proving Ground, Dahlgren, Virginia, 2 September 1919, NARA I, RG 72, Entry 25, General Correspondence, 1926-1943, File 3146, Document BB2; Indian Head Annual Report for 1923, 27 July 1923, PNAB, RG 181, Records of the 5th Naval District and Shore Establishments: Indian Head, MD, Naval Powder Factory, 1907-1925, Box 12, File 2103-1, Folder 3.

93.   Ibid. See footnote ** in McCollum, *Dahlgren*, 30.

94.   For a highly detailed and lengthy account of the "flying bomb" and radio-controlled aircraft projects quoting heavily from archived BUORD correspondence at NARA, see Rear Admiral Delmer S. Fahrney, *The History of Pilotless Aircraft and Guided Missiles*, undated but probably written from 1949 through 1958, 82-185, available at the Navy Department Library, Microfilm A-171, Washington Navy Yard; see also Captain L. S. Howeth, USN (retired), *History of Communications-Electronics in the United States Navy* (Washington, D.C.: Government Printing Office, 1963), 343-49. Howeth's history was distilled from Fahrney's cumbersome Navy bureaucratic prose. Interest in remote control technology can be traced to 1887, when Englishmen E. Wilson and C. J. Evans successfully controlled slow-moving boats by radio on the Thames River. Later, noted technologists such as Nikola Tesla and Admiral Bradley A. Fiske carried the concept further at the dawn of the twentieth century, but inventor John Hays Hammond Jr. has been credited for developing, from 1910 to 1912, the "automatic course stabilization principle" and a satisfactory method for "security of control," both of which were key to making remote control feasible.

95.   Memorandum for the Director of Air Service, War Department (Information Group), 14 November 1919, NARA I, RG 74, Entry 25A, Box 2, Folder 36463; Fahrney, *The History of Pilotless Aircraft and Guided Missiles*, 86-93; Howeth, *History of Communications-Electronics*, 343-44.

96.   Fahrney, *The History of Pilotless Aircraft and Guided Missiles*, 86-93; Howeth, *History of Communications-Electronics*, 343-44.

97.   Fahrney, *The History of Pilotless Aircraft and Guided Missiles*, 86-93, 186; Howeth, *History of Communications-Electronics*, 343-44.

98. Howeth, *History of Communications-Electronics*, 344-45.

99. Albert L. Pardini, *The Legendary Secret Norden Bombsight* (Atglen, Pa.: Schieffer Publishing Ltd., 1999), 43; Charles H. Leith, *Charles Candy Middlebrook Esq. and His Part in the Norden Bombsight of World War II* (Fredericksburg, Va.: Charles H. Leith, 1987), 12-13; Howeth, *History of Communications-Electronics*, 346-47.

100. Fahrney, *The History of Pilotless Aircraft and Guided Missiles*, 107-13; Howeth, *History of Communications-Electronics*, 345-46. Dahlgren legend recounts that the lost flying bomb was launched from the proving ground, but Fahrney, working from the original BUORD documents describing the test, plainly indicated that the aircraft flew away from Sperry's airfield at Amityville, New York.

101. Ibid.

102. Fahrney, *The History of Pilotless Aircraft and Guided Missiles*, 116-17; Howeth, *History of Communications-Electronics*, 346.

103. Fahrney, *The History of Pilotless Aircraft and Guided Missiles*, 117-19; Howeth, *History of Communications-Electronics*, 347.

104. Fahrney, *The History of Pilotless Aircraft and Guided Missiles*, 119-20; Howeth, *History of Communications-Electronics*, 347.

105. Fahrney, *The History of Pilotless Aircraft and Guided Missiles*, 120-22; Howeth, *History of Communications-Electronics*, 347.

106. Fahrney, *The History of Pilotless Aircraft and Guided Missiles*, 122-23, 173; Howeth, *History of Communications-Electronics*, 347.

107. Fahrney, *The History of Pilotless Aircraft and Guided Missiles*, 149-54; Howeth, *History of Communications-Electronics*, 348.

108. Fahrney, *The History of Pilotless Aircraft and Guided Missiles*, 152-56; Howeth, *History of Communications-Electronics*, xi-xv, 348.

109. Fahrney, *The History of Pilotless Aircraft and Guided Missiles*, 156-58, 160-63; Howeth, *History of Communications-Electronics*, 348.

110. Fahrney, *The History of Pilotless Aircraft and Guided Missiles*, 164-68; Howeth, *History of Communications-Electronics*, 349-50.

111. Fahrney, *The History of Pilotless Aircraft and Guided Missiles*, 170; Howeth, *History of Communications-Electronics*, 350.

112. Fahrney, *The History of Pilotless Aircraft and Guided Missiles*, 170-73.

113. The "pigeon" story entered oral legend and has been recounted in several different sources, including McCollum, *Dahlgren*, 39, and Christman/Hussey interview, 83, 85.

114. Christman/Hussey interview, 82-83, 85, James N. Payne, "Early Days of Dahlgren and Before," n.d., but ca. 1978, NSWCDD, Museum Historical Collection, 12-13.

115. Payne, "Early Days of Dahlgren and Before," 15; McCollum, *Dahlgren*, 109-17; BuOrd Annual Report, 1927, 16.

116. The total figure of thirty from Indian Head is noted in David Hedrick, *U.S. Naval Proving Ground Dahlgren Virginia History, April 1918-December 1945* (hereafter Hedrick History), 86, but no source for the data is shown; House Committee on Naval Affairs, *Hearings before the House Committee on Naval Affairs on Sundry Legislation Affecting the Naval Establishment*, No. 137, "Investigations of Appropriations and Expenditures at the Naval Proving Ground, Indian Head, Md., and Dahlgren, Va.," by a Special Committee of the Committee on Naval Affairs of the House of Representatives, 67th Cong., 1st sess. (25 July 1921), 707-10; Museum Collection, Weekly News Sheets, 14 October 1954, 1; see *Congressional Record* 62, pt. 6, 18 April 1922, 5676, for petitioners. At least two of those presenting the petition were apparently relatives of Roger Dement, including J. M. Dement and C. N. Dement. McCollum, *Dahlgren*, 55, 57, 66-67.

117. Inspector of Ordnance in Charge Captain G. L. Schuyler to Chief BUORD, Subject: Policy for Development, 16 October 1931, with attached Policy Statement for Naval Proving Ground, Dahlgren, Virginia, dated 15 October 1931, NARA I, RG 74, Entry 25A, Box 10, Folder H1-H16; Inspector of Ordnance in Charge Captain G. L. Schuyler to Chief BUORD, Subject: Regulations Governing U.S. Naval Proving Ground, Dahlgren, Virginia, and U.S. Naval Powder Factory, Indian Head, Maryland, 6 October 1931, with enclosed Changes in Regulations, dated 3 October 1931, NARA I, RG 74, Entry 25, Box 948, Folder NP-9/A 2-5 "S-8"; "General Regulations Governing the Naval Proving Ground, Dahlgren, VA, and Naval Powder Factory, Indian Head, MD," Volume I, 27 December 1928, NARA I, RG 74, Entry 25, Box 959, Folder NP9/S80-A2-5. The so-called "Hedrick History," compiled in 1945, indicated (p.14) that Accounting and Disbursing were established independently at Dahlgren on 1 January 1925. However, the 1928 organization chart in the official regulations shows the disbursing office at Dahlgren as a subsidiary office of Indian Head, and it was not until 1931 that Commander Schuyler's amendments to the regulations separated disbursing from the Indian Head organization. The Hedrick reference may be to the establishment of the subsidiary office at Dahlgren. Major General, Commandant Marines (B. H. Fuller?) to Chief BUORD, 27 July 1932, NARA I, RG 74, Entry 25, Box 751, Folder KP21 (3), requesting establishment of separate marine command "in view of the fact that the Proving Ground at Dahlgren and the Naval Powder Factory at Indian Head are now separate commands."

Chapter 3: *Dahlgren at War, 1932-1945*

1. Buford Rowland and William B. Boyd, *U.S. Navy Bureau of Ordnance in World War II* (Washington, D.C.: Bureau of Ordnance, Department of the Navy, ca. 1954), 1-2; William McBride, *Technological Change and the United States Navy, 1865-1945* (Baltimore, Md.: Johns Hopkins University Press, 2000), 161-64; George W. Baer, *One Hundred Years of Sea Power: The U.S. Navy, 1890-1990* (Stanford, Calif.: Stanford University Press, 1994), 113-18.

2. Albert B. Christman, *Sailors, Scientists, and Rockets: Origins of the Navy Rocket Program and of the Naval Ordnance Test Station, Inyokern*, vol. 1 of the *History of the Naval Weapons Center, China Lake, California* (Washington, D.C.: Naval History Division, 1971), 61-62; Rowland and Boyd, *Bureau of Ordnance*, 20; McBride, *Technological Change*, 164-65; Inspector of Ordnance in Charge Garrett L. Schuyler to Chief of the Bureau of Ordnance, 16 October 1931, with attached Policy Statement for the Naval Proving Ground, Dahlgren, Virginia, 15 October 1931, National Archives and Records Administration, Washington, D.C. (hereafter NARA I), Record Group (RG) 74, Entry 25A, Box 10, Folder A1-A16; Baer, One Hundred Years of Sea Power, 119-28; Vannevar Bush, *Modern Arms and Free Men: A Discussion of the Role of Science in Preserving Democracy* (New York: Simon and Schuster, 1949), 19, 25.

3. Admiral George Hussey, interview by A. B. Christman, at Navy Operational Archives, 1966, Navy Lab/Center Coordinating Group (NL/CCG), Box 1017, Folder "Interview of Vice Admiral George F. Hussey and Dr. L. T. E. Thompson" (hereafter Christman/Hussey Interview), 86; David Hedrick, *U.S. Naval Proving Ground Dahlgren Virginia History, April 1918-December 1945* (hereafter Hedrick History), 97, 317; Annual Report of the Chief of the Bureau of Ordnance to the Secretary of the Navy (hereafter BuOrd Annual Report), 1928, 7; BuOrd Annual Report, 1929, 9; BuOrd Annual Report, 1930 (included with the Annual Report of the Secretary of the Navy to the President (hereafter SecNav Annual Report), 1930), 295; BuOrd Annual Report, 1932 (included with the SecNav Annual Report, 1932), 277; Inspector of Ordnance in Charge Captain William R. Furlong to Chief of Bureau of Ordnance Rear Admiral Harold R. Stark, 24 December 1935, NARA I, RG 74, Entry 25, Box 959, Folder NP9/S80-A2-5; Vice Admiral George F. Hussey, who served as Proof Officer at Dahlgren in the 1930s, later described the involved nature of the Boulange screen velocity measurement process: "We fired the projectile through a wire screen and the breaking of the wire dropped a rod. The break of the second wire released a knife that nicked the falling rod. Measuring the distance between the zero mark and the point of the nick gave you the time of passage between the screens. With a prepared table you could tell what the velocity was and the correction factor to the muzzle." A new screen had to be hoisted down and up into a gun's line of fire between each round fired for proper use.

4. McBride, *Technological Change*, 164-71; SecNav Annual Report, 1933, 2-3; SecNav Annual Report, 1934, 2-3.

5. Memorandum from Bureau of Ordnance Sections D and E to the Assistant Chief of Bureau, 19 October 1933, NARA I, RG 74, Entry 25, Box 959, Folder NP9/S80-SG; Christman/Hussey Interview, 81.

6. Inspector of Ordnance in Charge Captain William R. Furlong to Chief of Bureau of Ordnance Rear Admiral Harold R. Stark, 24 December 1935, NARA I, RG 74, Entry 25, Box 959, Folder NP9/S80-A2-5.

7. Baer, *One Hundred Years of Sea Power*, 114-118, 131-135.

8. Regulations Governing the Naval Proving Ground, Dahlgren, Virginia, vol. 4: Proof Regulations, Approved by the Bureau of Ordnance, 2 November 1928 (with 1936 draft revisions), NARA I, RG 74, Entry 25, Box 959, Folder NP9/S80 (A2-5); Chief of the Bureau of Ordnance Rear Admiral Harold Stark to Inspector of Ordnance in Charge Captain C. R. Robinson, 7 August 1936, NARA I, RG 74, Entry 25 Box 99, Folder NP9/S80 (A2-5); Hedrick History, 29, 96-98, 140, 154, 317.

9. Kenneth G. McCollum, ed., *Dahlgren* (Dahlgren, Va.: Naval Surface Weapons Center, June 1977), 55-57; Charles Roble, "Dahlgren Career" (unpublished memoir), 11; Charles Roble, telephone interview by James P. Rife, 29 July 2003 (hereafter Roble Oral History), 1-10.

10. McCollum, *Dahlgren*, 57.

11. Ibid.; Roble, "Dahlgren Career," 12.

12. Roble, "Dahlgren Career," 2-7; Roble Oral History, 7-9.

13. Roble, "Dahlgren Career," 12-13.

14. Roble, "Dahlgren Career," 15-19; Roble Oral History, 3-4.

15. Christman, *Sailors, Scientists, and Rockets*, 72-73; Inspector of Ordnance in Charge William R. Furlong to Chief of the Bureau of Ordnance Rear Admiral Harold Stark, 11 June 1935, "Re: Necessity for and Value of Experimental Work at Naval Proving Ground," NARA I, RG 74, Entry 25, Box 948, Folder NP9/A1-1, "Projects"; L. Thompson and N. Riffolt, "Ballistic Engineering Problems: Experimental Development, *U.S. Naval Institute Proceedings*, March 1932, 383-87, and in "Notes on Ballistic Physics," *Science Leaflet*, Physics Section, 295-300, n.d., clipping in Navy Operational Archives, NL/CCG Records, Record Collection (RC) 12 (Personal Papers), Thompson, Dr. LTE, Box 1015, Folder L.T.E. Thompson Papers, 1915-1924.

16. Furlong to Stark, 11 June 1935, NARA I.

17. Ibid.

18. For the background and description of the bombing tests, see Archibald D. Turnbull and Clifford L. Lord, *History of United States Naval Aviation* (New

Haven: Yale University Press, 1949), 193-204; Rear Admiral Sherman E. Burroughs, interview by Albert Christman, 4 November 1966; Christman, Sailors, Scientists, and Rockets, 62-63.

19. Quoted in Christman, *Sailors, Scientists, and Rockets*, 62-63.

20. Louis Thompson, Ph.D., "Long-Range Gunnery and High-Altitude Bombing: A Comparison of Ballistic Merit," *Journal of the Franklin Institute* 215, no. 2 (February 1933): 119-20.

21. Dr. L. T. E. Thompson to Rear Admiral Harold R. Stark, Chief of the Bureau of Ordnance, 16 January 1935, NL/CCG Archives, RC 12 (Personal Papers), Thompson, Dr. LTE, Box 1015, Folder Correspondence 1923-1942.

22. Hedrick History, 29.

23. Loyd Searle, "The Bombsight War: Norden vs. Sperry," reprinted from *Spectrum* IEEE#0018-92235/89/0900-0060, http://world.std.com/~searle/IEEE1.htm. A detailed history of the Norden bombsight can be found in Robert Vance Brown, Lieutenant Commander, USNR, *The Navy's Mark 15 (Norden) Bomb Sight: Its Development and Procurement, 1920-1945*, 2 vols., April 1946, on microfilm in the Navy Department Library. See also Albert Pardini, *The Legendary Norden Bombsight* (Atglen, Pa.: Schiffer Military History, 1999).

24. Brown, *The Navy's Mark 15 Bomb Sight*, 1-9; Pardini, *Legendary Norden Bombsight*, 23.

25. Ralph Earle, ed., *Navy Ordnance Activities, World War, 1917-1918* (Washington, D.C.: Government Printing Office, 1920), 147-48; Seale, "The Bombsight War," 1-2; Brown, *The Navy's Mark 15 Bomb Sight*, 11-12.

26. Earle, *Navy Ordnance Activities*, 147-48.

27. Ibid.

28. Ibid.; Brown, *The Navy's Mark 15 Bomb Sight*, 14, 25-30.

29. Brown, *The Navy's Mark 15 Bomb Sight*, 31-45; Leith, *Charles Candy Middlebrook*, 22-23.

30. McCollum, *Dahlgren*, 36-38; Dr. L. T. E. Thompson, interview by Albert Christman, 10 November 1968 (hereafter Christman/Thompson Interview, Nov. 1968), 19-20; Brown, *The Navy's Mark 15 Bomb Sight*, 61; Lt. Schaeffer's Memorandum of 5 July 1927, Bureau of Ordnance Correspondence File, quoted in Brown, *The Navy's Mark 15 Bomb Sight*, 61-62.

31. Barth to Commander Picking, 26 June 1929, Bureau of Ordnance Correspondence File, quoted in Brown, *The Navy's Mark 15 Bomb Sight*, 66.

32. Pardini, *Legendary Norden Bombsight*, 67-69. Pardini gives a very careful technical description of how the synchronous Mark XV bombsight works.

33. Leith, *Charles Candy Middlebrook*, 25; Brown, *The Navy's Mark 15 Bomb Sight*, iii, 89-93.

34. Brown, *The Navy's Mark 15 Bomb Sight*, iii-iv, 102-10; Hedrick History, 240-41; Thompson to Stark, 16 January 1935, NL/CCG Archives.

35. Leith, *Charles Candy Middlebrook*, 27; Pardini, *Legendary Norden Bombsight*, 275.

36. McCollum, *Dahlgren*, 36; Leith, *Charles Candy Middlebrook*, 27.

37. Brown, *The Navy's Mark 15 Bomb Sight*, 320-21; Hedrick History, 240-41.

38. McCollum, *Dahlgren*, 50-52; Acting Secretary of the Navy William D. Leahy to Chairman of the Senate Committee on Appropriations The Honorable Carter Glass (Democrat from Virginia), 26 January 1939, and Leahy to Honorable Clifton A. Woodrum, House Deficiencies Subcommittee, 7 February 1939, NARA I, RG 80, Entry 22, Box 3310, Folder NP9/N1-13 (351007-1 to 4200228); Judge Advocate General to the Inspector of Ordnance in Charge, 6 June 1939, RG 80, Entry 13 (Spindle File), Box 210, Folder Dahlgren, VA 1935-1940; Chief of the Bureau of Ordnance Rear Admiral William R. Furlong to Senator Harry Flood Byrd, 7 August 1939, NARA I, RG 74, Entry 25, Box 943, Folder NP-9/N1-13 (1-81).

39. For an excellent biography of Parsons, see Albert Christman, *Target Hiroshima: Deak Parsons and the Creation of the Atomic Bomb* (Annapolis: Naval Institute Press, 1998); Dr. Charles C. Bramble, interview by Albert Christman, May 1969, NL/CCG Archives, Washington Navy Yard, Collection 0008 "Oral Histories," Box 870, Folder Dr. C. C. Bramble (hereafter Christman/Bramble Interview), 7; Dr. J. E. Henderson, interview by Albert Christman, April 1969, NL/CCG Archives, Washington Navy Yard, Collection 0008 "Oral Histories," Box 874, Folder Dr. J. E. Henderson (hereafter Christman/Henderson Interview), 17, 19.

40. Christman, *Target Hiroshima*, 38-39.

41. Christman/Henderson Interview, 14, 17-18; Admiral Horatio Rivero, interview by Albert Christman, May 1967, NL/CCG Archives, Washington Navy Yard, Collection 0008 "Oral Histories," Box 880, Folder Adm. Horatio Rivero (hereafter Christman/Rivero Interview), 3. The armor and sledgehammer story is recounted in Christman, *Target Hiroshima*, 64. See also Mrs. Robert Burroughs (the former Mrs. W. S. Parsons), interview by Albert Christman, 29 April 1966, NL/CCG Archives, Washington Navy Yard, Collection 0008 "Oral Histories," Box 870, Folder Mrs. Robert Burroughs (former Mrs. William S. Parsons) (hereafter Christman/Mrs. Burroughs Interview, April 1966), 58-59.

42. Christman/Mrs. Burroughs Interview, April 1966, 64-66; Curtis Youngblood, interview by Albert Christman, May 1967, NL/CCG Archives, Washington

Navy Yard, Collection 0008 "Oral Histories," Box 885, Folder Curtis Youngblood (hereafter Christman/Youngblood Interview), 6-7; Dr. L. T. E. Thompson, interview by Albert Christman, November 1965, NL/CCG Archives, Washington Navy Yard, Collection 0008 "Oral Histories," Box 882, Folder Dr. LTE Thompson, Nov. 1965 (hereafter Christman/Thompson Interview, Nov. 1965), 44.

43. Dr. L. T. E. Thompson, interview with Albert Christman, November 1967, NL/CCG Archives, Washington Navy Yard, Collection 0008 "Oral Histories," Box 883, Folder Dr. LTE Thompson, Nov. 1967 (hereafter Christman/Thompson Interview, Nov. 1967), 1-4; Thompson/ Christman Interview, July 1966, 57; Dr. L. T. E. Thompson, biographical sketch, NSWCDD Museum Historical Collection. For Thompson's "F" coefficient, see U.S. Naval Surface Warfare Center, Port Hueneme Division engineer and naval technology historian Nathan Okun's article "Major Historical Naval Armor Penetration Formulae" at http://www.warships1.com/index_nathan/index.htm (26 November 2001). A specialist in the history of warship armor and projectile design, Okun notes that Dr. Allen V. Hershey led the Naval Proving Ground's armor analysis effort during and after World War II and that Thompson's "F" coefficients remained the center of the formulae and test data sets developed by his department until the Navy shut down its Armor Program in 1955 and transferred it to the U.S. Army.

44. Christman, *Target Hiroshima*, 78.

45. Dr. Ralph A. Sawyer, interview by Albert Christman, May 1967, NL/CCG Archives, Washington Navy Yard, Collection 0008 "Oral Histories," Box 881, Folder Dr. Ralph A. Sawyer (hereafter Christman/Sawyer Interview), 1; Christman/Thompson Interview, Nov. 1968, 17; W. S. Parsons and L. Thompson, via Inspector of Ordnance in Charge, to the Chief of the Bureau of Ordnance Rear Admiral William R. Furlong, 3 September 1940, NARA I, RG 74, Entry 25, Box 727, Folder NP-9/A1-1.

46. Christman/Thompson Interview, July 1966, 57; Christman/Thompson Interview, Nov. 1968, 16-17; Dr. Leonard Loeb, interview by Albert Christman, March 1967, NL/CCG Archives, Washington Navy Yard, Collection 0008 "Oral Histories," Box 877, Folder Dr. Leonard Loeb (hereafter Christman/Loeb Interview), 5, 18; Christman/Henderson Interview, 11; Captain G. L. Schuyler to the Assistant Chief of the Bureau of Ordnance, 7 September 1940, NARA I, RG 74, Entry 25, Box 727, Folder NP-9/A1-1; Christman/Thompson Interview, Nov. 1967, 4.

47. Christman/Loeb Interview, 2-4, 12; Christman/Thompson Interview, Nov. 1967, 6-7; Inspector of Ordnance in Charge Captain David I. Hedrick to Chief of the Bureau of Ordnance William H. P. Blandy, 5 December 1942, National Archives and Records Administration, College Park, Md. (hereafter NARA II), RG 74, Entry 25, Box 212, Folder NP9/A9.

48. Christman/Loeb Interview, 5.

49. Biographical File, "Captain David Irvin Hedrick," Naval Historical Center Operational Archives Branch; Christman/Mrs. Burroughs Interview, April 1966, 37, 50; Christman/Sawyer Interview, 9; Christman/Loeb Interview, 9, 18; Captain Norm Scott, interview by Rodney P. Carlisle, Dahlgren, Virginia, 22 May 2003.

50. Christman, *Target Hiroshima*, 79; Christman/Youngblood Interview, 9-10.

51. Christman/Sawyer Interview, 11; for Hedrick's formal description of Dahlgren's mission (as he saw it), see Inspector of Ordnance in Charge Captain David I. Hedrick to Commandant, Potomac River Naval Command, 10 September 1942, NARA II, RG 74, Entry 25, Box 514, Folder NPG Dahlgren, VA-Projects.

52. Christman/Mrs. Burroughs Interview, April 1966, 48-49.

53. Christman/Loeb Interview, 4-6, 8-9. One of Schuyler's subordinates in the Bureau, Sam Shumaker, told Loeb that "Anything Dr. Tommy is for, I'm against."

54. Ibid., 17.

55. Christman/Thompson Interview, Nov. 1968, 20-21; Christman/Thompson Interview, July 1966, 55-56. The name Lukas-Harold was derived from Norden and Barth's middle names.

56. Christman/Loeb Interview, 8, 19.

57. McCollum, *Dahlgren*, 69-75; Christman/Loeb Interview, 11; Christman/Thompson Interview, Nov. 1967, 6-7; Christman/Thompson Interview, Nov. 1968, 17.

58. SecNav Annual Report, 1941, 1-6, 42-44; Inspector of Ordnance in Charge Captain J. S. Dowell to the Chief of the Bureau of Ordnance, 22 October 1940, NARA I, RG 74, Entry 25, Box 727, Folder NP-9/A1-1.

59. Inspector of Ordnance in Charge Captain J. S. Dowell to the Chief of the Bureau of Ordnance, 22 October 1940, NARA I, RG 74, Entry 25, Box 727, Folder NP-9/A1-1; Hedrick History, 15-16, 85-135; Bureau of Yards and Docks, *Building the Navy's Bases in World War II: History of the Bureau of Yards and Docks and the Civil Engineer Corps, 1940-1946*, vol. 1 (Washington, D.C.: U.S. Government Printing Office, 1947), 351; for Dahlgren's physical growth during World War II, see also R. Christopher Goodwin & Associates, Inc., *Final Report: Architectural Investigations of Dahlgren's Residential Area, Naval Surface Warfare Center, Dahlgren Laboratory, Dahlgren, Virginia*, 3 October 1994, maintained in the Office of the Department of Historic Resources, Commonwealth of Virginia, 2801 Kensington Avenue, Richmond, VA 23221, 27-35. The school was decommissioned in November 1945 per Bureau of Personnel Orders.

60. Hedrick History, 265, 280-82.

61. Ibid., 225-26; Chief of the Bureau of Ordnance to the Commanding Officer, 7 October 1944, NARA II, RG 74, Entry 25, Box 563, Folder NP-9/F41 (October 1, 1944); Commanding Officer, NPG, to Chief of the Bureau of Ordnance, 12 December 1944, NARA II, RG 74, Entry 25, Box 562, Folder NP-9/F41 (December 1, 1944); Chief of the Bureau of Ordnance to the Commanding Officer, 26 October 1944, NARA II, RG 74, Entry 25, Box 562, Folder NP-9/F41 (October 18, 1944); Chief of the Bureau of Ordnance to the Commanding Officer, NPG, 28 July 1944, NARA II, RG 74, Entry 25, Box 563, Folder NP-9/F41 (June 1, 1944); Chief of the Bureau of Ordnance to the Chief of the Bureau of Yards and Docks, 9 September 1944, NARA II, RG 74, Entry 25, Box 563, Folder NP-9/F41 (September 1, 1944); Chief of the Bureau of Ordnance to the Secretary of the Navy, 28 December 1944, NARA II, RG 74, Entry 25, Box 562, Folder NP-9/F41 (Dec. 1, 1944).

62. Hedrick History, 56-57. Compiled from testing reports and correspondence with the Bureau of Ordnance, Hedrick's history provides an extensively detailed account of the numerous testing activities undertaken at Dahlgren during the war, including numbers of rounds fired and bombs dropped, among other things.

63. Commanding Officer, NPG, to the Chief of the Bureau of Ordnance, 30 December 1943, NARA II, RG 74, Entry 25, Box 398, Folder NP-9/S78.

64. Christman/Thompson Interview, July 1966, 79. The Jean Bart story is recounted in Rowland and Boyd, *Bureau of Ordnance*, 55-57.

65. See, for example, NPG Report No. 9-43, *Examination of a Japanese 140 mm (5.5-Inch) Bombardment Projectile*, 26 April 1943, NSWCDD Technical Library; NPG Report No. 4-43, Examination of a Japanese 8-inch Common Projectile, 12 March 1943, NSWCDD Technical Library; NPG Report No. 6-43, *Examination of German 20mm Aircraft Ammunition*, 2 April 1943, NSWCDD Technical Library; NPG Report No. 7-45, *Examination of Italian Projectiles*, July 1945; NPG Report No. 5-43, *Examination of Japanese Equipment: 37mm Anti-Tank Ammunition, 3-inch A.A. Ammunition, 75mm Howitzer Ammunition, 50mm Mortar Projectile, 50mm Mortar Grenade*, 24 March 1943, NSWCDD Technical Library.

66. The story of the VT fuze is told in full detail in Ralph B. Baldwin, *The Deadly Fuze: The Secret Weapon of World War II* (San Rafael, Calif.: Presidio Press, 1980); Christman/Henderson Interview, 1, 10; Christman, *Target Hiroshima*, 74-75; Baldwin, The Deadly Fuze, 25.

67. Christman, *Target Hiroshima*, 73-74.

68. Baldwin, *The Deadly Fuze*, 25; Christman, *Target Hiroshima*, 74-75; Christman/Henderson Interview, 6, 11-12; Christman/Sawyer Interview, 10-12.

69. Baldwin, *The Deadly Fuze*, 62, 67-68; Christman/Henderson Interview, 6, 8.

70. Christman, *Target Hiroshima*, 80-81.

71. Baldwin, *The Deadly Fuze*, 87-89, 95.

72. Christman, *Target Hiroshima*, 82-83, 86-87; Christman/Henderson Interview, 11.

73. Baldwin, *The Deadly Fuze*, 97, 141; Christman, *Target Hiroshima*, 87-91; Chairman of "A" Division Richard C. Tolman to Dr. Vannevar Bush, Director of Scientific Research and Development, 31 March 1942, NARA II, RG 227, Entry 8, Box 6, Folder OSRD General; Vannevar Bush to Commander William S. Parsons, 19 June 1942, NARA II, RG 227, Entry 8, Box 6, Folder OSRD General.

74. Baldwin, *The Deadly Fuze*, 233-37.

75. Quoted in ibid., xxxi.

76. Ibid., 254-55, 265-67; Hedrick History, 212.

77. There are several good histories of the Manhattan Project available. A good start would be Leslie R. Groves, *Now It Can Be Told: The Story of the Manhattan Project* (New York: Harper & Brothers Publishers, 1962). See also Vincent C. Jones, *Manhattan: The Army and the Atomic Bomb*, in the U.S. Army's "Green Book" series (Washington, D.C.: Center of Military History, 1985) and Richard Rhodes, *The Making of the Atomic Bomb* (New York: Simon & Schuster, 1987).

78. Albert Christman Notes on Unrecorded Interview with Dr. N. E. Bradbury, 23 January 1967, NL/CCG Archives, Washington Navy Yard, Collection 0008 "Oral Histories," Box 870, Folder Dr. N. E. Bradbury; Albert Christman Interview with Dr. Norman Ramsey, 3 May 1966, NL/CCG Archives, Washington Navy Yard, Collection 0008 "Oral Histories," Box 880, Folder Ramsey, Dr. Norman, 2-5 (hereafter cited as Christman/Ramsey Interview); Jones, *Manhattan Project*, 508.

79. William S. Parsons to Dr. Norman Ramsey, 17 July 1943, NARA II, RG 77, Entry 5, Box 76, Folder 600.12 (Research); Christman/Ramsey Interview, 2-5.

80. Christman/Ramsey Interview, 2-5; Arthur Breslow, interview by Albert Christman, October 1966, NL/CCG Archives, Washington Navy Yard, Collection 0008 "Oral Histories," Box 870, Folder Arthur Breslow (hereafter Christman/Breslow Interview), 7-8.

81. William S. Parsons to Rear Admiral W. R. Purnell, 8 November 1943, NARA II, RG 77, Entry 5, Box 76, Folder 600.12 (Research); William S. Parsons to Brigadier General Leslie Groves, 24 December 1943, and attached "Program for Flight test of Dummy Bombs from B-29 Plane" and "Interoffice Memorandum" from W. S. Parsons on the subject of "Communication Chart, Ordnance Activities Originating from Project 'Y'," NARA II, RG 77, Entry 5, Box 77, Folder 600.913.

82. Christman/Rivero Interview, May 1967, 1-4, 10-15; Parsons to Ramsey, 17 July 1943, NARA II; William S. Parsons to Dr. L. T. E. Thompson, 30 May 1944, NL/

CCG Archives, Washington Navy Yard, Collection 0012 "Personal Papers," Box 1017, Folder Correspondence, 1942-1945; U.S. Naval Proving Ground News Sheet, Dahlgren, Va., 10 August 1945, NSWCDD Museum Historical Collection, Envelope: NPG Weekly News Sheets, 1945.

Chapter 4: *Numbers Over Guns, 1945-1959*

1. Kenneth G. McCollum, ed., *Dahlgren* (Dahlgren: Naval Surface Weapons Center, June 1977), 58; Norman Friedman, *U.S. Naval Weapons: Every Gun, Missile, Mine, and Torpedo Used by the U.S. Navy from 1883 to the Present Day* (Annapolis, Md.: Naval Institute Press, 1983), 187-88; Malcolm Muir Jr., *Black Shoes and Blue Water: Surface Warfare in the United States Navy, 1945-1975* (Washington, D.C.: Naval Historical Center, 1996), 5-34; Michael T. Isenberg, *Shield of the Republic: The United States Navy in an Era of Cold War and Violent Peace, 1945-1962* (New York: St. Martin's Press, 1993), 102-16, 213.

2. U.S. Naval Proving Ground *News Sheet*, Dahlgren, Virginia, 18 August 1945, NSWCDD Museum Historical Collection, Envelope: NPG Weekly News Sheets, 1945.

3. McCollum, *Dahlgren*, 83; U.S. Naval Proving Ground *News Sheet*, Dahlgren, Virginia, 21 September 1945, NSWCDD Museum Historical Collection, Envelope: NPG Weekly News Sheets, 1945; David Hedrick, *U.S. Naval Proving Ground Dahlgren Virginia History, April 1918-December 1945* (hereafter Hedrick History), 280-82, 317-31; Dr. Ralph A. Sawyer, interview by Albert Christman, May 1967, NL/CCG Archives, Washington Navy Yard, Collection 0008 "Oral Histories," Box 881, Folder Dr. Ralph A. Sawyer (hereafter Christman/ Sawyer Interview), 7-23; Buford Rowland and William B. Boyd, *U.S. Navy Bureau of Ordnance in World War II* (Washington, D.C.: Bureau of Ordnance, Department of the Navy, ca. 1954), 491-95; McCollum, *Dahlgren*, 62; U.S. Naval Proving Ground *News Sheet*, 23 November 1945, NSWCDD Museum Historical Collection, Envelope: NPG Weekly News Sheets, 1945; Biographical File, "Captain David Irvin Hedrick," Naval Historical Center Operational Archives Branch; *Admiral C. Turner Joy*, Navy Department Library, Special Collections, Senior Commander Biographies.

4. McCollum, *Dahlgren*, 62; Kenneth C. Baile, *Historical Perspective of NAVSWC/ Dahlgren's Organizational Culture*, NAVSWC MP 90-715 (Dahlgren: NSWC, March 1991), 17-18; Vannevar Bush, *Modern Arms and Free Men: A Discussion of the Role of Science in Preserving Democracy* (New York: Simon and Schuster, 1949). For more detailed treatments of the government's postwar policies relating to Navy Research, Development, Testing, and Evaluation, see Rodney P. Carlisle, *Navy RDT&E Planning in an Age of Transition: A Survey Guide to Contemporary Literature* (Washington, D.C.: Navy Laboratory/Center Coordinating Group [NL/CCG] and Naval Historical Center, 1997), 11-15, and Rodney P. Carlisle, *Management of the U.S. Navy Research and Development Centers During the Cold War: A Survey Guide to Reports* (Washington, D.C.: NL/CCG and Naval Historical Center, 1996), 1-5.

5. Muir, *Black Shoes and Blue Water*, 12; *Annual Report of the Secretary of the Navy for the Fiscal Year 1946* (Washington, D.C.: Government Printing Office, 1947) (hereafter SecNav Annual Report), 29-31; Rowland and Boyd, *Bureau of Ordnance*, 16-31, 510; SecNav Annual Report, 1947, 38; McCollum, *Dahlgren*, 89-90.

6. McCollum, *Dahlgren*, 82-85; Dr. Armido DiDonato, interview by James P. Rife, Dahlgren, Va., 13 August 2003 (hereafter DiDonato Oral History).

7. McCollum, *Dahlgren*, 99; NAVORD Report No. 1336/NPG Report No. 101, *Ballistic Test and Metallurgical Examination of German 6", 8", and 16" A.P. Projectiles*, undated but late 1947 or early 1948, NSWCDD Technical Library; NPG Report No. 289, *Final Report on Ballistic Test and Metallurgical Examination of Japanese 8", 14", 16" and 18" Armor Piercing Projectiles*, 18 May 1949, NSWCDD Technical Library; NPG Report No. 5-47, *Ballistic Tests and Metallurgical Examination of Japanese Heavy Armor Plate*, 16 December 1947, NSWCDD Technical Library; Experimental Department Memorandum No. 1244-47, *Jap Light Armor-Ballistic and Metallurgical Test of*, 1 July 1947, NSWCDD Technical Library.

8. McCollum, *Dahlgren*, 83; NPG Experimental Department Memorandum No. 1190-45, *Hypervelocity Guns*, 5 November 1945, NSWCDD Technical Library; SecNav Annual Report, 1946, 31; SecNav Annual Report, 1947, 35; Ian V. Hogg, *German Secret Weapons of the Second World War: The Missiles, Rockets, Weapons, and New Technology of the Third Reich* (Mechanicsburg, Pa.: Stackpole Books, 1999), 98-99, 148-52, 154-58; Muir, *Black Shoes and Blue Water*, 25-26; Rowland and Boyd, *Bureau of Ordnance*, 267, 517.

9. McCollum, *Dahlgren*, 58, 83; Muir, *Black Shoes and Blue Water*, 5-34; Isenberg, *Shield of the Republic*, 102-16, 213.

10. Raymond H. Hughey Jr., "History of Mathematics and Computing Technology at the Dahlgren Laboratory," *Technical Digest: Mathematics, Computing, and Information Science* (Dahlgren: NSWCDD, 1995), 13; "Nomination of Charles J. Cohen for the Department of Defense Distinguished Service Award," U.S. Naval Weapons Laboratory, Dahlgren, Va., 24 November 1967, NSWCDD Administrative Vault, Box 78-80, Folder 12000 Civilian Personnel; David R. Brown Jr., *The Navy's High Road to Geoballistics: Part I, The Founding Heritage: Exterior Ballistics*, 30 December 1986 (Dahlgren: NSWC, 1986), 13-18; McCollum, *Dahlgren*, 84.

11. McCollum, *Dahlgren*, 88.

12. Mark Bernard Schneider, "Nuclear Weapons and American Strategy, 1945-1953" (Ph.D. diss., University of Southern California, 1974), 183-84, 275; McCollum, *Dahlgren*, 96-108; William B. Anspacher et al., *The Legacy of the White Oak Laboratory* (Dahlgren: NSWCDD, 2000), 72; Friedman, *U.S. Naval Weapons*, 187-88.

13. Muir, *Black Shoes and Blue Water*, 7-12; SecNav Annual Report, 1946, 2-3; SecNav Annual Report, 1947, 3-4; Jeffrey G. Barlow, *Revolt of the Admirals: The Fight for*

Naval Aviation, 1945-1950 (Washington, D.C.: Naval Historical Center, 1994), 131-57; Isenberg, Shield of the Republic, 142-62; Friedman, *U.S. Naval Weapons*, 187-88; Dean C. Allard, "Interservice Differences in the United States, 1945-1950: A Naval Perspective," Airpower Journal (Winter 1989), available on-line at http://www.airpower.au.af.mil/airchronicles/apj/apj89/allard.html; Michael A. Palmer, *Origins of the Maritime Strategy: American Naval Strategy in the First Postwar Decade* (Washington, D.C.: Naval Historical Center, 1988); David A. Rosenberg, "American Postwar Air Doctrine and Organization: The Navy Experience," in *Air Power and Warfare*, edited by Alfred F. Hurley and Robert C. Ehrhart (Washington, D.C.: Office of Air Force History, 1979).

14. Steven L. Reardon, *History of the Office of the Secretary of Defense*, vol. 1, *The Formative Years*, 1947-1950 (Washington, D.C.: Office of the Secretary of Defense, 1984), 393-402.

15. Muir, *Black Shoes and Blue Water*, 17; Rowland and Boyd, *Bureau of Ordnance*, 519; Leland Johnson, *Sandia National Laboratories: A History of Exceptional Service in the National Interest* (Albuquerque, N.Mex., and Livermore, Calif., 1997), 41; Department of Defense Research and Development Board, "Military Objectives in the Field of Atomic Energy," 8 June 1953, NARA II, RG 330, Entry 341, Box 436, Folder 1, pp. 2-4; Anspacher et al., White Oak Laboratory, 72; Frank H. Shelton, *Reflections of a Nuclear Weaponeer* (Colorado Springs, Colo.: Shelton Enterprise Inc., 1988), p. 5-18.

16. Anspacher et al., *White Oak Laboratory*, 72; Johnson, *Sandia National Laboratories*, 5, 41; McCollum, Dahlgren, 96-103, 107.

17. McCollum, *Dahlgren*, 100, 102-4; Chief of the Bureau of Ordnance to Commanding Officer, Naval Proving Ground, 7 October 1948, Subj: Naval Proving Ground, Dahlgren, Virginia - Project ELSIE - Facilities for, NARA II, RG 74, Entry 1003A, Box 175, Folder 10/48-12/48; Chief of the Bureau of Ordnance to Facilities Review Board (Attn: Captain C. Phillips) and the Chief of Naval Operations, 8 October 1948, Subj: Naval Proving Ground, Dahlgren, Virginia – Facilities for another Government Agency, NARA II, RG 74, Entry 1003A, Box 175, Folder 10/48-12/48; Chief of the Bureau of Ordnance to the Chief of Naval Operations, 11 October 1948, Subj: Release and Shipment of Butler Type Hut to Naval Proving Ground, Dahlgren, Virginia, for Project ELSIE, NARA II, RG 74, Entry 1003A, Box 175, Folder 10/48-12/48; Commanding Officer, Naval Proving Ground, to the Chief of the Bureau of Yards and Docks, via the Commandant, Potomac River Naval Command, and the Chief of the Bureau of Ordnance, 12 October 1948, Subj: Location of Proposed New Laboratory, Request for Approval of, NARA II, RG 74, Entry 1003A, Box 175, Folder 10/48-12/48.

18. McCollum, *Dahlgren*, 99-100.

19. McCollum, *Dahlgren*, 100, 102-4, 106.

20. McCollum, *Dahlgren*, 102, 103, 106.

21. Shelton, *Reflections*, 1-22 - 1-23, 3-24 - 3-30, 5-14 - 5-21; Schneider, *Nuclear Weapons and American Strategy*, 183; Thomas B. Cochran, William M. Arkin, and Milton M. Hoenig, *Nuclear Weapons Databook*, vol. 1, U.S. *Nuclear Forces and Capabilities* (Cambridge, Mass.: Ballinger Publishing Company, 1984), 10; Friedman, *U.S. Naval Weapons*, 185-88; Muir, *Black Shoes and Blue Water*, 39-40; Anspacher et al., White Oak Laboratory, 92.

22. McCollum, *Dahlgren*, 105, 107; Cochran et al., *U.S. Nuclear Forces*, 7.

23. Hedrick History, 34-39; Ralph A. Niemann, *Dahlgren's Participation in the Development of Computer Technology* (Dahlgren: NSWC, 1982), 2-4.

24. Hedrick History, 33-39; Niemann, *Dahlgren's Participation*, 2-4. There is extensive literature on the history of computer technology. See, for example, Charles J. Bashe et al., *IBM's Early Computers* (Cambridge, Mass.: The MIT Press, 1986); David Ritchie, *The Computer Pioneers: The Making of the Modern Computer* (New York: Simon and Schuster, 1986); I. Bernard Cohen and Gregory W. Welch, eds., *Makin' Numbers: Howard Aiken and the Computer* (Cambridge, Mass.: The MIT Press, 1999); Michael R. Williams, *A History of Computing Technology*, 2nd ed. (Los Alamos, Calif.: IEEE Computer Society Press, 1997); and Raúl Rojas and Ulf Hashagen, *The First Computers: History and Architectures* (Cambridge, Mass.: The MIT Press, 2000); Chief of the Bureau of Ordnance to Inspector of Ordnance in Charge, NPG, Dahlgren, Va., 7 April 1943, Subj: Range Table Computations and Exterior Ballistic Program, NARA II, RG 74, Entry 25, Box 327, Folder 571-1(53); Director, Research and Development Division (Captain G. L. Schuyler), to Chief of the Bureau of Ordnance, 8 April 1942, NARA II, RG 74, Entry 25, Box 212, Folder NP9/S78; Commanding Officer, U.S. Naval Proving Ground, Dahlgren, Va., to the Chief of the Bureau of Ordnance, 22 September 1944, Subj: Differential Analyzer and Controlled Sequence Calculator-Preliminary Designs of, NARA II, RG 74, Entry 25, Box 1035, Folder NP9/A1-1, July 1, 1944.

25. "Biography of Dr. Charles C. Bramble," NSWCDD Headquarters Vault, Box 78-80, Folder 5512/1 L.T.E. Thompson Road and Bramble; McCollum, *Dahlgren*, 76-79; Baile, *Historical Perspective*, 18.

26. Memorandum to Re (Research and Development Desk), 1 May 1942, NARA II, RG 74, Entry 25, Box 212, Folder NP9/S78; Director, Research and Development Division (Captain G. L. Schuyler), to Chief of the Bureau of Ordnance, 8 April 1942, NARA II, RG 74, Entry 25, Box 212, Folder NP9/S78; Commanding Officer, U.S. Naval Proving Ground, Dahlgren, Va., to the Chief of the Bureau of Ordnance, 22 September 1944, Subj: Differential Analyzer and Controlled Sequence Calculator-Preliminary Designs of, NARA II, RG 74, Entry 25, Box 1035, Folder NP9/A1-1, July 1, 1944; McCollum, *Dahlgren*, 77, 79-80; Chief of the Bureau of Ordnance to Inspector of Ordnance in Charge, NPG, Dahlgren, Va., 7 April 1943, Subj: Range Table Computations and Exterior Ballistic Program, NARA II, RG 74, Entry 25, Box 327, Folder 571-1(53); Hedrick History, 33-34. The Naval Post Graduate School was located in Annapolis from its creation in

June 1909 until December 1951, when the Navy moved it across the country to Monterey, California, where it remains today.

27. McCollum, *Dahlgren*, 79; Niemann, *Dahlgren's Participation*, 3.

28. McCollum, *Dahlgren*, 79.   By January 1945, Bramble counted twenty naval officers, fifty-eight enlisted personnel (mostly WAVEs), and twenty-two civilians in his group (see Hedrick History, 33-34). Inspector of Ordnance in Charge Captain David I. Hedrick to the Chief of the Bureau of Ordnance, 11 January 1943, RG 74, NARA II, RG 74, Entry 25, Box 604, Folder NP-9/N36; NPG Commanding Officer Captain David I. Hedrick to the Chief of the Bureau of Ordnance, 4 December 1943, NARA II, RG 74, Entry 25, Box 604, Folder NP-9/N36; Chief of the Bureau of Ordnance to Inspector of Ordnance in Charge, NPG, Dahlgren, Va., 7 April 1943, Subj: Range Table Computations and Exterior Ballistic Program, NARA II, RG 74, Entry 25, Box 327, Folder 571-1(53); Commanding Officer Captain David Hedrick, U.S. Naval Proving Ground, Dahlgren, Va., to the Chief of the Bureau of Ordnance, 17 December 1943, Subj: Bombing Tables, NARA II, RG 74, Entry 25, Box 603, Folder NP9/F41; "Brief History of the Dahlgren Calculators," n.d. but early 1950s, NSWCDD Headquarters Vault, Cabinet 98, Drawer "46-53", Folder A-12 Historical, 3; Commanding Officer, U.S. Naval Proving Ground, Dahlgren, Virginia, to the Chief of the Bureau of Ordnance, 22 September 1944, Subj: Differential Analyzer and Controlled Sequence Calculator-Preliminary Designs of, NARA II, RG 74, Entry 25, Box 1035, Folder NP9/A1-1, July 1, 1944.

29. Chief of the Bureau of Ordnance Rear Admiral William H. P. Blandy to Dr. Warren Weaver, Chief of Applied Mathematics Panel, NDRC, 15 October 1943, NARA II, RG 74, Entry 25, Box 80, Folder A16 (CND) Oct. 12-Oct 24, 1943; "Final Report of Committee on Computing Aids for Naval Ballistics Laboratory," 28 April 1944, 1-5, NARA II, RG 74, Entry 25, Box 1035, Folder NP9/A1-1, July 1, 1944.

30. Commanding Officer, U.S. Naval Proving Ground, Dahlgren, Va., to the Chief of the Bureau of Ordnance, 22 September 1944, Subj: Differential Analyzer and Controlled Sequence Calculator-Preliminary Designs of, NARA II, RG 74, Entry 25, Box 1035, Folder NP9/A1-1, July 1, 1944; Cohen and Welch, *Makin' Numbers*, 1-5.

31. Cohen and Welch, *Makin' Numbers*, 2, 54-60; Bashe et al., *IBM's Early Computers*, 25-26; David Ritchie, *The Computer Pioneers: The Making of the Modern Computer* (New York: Simon and Schuster, 1986), 53-62; I. Bernard Cohen, *Howard Aiken: Portrait of a Computer Pioneer* (Cambridge, Mass.: The MIT Press, 1999), 109-14, 263-65.

32. Cohen, *Howard Aiken*, 156-58; Bashe et al., *IBM's Early Computers*, 25-26; Ritchie, *The Computer Pioneers*, 61.

33. Ritchie, *The Computer Pioneers*, 53-54; Cohen and Welch, *Makin' Numbers*, 59-60; Thomas Watson Jr., quoted in Ritchie, *The Computer Pioneers*, 59; Bashe et al., *IBM's Early Computers*, 33.

34. Chief of the Bureau of Ordnance to the Chief of the Bureau of Ships, 9 October 1944, Subj: Duties of Commander H. H. Aiken, USNR, Definition of, NARA II, RG 74, Entry 25, Box 1035, Folder NP9/A1-1, Oct. 1, 1944; Commanding Officer, U.S. Naval Proving Ground, Dahlgren, Va., to the Chief of the Bureau of Ordnance, 22 September 1944, Subj: Differential Analyzer and Controlled Sequence Calculator-Preliminary Designs of, NARA II, RG 74, Entry 25, Box 1035, Folder NP9/A1-1, July 1, 1944.

35. Chief of the Bureau of Ordnance to the Chief of the Bureau of Ships, 9 October 1944, Subj: Duties of Commander H. H. Aiken, USNR, Definition of, NARA II, RG 74, Entry 25, Box 1035, Folder NP9/A1-1, Oct. 1, 1944; Chief of the Bureau of Ordnance to the Commanding Officer, Naval Proving Ground, Dahlgren, Va., 9 October 1944, Subj: Design Contract for Bush Type Differential Analyzer, Authorization to Execute, NARA II, RG 74, Entry 25, Box 1035, Folder NP9/A1-1, Oct. 1, 1944; Navy Liaison Officer to the Chief of the Bureau of Ordnance, via Chief of the Bureau of Ships, 14 November 1944, Subj: "Proposed Automatic Calculator Dahlgren Proving Grounds," NARA II, RG 74, Entry 25, Box 1035, Folder NP9/A1-1, Oct. 1, 1944; Chief of the Bureau of Ordnance to the Commanding Officer, Naval Proving Ground, Dahlgren, Va., 28 November 1944, Subj: Design Contract for Controlled Sequence Calculator, Authorization to Execute, NARA II, RG 74, Entry 25, Box 1035, Folder NP9/A1-1, Oct. 1, 1944; Project Order No. 50281-Ord., Naval Proving Ground, Dahlgren, Va., Appropriation 1750702(1)-Ordnance and Ordnance Stores, Navy, 1945, NARA II, RG 74, Entry 25, Box 1035, Folder NP9/A1-1, Oct. 1, 1944; Cohen and Welch, *Makin' Numbers*, 59, 113.

36. McCollum, *Dahlgren*, 123.

37. Cohen, *Howard Aiken*, 263-65; "Brief History of the Dahlgren Calculators," 3; Ritchie, *The Computer Pioneers*, 60-61. For a highly detailed description of the physical and operational specifications of the Mark II Aiken Relay Calculator and its differences with the Mark I, see engineer and Aiken protégé Robert Campbell's recollections in Cohen and Welch, *Makin' Numbers*, 111-15, 118-22, 130. Henry Tropp, "Computer Report VII: The Effervescent Years: A Retrospective," IEEE Spectrum 11, no. 2 (1 February 1974): 73.

38. Tropp, *The Effervescent Years*, 73; Niemann, *Dahlgren's Participation*, 5.

39. Niemann, *Dahlgren's Participation*, 11-12; Ritchie, *The Computer Pioneers*, 60-61.

40. "Brief History of the Dahlgren Calculators," 4-5; Niemann, *Dahlgren's Participation*, 5-10; Tropp, "The Effervescent Years," 73.

41. Cohen and Welch, *Makin' Numbers*, 120; Niemann, *Dahlgren's Participation*, 9; "Biography of Dr. Charles C. Bramble," NSWCDD Headquarters Vault, Box 78-80, Folder 5512/1 L.T.E. Thompson Road and Bramble.

42. McCollum, *Dahlgren*, 126, 130; Niemann, *Dahlgren's Participation*, 18-19; Baile, *Historical Perspective*, footnote, p. 18.

43. McCollum, *Dahlgren*, 80, 130; "Brief History of the Dahlgren Calculators," 3, 7; Niemann, *Dahlgren's Participation*, 13-16; Cohen, *Howard Aiken*, 264.

44. Niemann, *Dahlgren's Participation*, 15-16.

45. Research and Development Board, Committee on Basic Physical Sciences, "Report of the Ad Hoc Panel on Electronic Digital Computers," 1 December 1949, NARA II, RG 330, Entry 341, Box 303, Folder 13 Comm. On Phy-Sc.; " Brief History of the Dahlgren Calculators," 4-5.

46. Ritchie, *The Computer Pioneers*, 61; Niemann, *Dahlgren's Participation*, 11, 16-18; "Brief History of the Dahlgren Calculators," 4-5; Williams, History of Computing Technology, 246-48. For the Mark III's full specifications, see Cohen and Welch, *Makin' Numbers*, 264-68; "Mark III Computer: Aiken Dahlgren Electronic Calculator," 9 September 1980, CID 546020, NSWCDD Technical Library.

47. "Biography of Dr. Charles C. Bramble," NSWCDD Headquarters Vault, Box 78-80, Folder 5512/1, L.T.E. Thompson Road and Bramble; McCollum, *Dahlgren*, 76.

48. Muir, *Black Shoes and Blue Water*, 35-72; McCollum, *Dahlgren*, 84, 87, 140; Rowland and Boyd, *Bureau of Ordnance*, 520-22.

49. Brown, *High Road to Geoballistics*, 9-10, 24; Charles Roble, telephone interview by James P. Rife, 29 July 2003, 16-17; "Minutes of the 11th Meeting of the U.S. Naval Proving Ground Advisory Council, Dahlgren, Va., 21-22 May 1959," 15 June 1959, NSWCDD Administrative Vault, Box 31-60, Folder 5050/7 Minutes of Advisory Council, 2; Commander, Naval Proving Ground, to Chief of the Bureau of Ordnance, 8 April 1958, Subj: NAF Dahlgren, retention of in partial maintenance status, NSWCDD Administrative Vault, Cabinet 98, Drawer "46-53," Folder A-12 Historical.

50. Ibid. Although the Naval Airfield was deactivated, a single aircraft, supported by one aviation officer and eight enlisted ratings, was maintained for administrative and VIP flights.

51. McCollum, *Dahlgren*, 84, 140, 142; "Navy Department Denies Report that NPG May Be Moved," *Free-Lance Star*, 29 April 1953; "Dahlgren NPG There to Stay, Navy Reiterates," *Free-Lance Star*, 30 April 1953; Harry F. Byrd to Horace T. Morrison, 30 April 1953, reprinted in *Free-Lance Star*, 30 April 1953 (all news items held in NSWCDD Museum Collection, 1952 Scrapbook); Muir, *Black Shoes and Blue Water*, 38-39; Isenberg, *Shield of the Republic*, 592-98.

52. McCollum, *Dahlgren*, 87-88.

53. Ibid; Christman/Bramble Interview, May 1969, 9-10, Navy RDT&E Management Archives, Operational Archives, Collection 0008, Oral Histories, Box 870, Folder Dr. C. C. Bramble; Chief of the Bureau of Ordnance to Commanding Officer C. Turner Joy, Naval Proving Ground, Dahlgren, Va., 14 July 1949, Subj: Loss of Scientific Talent to the Los Alamos Laboratory, NARA II, RG 74, Box 607, Folder 1/49-12/49; Commander, Naval Proving Ground, to Chief of the Bureau of Ordnance, 25 July 1949, Subj: Proselyting of Scientific Personnel by the Los Alamos Laboratory, NARA II, RG 74, Box 607, Folder 1/49-12/49.

54. McCollum, *Dahlgren*, 87.

55. Niemann, *Dahlgren's Participation*, 20-24; Christman/Bramble Interview, May 1969, 9-10, Navy RDT&E Management Archives, Operational Archives, Collection 0008, Oral Histories, Box 870, Folder Dr. C.C. Bramble.

56. Christman/Bramble Interview, May 1969, 9-10, Navy RDT&E Management Archives, Operational Archives, Collection 0008, Oral Histories, Box 870, Folder Dr. C.C. Bramble.

57. Christman/Bramble Interview, May 1969, 9-10; DiDonato Oral History; Albert Christman, *Target Hiroshima: Deak Parsons and the Creation of the Atomic Bomb* (Annapolis: Naval Institute Press, 1998), 232-55.

58. Niemann, *Dahlgren's Participation*, 27-30.

59. Ibid., 28-29; Bashe et al., *IBM's Early Computers*, 181. For a detailed description of NORC's specifications and operations, see Wallace J. Eckert and Rebecca Jones, *Faster, Faster: A Simple Description of a Giant Electronic Calculator and the Problems It Solves* (New York: International Business Machines Corporation, 1955).

60. McCollum, *Dahlgren*, 80.

61. Ibid., 175; Chief of the Bureau of Ordnance F. S. Withington to Commanding Officer Captain J. F. Byrne, Naval Proving Ground, Dahlgren, Va., 9 September 1955, NSWCDD Administrative Vault, Cabinet 98, Drawer "46-53," Folder A4-A4-3, Operation & Status of Vessels, Shore Stations, etc; Chief, Bureau of Ordnance, to Commander, U.S. Naval Proving Ground, 16 December 1955, Subj: Proposed Mission for NPG, Dahlgren, NSWCDD Administrative Vault, Cabinet 98, Drawer "46-53," Folder Unlabeled, 1955; "Minutes of the 8th Meeting of the U.S. Naval Proving Ground Advisory Council, 30 January 1958," NSWCDD Administrative Vault, Box 31-60, Folder 5050/7 Minutes of Advisory Council.

62. McCollum, *Dahlgren*, 81.

63. Ibid., 92-95.

64. Ibid; "Minutes of the 11th Meeting of the U.S. Naval Proving Ground Advisory Council, 15 June 1959," NSWCDD Administrative Vault, Box 31-60, Folder 5050/7 Minutes of Advisory Council; Anspacher, et al., *White Oak Laboratory,* 150-52. Between 1945 and 1970, the fleet recorded some twenty inadvertent ordnance detonations that were attributed to EMR; Robert L. Hudson, Final Review of Manuscript, 3 October 2005.

65. Ibid.; Robert L. Hudson, "Specific Contribution for History," NSWCDD Public Affairs Office, 17 June 2005.

66. Robert V. Gates, "Strategic Systems Fire Control," *Technical Digest: Mathematics, Computing, and Information Science* (Dahlgren: NSWCDD, 1995), 166-79; see also Graham Spinardi, *From Polaris to Trident: The Development of U.S. Fleet Ballistic Missile Technology* (Cambridge, England: Cambridge University Press, 1994), 25-34, and Robert A. Fuhrman, "The Fleet Ballistic Missile System: Polaris to Trident," Journal of Spacecraft 15, no. 5 (September-October 1978): 265-86; Isenberg, *Shield of the Republic,* 656-87.

67. Hughey, "History of Mathematics and Computing Technology," 13, 15-17; Gates, "Strategic Systems Fire Control," 167; Brown, *High Road to Geoballistics,* 30-32; Isenberg, *Shield of the Republic,* 676; David R. Brown Jr., "The Contributions of the U.S. Naval Weapons Laboratory in Geoballistics and Digital Fire Control to the Fleet Ballistic Missile Program: A Presentation to the Special Projects Office," 24 March 1964, NSWCDD Technical Library, #124084, 3; Donald T. Edwards and Doreen H. Daniels, *The SLBM Program: Brief History of NSWC's Involvement in the FBM/SLBM Fire Control Systems, 1956-1984* (Dahlgren: Naval Surface Weapons Center, November 1984); Robert Gates, interview by James P. Rife, Dahlgren, Va., 4 June 2003 (hereafter Gates Oral History), 9-11, 13-15; David R. Brown Jr. to Kenneth A. Boyd, 11 July 1992, Wayne Harman/David Brown Historical File, Folder Speeches, NSWCDD, 3; McCollum, *Dahlgren,* 90-91.

68. Hughey, "History of Mathematics and Computing Technology," 16-17; Gates Oral History, 14-15; Brown to Boyd, 11 July 1992, 3; Edwards and Daniels, *The SLBM Program,* 8-12; McCollum, *Dahlgren,* 90-91.

69. Edwards and Daniels, *The SLBM Program,* 8-11; Gates, "Strategic Systems Fire Control," 168-69; Isenberg, *Shield of the Republic,* 684-85.

70. McCollum, *Dahlgren,* 90-91; Gates, "Strategic Systems Fire Control," 166-67.

71. Isenberg, *Shield of the Republic,* 678-79; Gates, "Strategic Systems Fire Control," 166-67; Shelton, *Reflections,* 11-17 to 11-23; Fuhrman, "The Fleet Ballistic Missile System," 277-78; *FBM Facts/Chronology: Polaris-Poseidon-Trident* (Washington, D.C.: Navy Department, 1990), 28; Spinardi, *From Polaris to Trident,* 62-63; e-mail communication from Robert V. Gates to James P. Rife, 2 March 2004. Concerning the fate of FRIGATE BIRD's POLARIS target cards, Gates remembered that "the sad part of the story is that those target cards were at Dahlgren for a number of years after the test and, at some point in time, were thrown out when someone cleaned out old classified material rather than downgrading it. They obviously didn't know the significance of the cards."

72. Brown, *High Road to Geoballistics*, iv-vi; Brown, "Geoballistics and Digital Fire Control."

73. Paul B. Stares, *The Militarization of Space: U.S. Policy, 1945-1984* (Ithaca, N.Y.: Cornell University Press, 1985), 38-58.

74. Special Systems Department of the System Development Corporation, "SPASUR Automatic System Mark 1: Operating System Description," 15 August 1962, NSWCDD Technical Library, Number 546110, pp. 1-2 - 1-4; Norman Friedman, *Seapower and Space: From the Dawn of the Missile Age to Net-Centric Warfare* (London, England: Chatham Publishing, 2000), 126-28. Receiver stations were built at Fort Stewart, Georgia; Silver Lake, Mississippi; Elephant Butte, New Mexico; and Brown Field, San Diego, California; transmitter stations were built at Jordan Lake, Alabama; Lake Kickapoo, Archer City, Texas; and Gila River, Arizona.

75. "Minutes of the 9th Meeting of the U.S. Naval Proving Ground Advisory Council, Dahlgren, Virginia," 4 June 1958, "Comments of the Advisory Council," 7, NSWCDD Administrative Vault, Box 31-60, Folder 5050/7 Minutes of Advisory Council; Niemann, *Dahlgren's Participation*, 29; Hughey, "History of Mathematics and Computing Technology," 18; Special Systems Department of the System Development Corporation, "SPASUR Automatic System Mark 1: Operating System Description," 15 August 1962, NSWCDD Technical Library, Number 546110, pp. 1-2 - 1-4; Commanding Officer, U.S. Naval Space Surveillance System, Dahlgren, Va., to Chief of Naval Operations, 25 July 1962 (with Enclosed "Background Data on the U.S. Naval Space Surveillance System" and "History of U.S. Naval Space Surveillance System from 20 June 1958 to 31 December 1961"), NSWCDD Administrative Vault, Cabinet 3/97, Drawer 5, Folder 5750-History.

76. Special Systems Department of the System Development Corporation, "SPASUR Automatic System Mark 1: Operating System Description," 15 August 1962, NSWCDD Technical Library, Number 546110, pp. 1-2 - 1-4; Commanding Officer, U.S. Naval Space Surveillance System, Dahlgren, Va., to Chief of Naval Operations, 25 July 1962 (with Enclosed "Background Data on the U.S. Naval Space Surveillance System" and "History of U.S. Naval Space Surveillance System from 20 June 1958 to 31 December 1961"), NSWCDD Administrative Vault, Cabinet 3/97, Drawer 5, Folder 5750-History; Chief, Bureau of Naval Weapons, to Distribution List, 13 February 1961, Subj: Command Relationships and Management Control Coordination of U.S. Naval Space Surveillance System (NAVSPASUR), Dahlgren, Virginia, assignment of, NARA II, RG 402, A1/Entry 10, Box 6, Folder BuWeps Notice 5450-13 February 1961, canceled, 31 December 1961-obsolete.

77. Chief of the Bureau of Ordnance M.F. Schoeffel to Commander, U.S. Naval Proving Ground, Dahlgren, Va., 24 April 1952, Subj: Creation of an Advisory Board, NSWCDD Administrative Vault, Cabinet 3(96), Drawer 3, Folder 5420/2 NWL Advisory Council (Naval Ordnance Adv. Bd.); "Tentative Minutes of the

First Meeting of the Naval Proving Ground Advisory Council," 19 May 1953, NSWCDD Administrative Vault, Cabinet 98, Drawer "46-53," Folder NPG Advisory Council Conferences; "Minutes of the 9th Meeting of the U.S. Naval Proving Ground Advisory Council, Dahlgren, Virginia," 4 June 1958, 3, NSWCDD Administrative Vault, Box 31-60, Folder 5050/7 Minutes of Advisory Council; "Minutes of the 11th Meeting of the U.S. Naval Proving Ground Advisory Council, Dahlgren, Virginia," 15 June 1959, "Comments of the Advisory Council," 1, 6, NSWCDD Administrative Vault, Box 31-60, Folder 5050/7 Minutes of Advisory Council. For a thorough discussion of the evolution of Dahlgren's management culture, see Baile, *Historical Perspective*; McCollum, *Dahlgren*, 95.

78. McCollum, *Dahlgren*, 95; "Minutes of the 10th Meeting of the U.S. Naval Proving Ground Advisory Council, Dahlgren, Virginia," 3 December 1958, 1, 5-6, NSWCDD Administrative Vault, Box 31-60, Folder 5050/7 Minutes of Advisory Council; "Minutes of the 11th Meeting of the U.S. Naval Proving Ground Advisory Council, Dahlgren, Virginia," 15 June 1959, "Comments of the Advisory Council," 1, 6, NSWCDD Administrative Vault, Box 31-60, Folder 5050/7 Minutes of Advisory Council.

79. "Minutes of the 9th Meeting of the U.S. Naval Proving Ground Advisory Council, Dahlgren, Virginia," 4 June 1958, "Comments of the Advisory Council," 7, NSWCDD Administrative Vault, Box 31-60, Folder 5050/7 Minutes of Advisory Council; "Minutes of the 11th Meeting of the U.S. Naval Proving Ground Advisory Council, Dahlgren, Virginia," 15 June 1959, "Comments of the Advisory Council," 1, 6, NSWCDD Administrative Vault, Box 31-60, Folder 5050/7 Minutes of Advisory Council; McCollum, *Dahlgren*, 92-93.

80. "Minutes of the 11th Meeting of the U.S. Naval Proving Ground Advisory Council, Dahlgren, Virginia," 15 June 1959, "Comments of the Advisory Council," 1, NSWCDD Administrative Vault, Box 31-60, Folder 505017 Minutes of Advisory Council.

Chapter 5: *Rebels and Revolution, 1959-1973*

1. "Minutes of the 10th Meeting of the U.S. Naval Proving Ground Advisory Council, Dahlgren, Virginia," 3 December 1958, 5, NSWCDD Administrative Vault, Box 31-60, Folder 5050/7 Minutes of Advisory Council.

2. "Minutes of the 12th Meeting of the U.S. Naval Weapons Laboratory Advisory Council, Dahlgren, Virginia," 30 November 1959, 2-3, including "Comments and Recommendations," 1-5, NSWCDD Administrative Vault, Box 31-60, Folder 5050/7 Minutes of Advisory Council.

3. Ibid.

4. Ibid.; Michael R. Williams, *A History of Computing Technology*, 2nd ed. (Los Alamos, Calif.: IEEE Computer Society Press, 1997), 385-94. For the history of

STRETCH's design and development, see Charles J. Bashe et al., *IBM's Early Computers* (Cambridge, Mass.: The MIT Press, 1986), 416-58; Ralph A. Niemann, *Dahlgren's Participation in the Development of Computer Technology* (Dahlgren: NSWC, 1982), 30-31.

5. Niemann, *Dahlgren's Participation*, 30-31; "History of Command from 1 January 1960 to 31 December 1960," NSWCDD Administrative Vault, Cabinet 3/97, Drawer 5, Folder 5750 Historical Matters; Robert V. Gates, "Strategic Systems Fire Control," *Technical Digest: Mathematics, Computing, and Information Science* (Dahlgren: NSWCDD, 1995), 170.

6. Niemann, *Dahlgren's Participation*, 31.

7. Ibid.

8. Ibid., 31-32; Raymond H. Hughey Jr., "History of Mathematics and Computing Technology at the Dahlgren Laboratory," *Technical Digest: Mathematics, Computing, and Information Science* (Dahlgren: NSWCDD, 1995), 15-18; Kenneth G. McCollum, ed., *Dahlgren* (Dahlgren: Naval Surface Weapons Center, June 1977), 135-36.

9. "History of Command from 1 January 1960 to 31 December 1960," NSWCDD Administrative Vault, Cabinet 3/97, Drawer 5, Folder 5750 Historical Matters; Naval Research Advisory Committee, "Study of Bureau of Naval Weapons Research and Development Facility Requirements, Phase II: Station Survey," 20 July 1960, 7, Naval Historical Center, Operational Archives Branch, NL/CCG Archives, Record Collection 3, Box 102, Folder NRAC Study of BUWEPS R&D Facility Requirements, Phase II, Station Survey, Naval Weapons Laboratory, Dahlgren, VA, 1960 (hereafter cited as "1960 NRAC Survey of NWL"); Dr. W. D. Lewis, "Report on Visit to Naval Weapons Laboratory, Dahlgren, Virginia," 25 to 27 January 1960, 2-8, Naval Historical Center, Operational Archives Branch, NL/CCG Archives, Record Collection 3, Box 102, Folder NRAC Study of BUWEPS R&D Facility Requirements, Phase II, Station Survey, Naval Weapons Laboratory, Dahlgren, VA, 1960 (hereafter cited as "1960 Lewis Report on NWL Visit"; Commander, U.S. Naval Weapons Laboratory Captain Robert F. Sellars to Chief of Naval Operations, 19 June 1964, "History of Command from 1 January 1963 to 31 December 1963," NSWCDD Administrative Vault, Cabinet 3/97, Drawer 5, Folder 5750-History (hereafter "1963 History of Command").

10. Robert L. Hudson, "Specific Contribution for History," NSWCDD Public Affairs Office, 17 June 2005 (hereafter cited as "Hudson Contribution"); David B. Colby listing in "Senior Scientific and Engineering Personnel," NSWC/DL MP-24/77, December 1977, NSWCDD Administrative Vault, Shelf Materials; David B. Colby, telephone interview by James Rife, 13 September 2005 (hereafter cited as Colby/Rife Interview).

11. Ibid; "1960 NRAC Survey of NWL," 1-27.

12. "1960 Lewis Report on NWL Visit," 1; McCollum, *Dahlgren*, 129; *Key Scientific and Engineering Personnel*, 1960, U.S. Naval Weapons Laboratory, NSWCDD, Museum Historical Collection; David R. Brown Jr., "Highlights of NSWC/DL Contributions to Strategic Missile Systems," 1973, Robert Gates Historical File, NSWCDD (hereafter cited as "Highlights"); Hughey, "History of Mathematics and Computing Technology," 15-18; Gates, "Strategic Systems Fire Control," 170.

13. Brown, "Highlights"; Gates, "Strategic Systems Fire Control," 168-70; Donald T. Edwards and Doreen H. Daniels, *The SLBM Program: Brief History of NSWC's Involvement in the FBM/SLBM Fire Control Systems, 1956-1984* (Dahlgren: Naval Surface Weapons Center, November 1984), 8-12.

14. Gates, "Strategic Systems Fire Control," 168-70; Edwards and Daniels, *The SLBM Program*, 8-12; Norman Friedman, *U.S. Naval Weapons: Every Gun, Missile, Mine, and Torpedo Used by the U.S. Navy from 1883 to the Present Day* (Annapolis, Md.: Naval Institute Press, 1983), 220-23.

15. Strategic Systems Programs, *FBM Facts/Chronology: Polaris-Poseidon-Trident* (Washington, D.C.: Navy Department, 1990), 6; Brown, "Highlights"; Gates, "Strategic Systems Fire Control," 169-72; Edwards and Daniels, *The SLBM Program*, 12-15; Friedman, *U.S. Naval Weapons*, 223-24. For a history of the POSEIDON program, see Graham Spinardi, *From Polaris to Trident: The Development of U.S. Fleet Ballistic Missile Technology* (Cambridge, England: Cambridge University Press, 1994), 86-112; e-mail communication from Robert Gates to James P. Rife, 12 August 2003.

16. Gates, "Strategic Systems Fire Control," 169-72; Edwards and Daniels, *The SLBM Program*, 12-15; Friedman, *U.S. Naval Weapons*, 223-24; Strategic Systems Programs, *FBM Facts/Chronology*, 6.

17. "1960 NRAC Survey of NWL," 1, 8, 12-14, 25-27; "1960 Lewis Report on NWL Visit," 2. The Naval Research Advisory Council is an independent civilian scientific group founded in 1946 that advises the Office of Naval Research on matters of RDT&E policy and management.

18. "1960 NRAC Survey of NWL," 1, 8, 12-14, 25-27.

19. McCollum, *Dahlgren*, 90, 131-32; Dr. Armido DiDonato, interview by James P. Rife, Dahlgren, Va., 13 August 2003 (hereafter DiDonato Oral History), 16-17.

20. McCollum, *Dahlgren*, 90.

21. Ibid.; DiDonato Oral History, 12, 31; "1963 History of Command," Enclosure 1, "Administrative History"; DiDonato Oral History, 31.

22. McCollum, *Dahlgren*, 131-32.

23. "1960 Lewis Report on NWL Visit," 3; "Minutes of the 12th Meeting of the U.S. Naval Weapons Laboratory Advisory Council, Dahlgren, Virginia," 30 November 1959, 1, NSWCDD Administrative Vault, Box 31-60, Folder 5050/7 Minutes of Advisory Council; McCollum, *Dahlgren*, 117-22; "Report of the 22nd Meeting of the U.S. Naval Weapons Laboratory Advisory Council," Dahlgren, Virginia, 15 June 1966, 2, NSWCDD Administrative Vault, Box 78-80, Folder 5050/7 NWL Advisory Council Minutes (hereafter "Report of the 22nd Meeting of the NWL Advisory Council").

24. "1963 History of Command," Enclosure 1, "Administrative History"; McCollum, *Dahlgren*, 82; DiDonato Oral History, 23; Bernard Smith, *Looking Ahead from Way Back: An Autobiography by Bernard Smith* (Richmond, Ind.: Prinit Press, 1999), 69-73; Oral History Interview with Barney Smith, circa 1974, 21, Naval Historical Center Operational Archives Branch, NL/CCG Archives, Record Collection 8, "Oral Histories," Box 882, Folder Barney Smith, ca. 1974; "1963 History of Command," Enclosure 3, "Brief Biography of Officer in Command-December 1962, Captain Robert F. (Mike) Sellars," NSWCDD Administrative Vault, Cabinet 3/97, Drawer 5, Folder 5750-History.

25. Smith, *Looking Ahead*, 1-59, 64-69; McCollum, *Dahlgren*, 151-53. For Thompson's research philosophy and tenure at NOTS Inyokern, California, see Christman, *Sailors, Scientists, and Rockets: Origins of the Navy Rocket Program and of the Naval Ordnance Test Station, Inyokern* (Washington, D.C.: Naval History Division, 1971); J. D. Gerrard-Gough and A. B. Christman, *The Grand Experiment at Inyokern, 1944-1948* (Washington, D.C.: Naval History Division, 1978).

26. McCollum, *Dahlgren*, 153; Smith, Looking Ahead, 71-72, 75.

27. Ibid. For a full discussion of McNamara's Navy RDT&E management reforms and policies, see Rodney P. Carlisle, *Management of the U.S. Navy Research and Development Centers During the Cold War: A Survey Guide to Reports* (Washington, D.C.: NL/CCG and Naval Historical Center, 1996), 18-43; Cynthia Rouse Interview with James E. Colvard, 10 December 1976, 8-9, Naval Historical Center, Operational Archives Branch, NL/CCG Archives, Record Collection 8, Box 871, Folder Colvard, James E., "The Center Viewpoint," 10 Dec. 1976, Dahlgren, Virginia (hereafter Rouse/Colvard Interview); Dr. Vincent Ponko Jr. Interview with Mr. James E. Colvard, 3 November 1980, 28-29, Naval Historical Center, Operational Archives Branch, NL/CCG Archives, Record Collection 8, Box 871, Folder NL-T28, Mr. James E. Colvard, Final Transcript of Interview (hereafter Ponko/Colvard Interview).

28. Smith, *Looking Ahead*, 73-75.

29. McCollum, *Dahlgren*, 154, Rouse/Colvard Interview, 8-9; Susan Frutkin and Dr. Peter Bruton (Booz-Allen and Hamilton) Interview with Barney Smith, circa 1974, 2, Naval Historical Center, Operational Archives Branch, NL/CCG Archives, Record Collection 8, Box 882, Folder Barney Smith, ca. 1974 (hereafter Frutkin/Bruton/Smith Interview); McCollum, Dahlgren, 161; Technical Director

Bernard Smith and Captain William A. Hasler Jr., "Recommendations for the Development of the Naval Weapons Laboratory," 5, NSWCDD Administrative Vault, Shelf Materials, Folder Loose.

30. "Recommendations for the Development of the Naval Weapons Laboratory," 1-7; Smith, *Looking Ahead*, 74-75; Frutkin/Bruton/Smith Interview, 15.

31. Ibid.

32. Frutkin/Bruton/Smith Interview, 87; Smith, *Looking Ahead*, 75; NWL Administrative Report No. AR-103, "Recommendations for the Development of the Naval Weapons Laboratory: Mission, Organization, Program," May 1968, 3-4, NSWCDD Administrative Vault, Shelf Materials, Folder Loose.

33. Frutkin/Bruton/Smith Interview, 112-13.

34. Ibid.; McCollum, *Dahlgren*, 163.

35. McCollum, *Dahlgren*, 163.

36. McCollum, *Dahlgren*, 163.

37. Smith, *Looking Ahead*, 69, 73-74; Biography, "Captain William August Hasler, Jr., United States Navy," NSWCDD Administrative Vault, Box 78-80, Folder 5512-Identification.

38. Smith, Looking Ahead, 69, 73-74; Captain Steven N. Anastasion to Dr. William F. Whitmore, 31 October 1969, NSWCDD Administrative Vault, Cabinet 3(96), Drawer 3, Folder 5420/1 Naval Ordnance Advisory Board, 1969-1970.

39. Smith, *Looking Ahead*, 69, 73-74; Frutkin/Bruton/Smith Interview, 55; McCollum, *Dahlgren*, 157-58.

40. Carlisle, *Management of the U.S. Navy Research and Development Centers*, 35-37; Smith, *Looking Ahead*, 69; see also William B. Anspacher et al., *The Legacy of the White Oak Laboratory* (Dahlgren: NSWCDD, 2000), 383-86. Jim Colvard discusses the debate over who should control Navy R&D Centers, ASN(R&D) or the SYSCOMs, in his interview with Dr. Vincent Ponko Jr. Colvard felt that the compromise was a rational decision (Ponko/Colvard Interview, 7-8).

41. Ibid.

42. Carlisle, *Management of the U.S. Navy Research and Development Centers*, 41-43; Anspacher et al., *White Oak Laboratory*, 389-94.

43. Carlisle, *Management of the U.S. Navy Research and Development Centers*, 45-48; Anspacher et al., *White Oak Laboratory*, 389-94; Deputy Director of Navy Laboratories B. H. Andrews to Captain J. E. Rawls, NWL, Dahlgren, Virginia, Attached Paper "Thoughts on Centers," 20 May 1968, NSWCDD Administrative Vault, Box 78-80, Folder 5450/5713 Missions-Centers; Assistant Secretary of the

Navy for Research and Development Robert A. Frosch to the Chief of Naval Material, 21 March 1967, NSWCDD Administrative Vault, Box 78-80, Folder Merging of Laboratories.

44. Frosch to CNM, 21 March 1967; Anspacher et al., *White Oak Laboratory*, 359-60, 390-91; NSWC *Command History* for 1974-1977, Dahlgren, Virginia, NSWCDD Administrative Vault, Shelf Materials, Folder: Loose.

45. Captain William A. Hasler Jr. to Dr. L. T. E. Thompson, 22 December 1966, NSWCDD Administrative Vault, Box 78-80, Folder Merging of Laboratories; "Report of the Naval Weapons Laboratory Advisory Council Meeting, May 4-6, 1967," 1-7, NSWCDD Administrative Vault, Box 78-80, Folder 5050/7 NWL Advisory Council Minutes.

46. "Report of the Naval Weapons Laboratory Advisory Council Meeting, 4-6 May 1967," attachment, "Warfare Centers and the Amalgamation Problem," 5-7, NSWCDD Administrative Vault, Box 78-80, Folder 5050/7 NWL Advisory Council Minutes (hereafter "Report of the NWL Advisory Council, 4-6 May 1967").

47. Smith, *Looking Ahead*, 72.

48. NWL Technical Director Bernard Smith to Assistant Secretary of the Navy (Research and Development) Robert A. Frosch, Subject: Formation of Warfare Centers, 20 April 1967, NSWCDD Administrative Vault, Box 78-80, Folder Merging of Laboratories.

49. Ibid.

50. Ibid.

51. Captain William A. Hasler Jr. to Rear Admiral Frederic S. Withington, USN (Retired), 4 January 1968, NSWCDD Administrative Vault, Box 78-80, Folder 5420/1 NWL Advisory Council; NSWC *Command History* for 1974-1977; Anspacher et al., *White Oak Laboratory*, 360-61, 390-94. For a detailed listing of all the mergers, see Carlisle, *Management of the U.S. Navy Research and Development Centers*, 45-47.

52. Ibid.

53. McCollum, *Dahlgren*, 140-43; Smith, *Looking Ahead*, 78-79.

54. McCollum, *Dahlgren*, 140-41.

55. "Report of the 22nd Meeting of the NWL Advisory Council," 4; "Report of the NWL Advisory Council, 4-6 May 1967," 1, 3.

56. Lt. R. A. Green and D. W. Harris, NWL Administration Note AN-E/1-71, "Review of Quick Reaction Programs at NWL from April 1967 through August 1971: The Vietnam Laboratory Assistance Program and the Navy Science Assistance Program," August 1971, U.S. Naval Weapons Laboratory, Dahlgren, Virginia

(hereafter "Quick Reaction Programs at NWL"), ii, 1-3, 9-18; *Senior Scientific and Engineering Personnel*, "Donald H. George," January 1983, NSWC MP 82-247.

57. "Review of Quick Reaction Programs," ii, 8, 18-23; Smith, *Looking Ahead*, 80; Frutkin/Bruton/Smith Interview, 45.

58. E-mail from Wayne L. Harman to Russell L. Coons, 18 March 2004, NSWCDD, Dahlgren, Virginia; Green and Harris, "Quick Reaction Programs at NWL," 34-40.

59. J. E. Colvard to Distribution, 9 June 1970, 4, NSWCDD Administrative Vault, Box 78-80, Folder Camp A.P. Hill Retreat, 5420; Captain Steven N. Anastasion to Dr. William F. Whitmore, with attached "Status Report, U.S. Naval Weapons Laboratory," 13 April 1971, NSWCDD Administrative Vault, Cabinet 3(96), Drawer 3, Folder Laboratory Advisory Board for Ordnance, 1971; Tommy Tschirn, interview by James P. Rife, Dahlgren, Va., 25 June 2003 (hereafter Tschirn Oral History) (Unrecorded Postscript to Interview).

60. Hudson Contribution; Robert L. Hudson listing in "Senior Scientific and Engineering Personnel," NSWC/DL MP-24/77, December 1977, NSWCDD Administrative Vault, Shelf Materials; Robert L. Hudson, Final Review of Manuscript, 3 October 2005, hereafter cited as "Final Review"; Charles E. "Gene" Gallaher, interview by James P. Rife, Dahlgren, Va., 14 May 2004, 1-13.

61. Hudson, "Final Review."

62. Ibid. The P-19 anechoic chamber is still functional in Building 150 at Dahlgren today.

63. Ibid.

64. Naval Electronic Systems Command, "Draft Project Master Plan for the Electromagnetic Performance of Aircraft and Ship Systems (EMPASS)," 22 January 1979, located in the files of Robin R. Staton, NSWCDD, Dahlgren, Virginia; Colby/Rife Interview; "Empass Aircraft EP-3A departs for VX-1, Patuxent River," *The Laboratory Log*, 18 July 1975.

65. Ibid.

66. Ibid.

67. Defense Threat Reduction Agency, *Defense's Nuclear Agency, 1947-1997* (Washington, D.C.: Department of Defense, 2002), 238-42. Although no author is credited except the Defense Threat Reduction Agency, Dr. Richard G. Hewlett, Dr. Philip L. Cantelon, and Dr. Rodney P. Carlisle of History Associates Incorporated all contributed significantly to the work. See also *The NWL Story (1918-1971): Proving Ground to Weapons Laboratory* (Dahlgren, Va.: Naval Weapons Laboratory, 1971), Appendix B, 8, held in NSWCDD Administrative Vault, Shelf Materials, Folder Loose; "Report of the 22nd Meeting of the NWL Advisory Council," 3; "Current and Projected Programs for NWL, Dahlgren,"

Naval Ordnance Systems Command, April 1967, 2-3, Appendix III; *Washington Post Magazine*, "Backlight," 4 January 2004, 4.

68. "4th Report of the Naval Research Advisory Committee, Laboratory Advisory Board for Ordnance," 1 December 1969, 3, NSWCDD Administrative Vault, Cabinet 3(96), Drawer 3, Naval Ordnance Advisory Board, Folder Naval Ordnance Advisory Board, 1969-1972 (1).

69. "Report of the NWL Advisory Council, 4-6 May 1967," 3; McCollum, *Dahlgren*, 143-46; Captain Steven Anastasion to Dr. William F. Whitmore, 13 April 1971, NSWCDD Administrative Vault, Cabinet 3(96), Drawer 3, Folder Laboratory Advisory Board for Ordnance, 1971; Tschirn Oral History, 1-4, 15, 29.

70. Ibid.

71. Ian V. Hogg, *German Secret Weapons of the Second World War: The Missiles, Rockets, Weapons, and New Technology of the Third Reich* (Mechanicsburg, Pa.: Stackpole Books, 1999), 148-52; Captain R. F. Schniedwind to Commander, Naval Ordnance Systems Command, 28 September 1972, Subject: Space Research Corporation; profile of, NSWCDD Administrative Vault, Box 31-60, Folder SRC Claim; for Bull's interest in the history of German "big gun" technology and its applications to modern artillery and space research, see Gerald V. Bull and Charles H. Murphy, *Paris Kanonen - The Paris Guns (Wilhelmgeschutze) and Project HARP: The Application of Major Calibre Guns to Atmospheric and Space Research* (Herford, Germany: E.S. Mittler, ca.1988).

72. McCollum, *Dahlgren*, 144-45; Schniedwind to NAVORD, 28 September 1972; SRC Appeal Before the Armed Services Board of Contract Appeals, ASBCA No. 16241, Contract No. N00178-71-C-0002, 12 July 1971; Sterling Cole to Rear Admiral Thomas J. Christman, USN, 20 September 1971; Robert C. Stacey to Major General J. R. Guthrie, 27 January 1971; Commander, NWL, to Commander, Naval Ordnance Systems Command, 28 January 1972; Dr. Gerald V. Bull to Bernard Smith, 1 March 1972; Sterling Cole to Admiral M. W. Wood, 19 January 1972; Sterling Cole to Secretary of the Navy, 13 April 1972; Sterling Cole to Admiral Isaac Kidd Jr., 26 May 1972; Captain R. F. Schniedwind, Memorandum for the Record, 23 August 1972; Sterling Cole to Captain R. F. Schniedwind, 22 November 1972; and James E. Colvard, Memorandum for the Record, 20 August 1973, all in NSWCDD Administrative Vault, Box 31-60, Folder SRC Claim. After his early work with the U.S. and Canadian defense establishments in the 1960s, Bull's fortunes sharply declined. In the 1970s Bull struggled to keep the SRC viable by becoming an international arms designer and dealer, and he served a brief stint in federal prison for violating President Jimmy Carter's arms embargo against South Africa. In the 1980s Bull went to work for Sadaam Hussein and designed a "super" gun capable of striking Israel from the western Iraqi desert. Bull was assassinated in Brussels in March 1990 by unknown agents. For Bull's life, work, and death, see James Adams, *Bull's Eye: The Assassination and Life of Supergun Inventor Gerald Bull* (New York: Times Books, 1992).

73. Ibid.

74. Ibid.

75. Smith, *Looking Ahead*, 78-79.

76. Ibid., 77; Ponko/Colvard Interview, 17.

77. "Recommendations for the Development of the Naval Weapons Laboratory," 11.

78. Ibid., 7.

79. Ibid., 18; Captain William A. Hasler Jr. to Rear Admiral Frederic S. Withington, USN (Ret.), 2 July 1968, NSWCDD Administrative Vault, Box 78-80, Folder 5450/5713 Missions-Centers.

80. McCollum, *Dahlgren*, 154; Smith, *Looking Ahead*, 76; Ponko/Colvard Interview, 17; "Recommendations for the Development of the Naval Weapons Laboratory," 24-33; Commander, Naval Weapons Laboratory, Dahlgren, Virginia, to Commanding General, U.S. Army Material Command, 15 July 1970, Subject: Naval Weapons Laboratory Test and Evaluation Capability, information concerning, 14, NSWCDD Administrative Vault, Box 78-80, Folder NWL Test & Evaluation Capability 5000; Tschirn Oral History, 3, 14, 16.

81. NWL Commander to Commanding General, U.S. Army Material Command, 15 July 1970, 14; McCollum, *Dahlgren*, 154-55; Smith, *Looking Ahead*, 77; Rouse/Colvard Interview, 7; Ponko/Colvard Interview, 28-29; Frutkin/Bruton/Smith Interview, 89-90; Carlisle, *Management of the U.S. Navy Research and Development Centers*, 52-53.

82. Smith, *Looking Ahead*, 81-82; Anspacher et al., *White Oak Laboratory*, 361; Rouse/Colvard Interview, 18.

83. Smith, *Looking Ahead*, 77-78; Rouse/Colvard Interview, 18; Ponko/Colvard Interview, 15.

Chapter 6: On the Surface, 1973-1987

1. James E. Colvard, interview by Dr. Vincent Ponko, 3 November 1980, Naval Historical Center, Operational Archives Branch, NL/CCG Archives, Record Collection 8, Box 871, Folder NL-T28, Mr. James E. Colvard, Final Transcript of Interview (hereafter Ponko/Colvard Interview), p. 14.

2. James Colvard, interview by Cynthia Rouse, 10 December 1976, Naval Historical Center Operational Archives Branch, NL/CCG Archives, Record Collection 8, Box 871, Folder Colvard (hereafter Rouse-Colvard Interview), p. 1; Ponko-Colvard Interview, p. 14.

3. Rouse-Colvard Interview, pp. 1-2; Ponko-Colvard Interview, p. 15.

4. Ibid., p. 3; James E. Colvard, interview by Joseph Marchese, 2 March 1990, Naval Historical Center, Operational Archives Branch, NL/CCG Archives, Record Collection 8, Box 871, Folder Colvard, summary vita.

5. Ponko-Colvard Interview, pp. 16-17.

6. Ibid., p. 17.

7. Smith in *On the Surface*, 21 March 1980, quoting *Aeronautics and Astronautics* interview from September 1979.

8. Ponko-Colvard Interview, p. 15.

9. Ibid., p. 17; Robert Gates, interview by James P. Rife, Dahlgren, Va., 4 June 2003 (hereafter Gates Oral History), p. 49.

10. Ponko-Colvard Interview, p. 18-19; Rouse/Colvard Interview, pp. 2-3.

11. Ponko-Colvard Interview, p. 18-19; Rouse/Colvard Interview, pp. 2-3; William Anspacher, et al., *The Legacy of the White Oak Laboratory* (Dahlgren: NSWCDD, 2000), p. 400.

12. Anspacher, *The Legacy of the White Oak Laboratory*, p. 371; Ponko-Colvard Interview, pp. 19-20.

13. Anspacher, *The Legacy of the White Oak Laboratory*, p. 400; Rouse/Colvard Interview, pp. 2-3; Memorandum for Vice Admiral J. H. Doyle, Jr., 18 April 1977, NSWCDD Administrative Vault, Box 31-60, Folder: Headquarters-Corres.; Smith, et al., *NOL-NWL Consolidation Study*, p. 13.

14. Anspacher, *The Legacy of the White Oak Laboratory*, pp. 410-412; Bernard Smith (NAVMAT), William Wineland (WOL), Charles Cohen (NWL), and Bernard Connolly (NAVMAT), *NOL-NWL Consolidation Study*, Prepared for the Deputy Chief of Naval Material (Development), 12 October 1973, NSWCDD Administrative Vault, Shelf Materials, Folder: Green Binder "Sensitive Material-Hold Close"; Commander, Naval Ordnance Laboratory to Chief of Naval Material, 29 April 1974, Subject: "A Study of the Feasibility of Consolidating the Naval Ordnance and Naval Weapons Laboratories"; comments on, NSWCDD Administrative Vault, Shelf Materials, Folder: Green Binder "Sensitive Material-Hold Close."

15. Anspacher, *The Legacy of the White Oak Laboratory*, pp. 410-412; Bernard Smith (NAVMAT), William Wineland (WOL), Charles Cohen (NWL), and Bernard Connolly (NAVMAT), *NOL-NWL Consolidation Study*, Prepared for the Deputy Chief of Naval Material (Development), 12 October 1973, NSWCDD Administrative Vault, Shelf Materials, Folder: Green Binder "Sensitive Material-Hold Close"; Commander, Naval Ordnance Laboratory to Chief of Naval Material, 29 April 1974, Subject: "A Study of the Feasibility of Consolidating the

Naval Ordnance and Naval Weapons Laboratories"; comments on, NSWCDD Administrative Vault, Shelf Materials, Folder: Green Binder "Sensitive Material-Hold Close."

16. OPNAV Report 5750-1, "Naval Surface Weapons Center Command History, 1974-1977," NSWCDD Administrative Vault, Shelf Materials; Ponko-Colvard Interview, p. 19; James H. Probus, by direction of Chief of Naval Material to Commander, Naval Surface Weapons Center, 20 September 1974, NSWCDD Administrative Vault, Shelf Materials, Folder: Looseleaf (Merger); Dr. Joseph Marchese Interview with Dr. James Colvard, 2 March 1990, p. 1, NL/CCG Archives, Record Collection 8, Box 871, Folder: Colvard, Dr. James, 2 March 1990; Anspacher, *The Legacy of the White Oak Laboratory*, pp. 364-365.

17. Ponko-Colvard Interview, 19.

18. Ibid., 20.

19. Ibid., 19-22; *On the Surface*, 29 Sept 1978.

20. *On the Surface*, 16 March 1979.

21. Ponko-Colvard Interview, 23-24.

22. Rouse-Colvard Interview, 6; Memorandum from Captain C. J. Rorie to OICs/ATDs/Department Heads/Division Heads/Branch Heads, Subject: The Role of the Surface Weapons Center, 3 August 1976, NSWCDD Administrative Vault, Box 31-60, Folder: Lttrs & Memos, Capt. Rorie signed.

23. Lemmuel Hill, later TD at NSWC, believed the intention of placing the headquarters address at White Oak was to allay some of the concern about being absorbed. Lemmuel Hill, interview by Rodney Carlisle, Bowie, Md., 9 May 2003 (hereafter Hill Oral History), 40.

24. NSWCDD Administrative Vault, Box 31-60, Folder Headquarters-Corr. This whole folder is instructive, but see especially, Rear Admiral W. R. Smedberg to Captain C. J. Rorie, (n.d., 1974); Chief of Naval Material to Chief of Naval Operations, 19 November 1975, Subject: "Change to Manpower Authorization; request for"; Chief of Naval Material to Chief of Naval Operations, 8 November 1976, Subject: "Headquarters Location of the Naval Surface Warfare Center; request for re-designation of"; Memorandum for Vice Admiral J. H. Doyle, Jr., 18 April 1977, Subject: "Surface Weapons Center, White Oak Laboratory; closure," all in NSWCDD Administrative Vault, Box 31-60, Folder Headquarters-Corr.

25. Paul Anderson, interview with Rodney Carlisle, King George, Va., 26 September 2003 (hereafter Anderson Oral History), 40.

26. Ray Shank, "Protection Systems Department Accomplishments," 87-88, and Robert T. Ryland Jr., "Protection System Department Assessment," 89-91, both

in Naval Surface Weapons Center, *Revolution at Sea Starts Here, A 1987 History of the Naval Surface Warfare Center* (Washington, D.C.: Government Printing Office, 1 March 1988); Capt. Paul Anderson to Prof. Jerome Smith, 1 November 1977, NSWCDD Administrative Vault, Cabinet 3, Drawer 3, Naval Ord. Advisory Bd, Folder 5420/1 Adv. Council.

27. Commander, Naval Surface Weapons Center, to All Hands, Subject: Naval Surface Weapons Center Organization, 20 October 1977, NSWCDD Administrative Vault, Box 31-60, Folder: Looseleaf Material; Captain Paul Anderson to All Hands, 8 December 1977, NSWCDD, Karen Melichar File Collection, Folder "Merger & Hdqtrs Change, October 1978.

28. Paul Anderson interview, 95-96; *On the Surface*, 5 May 1978.

29. On the Surface, 14 July 1978; *On the Surface*, 29 September 1978.

30. Gene Gallaher, interview by Rodney Carlisle, Dahlgren, Va., 2 May 2003, 20-21.

31. Gates Oral History, 45.

32. Robert V. Gates, "Strategic Systems Fire Control," *Technical Digest: Mathematics, Computing, and Information Science* (Dahlgren: NSWCDD, 1995), 172.

33. Strategic Systems Programs, *FBM Facts/Chronology: Polaris-Poseidon-Trident* (Washington, D.C.: Navy Department, 1990), 7, 10, 13; Gates, "Strategic Systems Fire Control," 172.

34. Gates, "Strategic Systems Fire Control," 173.

35. Ibid., 173.

36. Ibid., 173-75; Email from Robert V. Gates to Russell L. Coons, Janice R. Miller, and James Rife, 12 December 2005.

37. Graham Spinardi, *From Polaris to Trident: The Development of U.S. Fleet Ballistic Missile Technology* (Cambridge, England: Cambridge University Press, 1994), 148 ff.

38. George W. Baer, *One Hundred Years of Sea Power* (Stanford, Calif.: Stanford University Press, 1994), 394 ff; Robert Waring Herrick, Soviet Naval Doctrine and Policy, 1956-1986, *Book 3. Studies in Russian History, Volume 8c* (Lewiston, NY: The Edwin Mellen Press, 2003).

39. Ibid., 397, 399, 411. The puzzle of Soviet naval intentions in the 1970s was detailed in Paul Nitze et al., *Securing the Seas: The Soviet Naval Challenge and Western Alliance Options* (Boulder, CO.: Westview Press, 1979), 53-56.

40. Baer, *One Hundred Years of Sea Power*, 407-8.

41. Ibid., 408; Desmond Wilson and Nicholas Brown, "Warfare at Sea: Threat of the Seventies," Professional Paper No. 79, 4 November 1971, Center for Naval Analyses, 10-11.

42. Baer, *One Hundred Years of Sea Power*, 410; *On the Surface*, 29 February 1980.

43. *On the Surface*, 29 Oct 1978 and 18 May 1979.

44. *On the Surface*, 19 January 1979.

45. Ibid.

46. Jerome Smith, "Report of the NRAC Laboratory Advisory Board for Ordnance," 18th Meeting, NSWC, 24-26 May 1976, 4.

47. *On the Surface*, 9 May 1980; Walter Hoye, "Dahlgren Division NAVSWC Surges Ahead in Missile Warhead Development, Application," *On the Surface*, 24 January 1992.

48. Ibid.

49. Ibid.

50. Hill Oral History, 10, 46-47. Under Defense Secretary Robert McNamara's reforms of the mid-1960s, Navy funding was classified according to program and budget element. Program 6, RDT&E, was divided into six budget categories: 6.1-Basic Research, 6.2-Exploratory Development, 6.3-Advanced Development, 6.4-Engineering Development, 6.5-Management and Support, 6.6-Operational Development (see Rodney P. Carlisle and William M. Ellsworth, *Shaping the Invisible: Development of the Advanced Enclosed Mast/Sensor System, 1992-1999* (West Bethesda, Md.: Naval Surface Warfare Center, Carderock Division, 2000), 9).

51. Hill Oral History, 15, 16.

52. Ibid., 34.

53. Dr. Joseph Marchese Interview with Rear Admiral Wayne Meyer, 30 April 1990, NL/CCG Archives, Record Collection 8, Box 878, Folder: Meyer, Rear Admiral Wayne, 30 April 1990, 1-3 (hereafter cited as Marchese/Meyer Interview).

54. Ibid.

55. Hill Oral History, 33.

56. Marchese/Meyer Interview, 6-10; see also Thomas Tschirn, interview by James P. Rife, Dahlgren, Virginia, 25 June 2003 (hereafter Tschirn Oral History), 4-5.

57. Ibid., 11-12.

58. Radio Corporation of America, *AEGIS: Shield of the Fleet*, (Moorestown, NJ: RCA, Inc., 1972); 1981 *Command History*, NSWCDD Administrative Vault, Shelf Materials, Folder: Looseleaf, 1978, p. 6; Anderson Oral History, 40-41.

59. Ibid.; *On the Surface*, 25 July 1980.

60. *1978 Command History*, NSWCDD Administrative Vault, Shelf Materials, Folder: Looseleaf, 1978), p. 8; 1981 *Command History*, NSWCDD Administrative Vault, Shelf Materials, Folder: Looseleaf, 1978, p. 6.

61. David Miller and Chris Miller, *Modern Naval Combat* (New York: Crescent Books, 1986), 58-63, 156-57; Hill Oral History, 20-21.

62. Miller and Miller, 20-21.

63. *On the Surface*, 20 February 1987; *Free Lance-Star*, 27 June 1988.

64. *On the Surface*, 25 July 1980; Hill Oral History, 15.

65. The Falklands War has generated an impressive number of analytical studies within academia and professional military circles. For perhaps the best written and most authoritative work on the naval war, see retired Head of the (Royal) Naval Historical Branch David R. Brown's *The Royal Navy and the Falklands War* (Annapolis, Md.: Naval Institute Press, 1987); see also Rodney A. Burden et al., *Falklands: The Air War* (British Aviation Research Group, 1986); for more recent treatment, see Chris Hobson and Andrew Noble, *Falklands: Air War* (Aerofax Midland Pub Ltd., 2002). The literature on the *Stark* incident is less impressive, but for basic information, see Jeffrey L. Levinson and Randy L. Edwards, *Missile Inbound: The Attack on the Stark in the Persian Gulf* (Annapolis, Md.: Naval Institute Press, 1997).

66. House Committee on Armed Services, *The July 3, 1988 Attack by the* Vincennes *on an Iranian Aircraft: Hearing Before the Investigations Subcommittee and the Defense Policy Panel of the Committee on Armed Services*, 102nd Cong., 2nd sess., 21 July 1992, HASC No. 102-77, 4-29; Senate Committee on Armed Services, *Investigation into the Downing of an Iranian Airliner by the U.S.S. "Vincennes": Hearing Before the Committee on Armed Services*, 100th Cong., 2nd sess., 8 September 1988, S. Hrg. 100-1035, pp. 9-13.

67. House Armed Services Committee, *The July 3, 1988 Attack by the* Vincennes, 25; Senate Armed Services Committee, *Investigation into the Downing of an Iranian Airliner by the U.S.S. "Vincennes": Hearing Before the Committee on Armed Services*, 100th Cong., 2nd sess., 8 September 1988, S. Hrg. 100-1035, pp. 9-13; *Free Lance-Star*, 8 July 1988; *Journal*, 12 July 1988 (both articles maintained in NSWCDD Museum Historical Collection, Folder: Newspaper Articles, May 1986-December 1990.

68. Ibid.

69. Richard L. Schwoebel, *Explosion Aboard the Iowa* (Annapolis, Md.: Naval Institute Press, 1999), xviii.

70. Ibid., 145-47, 181, 188-90, 223.

71. Ibid., 183-84.

72. Marchese/Meyer Interview, 9; "Independent Research and Independent Exploratory Development Annual Report for Fiscal Year 1970," 16-21, NL/CCG Archives, Record Collection 3, Box 134, Folder Naval Weapons Laboratory IR/IED Annual Report for FY 1970, October 1970; Lt. Commander Stuart A. Borland, Royal Navy, "BOGHAMMER-Make My Day! Phalanx Enters the Surface Warfare Arena," in NSWCDD *Technical Digest*, September 1994, (Dahlgren, Va.: NSWCDD, 1994), pp. 100-107.

73. Commander, Naval Sea Systems Command, to Commander, Naval Surface Warfare Center, "SEATASK NAVSEA 3900/2 (REV 9-78)," 3 October 1988, NSWCDD, Building 183 "Wine Cellar" Collection, Cabinet 1, Drawer 3, Folder PHALANX Seatask, FY89 G Dept., 2.

74. Borland, "BOGHAMMER," 100.

75. Ibid., 101.

76. Commander, Naval Sea Systems Command, to Commander, Naval Surface Warfare Center, "SEATASK NAVSEA 3900/2 (REV 9-78)," 3 October 1988, NSWCDD, Building 183 "Wine Cellar" Collection, Cabinet 1, Drawer 3, Folder PHALANX Seatask, FY89 G Dept., p. 6 and budget appendix.

77. "F" Department technicians actually reconstructed the event from available information by replaying the situation using computer simulations and some of the real equipment. When the missile hit, the computer had gone down, and as soon as the crewmen could, they reloaded the computer over the old core, which held the memory that was present just before the strike. As Lemmuel Hill reported, "We tried to analyze what happened by getting at that valuable information. Security people have been telling us for years that to obliterate information on hard disk, you should write over it many times. If the same logic is applied in reverse, maybe the original information can still be found. It was our understanding that the crew tried to reload the system three times. We tried to peel back that data, loaded over three times, to figure out what was underneath." Hill further reported that "I am very proud of this organization's ability to respond and literally work around the clock to get things done. If you had gone into the SLQ-32 room after the *Stark* incident, you would have found people sleeping in the corner . . . people who had been working 24 hours a day, without sleep, trying to solve problems. They really showed tremendous commitment and made an outstanding effort." The technicians were ultimately successful in identifying several key problems, and as a result of their findings, the NSWC initiated some important changes to some of the ship's systems as well as changes in methodology for gun operation and anti-cruise missile

warfare. See *Revolution at Sea Starts Here: A 1987 History of the Naval Surface Warfare Center* (Dahlgren, Va.: NSWCDD, 1988), 43-44, 63-64; Committee on Armed Services, House of Representatives, 100th Cong., 1ˢᵗ sess., *Report on the Staff Investigation into the Iraqi Attack on the USS Stark,* (Washington, D.C.: Government Printing Office, 1987).

78. Borland, "BOGHAMMER," 101-6.

79. Ibid.

80. For a detailed history of the TOMAHAWK Program, see Nigel Macknight, TOMAHAWK Cruise Missile, (Osceola, Wis.: Motorbooks International, 1995); for the history of early guided missiles, see Norman Friedman, *U.S. Naval Weapons: Every Gun, Missile, Mine and Torpedo Used by the U.S. Navy from 1883 to the Present Day* (Annapolis, Md.: Naval Institute Press, 1982), 215-16; Captain Linwood S. Howeth, USN (retired), *History of Communications-Electronics in the United States Navy* (Washington, D.C.: Bureau of Ships and Office of Naval History, 1963), 479-93; see also RADM D. S. Fahrney, *The History of Pilotless Aircraft and Guided Missiles*, Microfilm A-171 (Washington, D.C.: Navy Department Library, n.d. but probably 1949 through 1958).

81. Friedman, *U.S. Naval Weapons*, 215-16; Howeth, *History of Communications-Electronics*, 479-93; Fahrney, *The History of Pilotless Aircraft and Guided Missiles.*

82. William B. Boyd and Buford Rowland, *U.S. Navy Bureau of Ordnance in World War II* (Washington, D.C.: Government Printing Office, 1954), 340-44.

83. Friedman, *U.S. Naval Weapons*, 216-20; Macknight, *TOMAHAWK*, 29-31; Tschirn Oral History, 4-5.

84. Friedman, *U.S. Naval Weapons*, 225-26; Macknight, *TOMAHAWK*, 38-39.

85. Friedman, *U.S. Naval Weapons*, 225-26; Macknight, *TOMAHAWK*, 42-45; Wayne L. Harman comments, 8 March 2004, NSWCDD, Dahlgren, Virginia.

86. Friedman, *U.S. Naval Weapons*, 225-26; Macknight, *TOMAHAWK*, 42-45.

87. On the Surface, 20 February 1987; Wayne L. Harman comments, 8 March 2004, NSWCDD, Dahlgren, Virginia.

88. N Department, "TOMAHAWK Project Plan 27 May 1988," NSWCDD, Building 183 "Wine Cellar" Collection, Cabinet 1, Drawer 3, Folder Project Plan TOMAHAWK, 7/28/88, 11-19, and appendices.

89. Ibid., 17-18.

90. Ibid., appendices; *King George Journal*, 20 June 1989.

91. NSWCDD "Wine Cellar" Collection, Cabinet 1, Drawer 3, Folder Program Plan, Standard Missile G Department, 3/13/90, 20-21.

92. Each section detailed its responsibilities for adjusting to modifications through the whole ninety-two pages of the Standard Missile Program Plan.

93. Quoted in "AEGIS: 'It worked perfectly . . . unfortunately," Fredericksburg *Free Lance-Star*, 8 July 1988.

94. Tommy Tschirn, interview by James P. Rife, Dahlgren, Va., 25 June 2003, 6, 30.

95. *On the Surface*, 19 October 1979.

Chapter 7: *A New World Order, 1987-1995*

1. Rodney P. Carlisle, *Navy RDT&E Planning in an Age of Transition: A Survey Guide to Contemporary Literature* (Washington, D.C.: NL/CCG and Naval Historical Center, 1997), 27-42; Rodney P. Carlisle, *Management of the U.S. Navy Research and Development Centers During the Cold War: A Survey Guide to Reports* (Washington, D.C.: NL/CCG and Naval Historical Center, 1996), 67-85; Captain Charles E. Biele to all Center COs and TDs, Subject: Disestablishment of NAVMAT, 9 April 1985, NSWCDD, Karen Melichar File Collection, Folder Disestablishment of NAVMAT, 5450, (1985); *On the Surface*, undated but April 1985, NSWCDD, Karen Melichar File Collection, Folder NAVMAT Transition (Disestablishment), Effective 6 May 85; Dr. Tom Clare, interview by Rodney Carlisle, Dahlgren, Va., 1 May 2003 (hereafter Clare Oral History), 8-9.

2. Commander, Naval Surface Weapons Center, Captain Carl A. Anderson to Chief of Naval Operations (OP-09B), via Commander, Space and Naval Warfare Systems Command, Subject: Request for Name Change, 22 May 1987, NSWCDD, Karen Melichar File Collection, Folder NSWC Name Change '87; *On the Surface*, 18 September 1987, NSWCDD, Karen Melichar File Collection, Folder NSWC Name Change '87.

3. John E. Holmes e-mail to D21, Forwarded to C1, D1, and D2, Subject: NSWC Name Change, 12 June 1987, NSWCDD, Karen Melichar File Collection, Folder NSWC Name Change '87.

4. Dave Colby e-mail to C and D, 12 June 1987, Subject: Thinking About the Meaning of Name Changes, NSWCDD, Karen Melichar File Collection, Folder NSWC Name Change, '87.

5. Commander, Naval Surface Warfare Center, to CNO, 11 September 1987, Subject: Change in Activity Title, NSWCDD, Karen Melichar File Collection, Folder NSWC, Name Change, '87.

6. Lemmuel L. Hill and Captain Carl A. Anderson, *A Strategic Perspective on the Future of the Naval Surface Warfare Center: Today's Commitments, Tomorrow's Challenges* (Dahlgren, Va.: NSWC, May 1988), 1-2, 5-6.

7. Ibid.

8. Ibid., 5-7, 14-15.

9.   *1990 Report to the Community* (Dahlgren, Va.: NSWC, 26 March 1990), 5; *Spawarrior*, May 1989, 6; Clare Oral History, 1-3, 44-49.

10.   Clare Oral History, 1-3, 44-49.

11.   *1990 Report to the Community*, 5; Clare Oral History, 3-10; Lemmuel Hill, interview by Rodney Carlisle, Bowie, Md., 9 May 2003 (hereafter Hill Oral History), 34.

12.   *Command History for 1990 Calendar Year*, NSWC, Dahlgren, Robert Gates Historical File, Folder Command History, 2-3; *1990 Report to the Community*, 12-16.

13.   Memorandum for Correspondents, "Naval Space Command to be Former," 15 June 1983, NSWCDD, Karen Melichar File Collection, Folder Naval Space Command; Chief of Naval Operations to Commander, Naval Surface Weapons Center, Dahlgren, via Chief of Naval Material, Subject: Establishment of the Naval Space Command, Dahlgren, 23 June 1983, NSWCDD, Karen Melichar File Collection, Folder Naval Space Command; 1990 Report to the Community, 17-25.

14.   Ibid.

15.   Ibid.

16.   Carlisle, *Navy RDT&E*, 43.

17.   Victor A. Meyer, "A Vision of Naval Surface Force Structure in 2030," *Technical Digest*, September 1991 (Dahlgren, Va.: NSWC, 1991), 8-9; Carlisle, Navy RDT&E, 42-49; Rodney P. Carlisle, *Where the Fleet Begins: A History of the David Taylor Research Center* (Washington, D.C.: Naval Historical Center and Department of the Navy, 1998), 421-61; President George W. Bush Speech to Congress, 6 March 1991.

18.   For the impact of the Goldwater-Nichols Act on defense acquisition and the Navy's response, see Leslie Lewis, Roger Allen Brown, and C. Robert Roll, *Service Responses to the Emergence of Joint Decision Making*, Publication MR-1438-AF (RAND Corporation, 2001), 13-19, 49-60, available on-line at http://www.rand.org/publications/MR/MR1438; Clare Oral History, 13-15, 37-38.

19.   Clare Oral History, 13-15, 37-38.

20.   Clare Oral History, 37-38; Dr. Thomas A. Clare, "Identity of the Navy R&D Centers: Known or Unknown?" February 1990, NSWCDD, Executive Director Signature File, Folder Clare, 1990; Director of Naval Laboratories Gerald R. Schiefer to Dr. Thomas Clare, 9 February 1990, NSWCDD, Executive Director Signature File, Folder Clare, 1990.

21.   Clare, "Identity of the Navy R&D Centers," 4-7.

22.   Ibid., 7-8.

23. Director of Navy Laboratories to Technical Director, Naval Surface Warfare Center, 9 February 1990, NSWCDD, Executive Director Signature File, Folder Clare, 1990. For a detailed discussion concerning the numerous studies that Clare and Schiefer alluded to, see Carlisle, *Navy RDT&E Planning*, 43-54.

24. Carlisle, *Navy RDT&E Planning*, 45-47.

25. Ibid., 47.

26. Ibid., 51-52.

27. Ibid., 56; William B. Anspacher, et al., *The Legacy of the White Oak Laboratory* (Dahlgren, Va.: NSWCDD, 2000), 365-67, 418-21.

28. Carlisle, *Navy RDT&E Planning*, 56-58; Department of the Navy Base Structure Committee, "Department of the Navy Base Closure and Realignment Recommendations, Detailed Analysis," April 1991, 13-15, 43-48.

29. "Navy Puts 94 Bases on 'Hit List,'" *Washington Post*, 25 April 1990; "Dahlgren May See Changes," Fredericksburg *Free Lance-Star*, 27 April 1990.

30. "Military Construction Freeze Extended," *Washington Post*, 2 May 1990; "Dahlgren Preschool on List of Defense Cuts," Fredericksburg *Free Lance-Star*, 2 May 1990; Congressman Charles E. Bennett and Congresswoman Pat Schroeder to Secretary of Defense Richard B. Cheney, 1 May 1990, NSWCDD Museum Historical Collection, Newspaper Articles, May 1986-December 1990.

31. Memorandum from SPAWAR, A-E Norton to Distribution List, Subject: Navy Base/Facility Realignment Proposals, 25 April 1990, NSWCDD Museum Historical Collection, Newspaper Articles, May 1986-December 1990; Memorandum from Commander and Technical Director to All Hands, Subject: Consolidation Issues, 10 May 1991, NSWCDD Executive Director Signature File, Folder Clare, 1991.

32. Ibid.

33. Clare Oral History, 18-19; Robert Gates, interview by James Rife, Dahlgren, Va., 4 June 2003 (hereafter Gates Oral History), 27.

34. Clare Oral History, 12, 15, 17-18, 20, 22-28.

35. Ibid., 24-25; Commander and Technical Director, Naval Surface Warfare Center, to Chief of Naval Operations, Subject: Vision for the Dahlgren Division of the Naval Surface Warfare Center, 23 December 1991, NSWCDD, Executive Director Signature File, Folder Clare, 1991; Executive Director to Dahlgren Laboratory and Technical Department, Division, and Branch Heads, and Program Managers, Subject: Research and Technology (R&T) at the Dahlgren Laboratory, 6 July 1992, NSWCDD, Executive Director Signature File, Folder Clare, 1992; Executive Director, Dahlgren Division, to Deputy Executive Director, Dahlgren Division, and Executive Director, Coastal Systems Station Heads of Technical

Departments, Dahlgren Division, Subject: Performance Expectations and Evaluation, 7 December 1992, NSWCDD, Executive Director Signature File, Folder Clare, 1992; *Strategic Plan: Naval Surface Warfare Center-Dahlgren Division*, January 1993, NSWCDD Administrative Vault, Cabinet 3/97, Drawer 4, Folder Strategic Planning.

36. Clare Oral History, 27-28.

37. Carlisle, *Navy RDT&E Planning*, 58-59; Clare Oral History, 10, 27-28; OPNAV NOTICE 5450, From Chief of Naval Operations, Subject: Establishment of Naval Surface and Undersea Warfare Centers, Modification of Title and Disestablishment of Shore Establishment Activities and Detachments, 23 December 1991, NSWCDD "Wine Cellar" Collection, Box H-109, Folder MEGACENTER: Name, Charter, Etc.; Anspacher, et al., *The Legacy of the White Oak Laboratory*, 366-67, 418-21; "Testimony of the Honorable Constance A. Morella Before the Base Closure and Realignment Commission," 22 May 1991, NSWCDD "Wine Cellar" Collection, Box H-110, Loose Materials.

38. Commander, Naval Surface Warfare Center, to Division COs, Subject: Stand-Up of Naval Surface Warfare Center Divisions, 2 January 1992, NSWCDD "Wine Cellar" Collection, Box H-109, Folder MEGACENTER: Name, Charter, Etc.; NSWC-01 to NSWC Division Commanders, Subject: Stand-up, 2 January 1992, NSWCDD "Wine Cellar" Collection, Box H-109, Folder MEGACENTER: Name, Charter, Etc.; Naval Surface Warfare Center, Dahlgren Division Notice 5450, Commander to Distribution Lists A and T, and Coastal Systems Station, Subject: Organization and Administration of the Naval Surface Warfare Center, Dahlgren Division, 8 January 1992, NSWCDD "Wine Cellar" Collection, Box H-109, Folder MEGACENTER: Name, Charter, Etc.; OPNAV NOTICE 5450, From Chief of Naval Operations, Subject: Establishment of Naval Surface and Undersea Warfare Centers, Modification of Title and Disestablishment of Shore Establishment Activities and Detachments, 23 December 1991, NSWCDD "Wine Cellar" Collection, Box H-109, Folder MEGACENTER: Name, Charter, Etc., 10; Commander and Executive Director to Division Council, Division Heads, and Branch Heads, Subject: NSWC, Dahlgren Division Reorganization, 19 August 1992, NSWCDD, Karen Melichar File Collection, Folder Org. Stuff "Standup, etc."; Captain Norman S. Scott, interview by Rodney Carlisle, Dahlgren, Va., 22 May 2003 (hereafter Scott Oral History), 7-9; Clare Oral History, 10-12.

39. Clare Oral History, 10-12, Scott Oral History, 7-9.

40. Scott Oral History, 28-31, 42-43; *Revolution at Sea Starts Here: A 1987 History of the Naval Surface Warfare Center*, 1 March 1988, NSWCDD, Robert Gates Historical File, 44.

41. Scott Oral History, 30-31; Director of Congressional and Public Affairs to Commander, Naval Sea Systems Command, Subject: White Oak Congressional and Public Affairs Plan, 21 October 1992, with attached "Summary of the Investigation into the Circumstances Surrounding the 28 June 1992

Explosion at the Naval Surface Warfare Center, Dahlgren Division, White Oak Detachment, White Oak, Maryland," NSWCDD Administrative Vault, Cabinet 3/97, Drawer 4, Folder W.O.; Navy Ordnance Environmental Support Office, "Environmental Program Management Review (DRAFT)," White Oak Detachment, Naval Surface Warfare Center, Dahlgren Division, October 1992, NSWCDD Administrative Vault, Cabinet 3/97, Drawer 4, Folder ESB/ECE; Anspacher, et al., *The Legacy of the White Oak Laboratory*, 366-67, 419-20.

42. Anspacher, et al., *The Legacy of the White Oak Laboratory*, 366-67, 419-20; *DoD Base Closure and Realignment Report to the Commission, Department of the Navy Analyses and Recommendations*, vol. 4, Attachment X-21, "Recommendation for Closure: Naval Surface Warfare Center, Dahlgren Division, Detachment, White Oak, Maryland, X-53 and X-54" (Washington, D.C.: Department of the Navy, March 1995); Office of Assistant Secretary of Defense (Public Affairs), No. 095-95, "Secretary Perry Recommends Closing, Realigning 146 Bases, 28 February 1995, NSWCDD "Wine Cellar" Collection, Box H-118, Binder BRAC; "Md. Expects to Escape Military Base Closings," *Baltimore Sun*, 28 February 1995; "Sarbanes to Oppose Base Closing," Washington Post, 27 February 1995; "Base Closings Affect Area," *Washington Times*, 28 February 1995; *Silver Spring Gazette*, "White Oak on Hit List; Community Concerned," 1 March 1995.

43. Commander, Dahlgren Division, Naval Surface Warfare Center, to Commander, Naval Surface Warfare Center, Subject: Disestablishment of White Oak Detachment, Dahlgren, Division, Naval Surface Warfare Center, 11 August 1997, and enclosed "Fact and Justification Sheet," NSWCDD Administrative Vault, Cabinet 3/97, Drawer 2, Folder White Oak; Anspacher, et al., *The Legacy of the White Oak Laboratory*, 366-67, 419-20; *Baltimore Sun*, 2 March 1995, "Keep White Oak Wind Tunnel, Top General Urges."

44. Clare Oral History, 18; Scott Oral History, 28-29.

45. Naval Surface Warfare Center, Dahlgren, Virginia, Superfund Program Site Fact Sheet; U.S. Environmental Protection Agency: National Priorities List, "NPL Site Narrative for Naval Surface Warfare Center-Dahlgren," Federal Register Notice, 14 October 1992, can be found at http://www.epa.gov/superfund/sites/npl/nar1326.htm; "Superfund Cleanup Goes On and On," Fredericksburg *Free Lance-Star*, 9 November 2003.

46. *Revolution at Sea Starts Here: A 1987 History of the Naval Surface Warfare Center*, 1 March 1988, NSWCDD, Robert Gates Historical File, 44; Dahlgren Superfund Program Site Fact Sheet, "NPL Site Narrative"; "Dahlgren Wages War on Hazardous Wastes," Fredericksburg *Free Lance-Star*, 27 March 1990.

47. "EPA Raps Dahlgren Base on Sewage," Fredericksburg *Free Lance-Star*, 18 September 1990; Scott Oral History, 38-39.

48. *Revolution at Sea Starts Here: A 1987 History of the Naval Surface Warfare Center*, 1 March 1988, NSWCDD, Robert Gates Historical File, 44; Dahlgren Superfund Program Site Fact Sheet, "NPL Site Narrative"; "Superfund Cleanup Goes On

and On," Fredericksburg Free Lance-Star, 9 November 2003; Scott Oral History, 38-39.

49. "Superfund Cleanup Goes On and On," Fredericksburg *Free Lance-Star*, 9 November 2003.

50. Archester Houston and Steven L. Dockstader, *Total Quality Leadership: A Primer,* (Washington, D.C.: Department of the Navy Total Quality Leadership Office, n.d.), 5-16; CNO Admiral Frank B. Kelso, *Navy Force 2001: Readiness 1994* (Washington, D.C.: Department of the Navy, 1993), 6.

51. Memorandum from Paul Credle to NSWCDD Board of Directors, Subject: Declare Victory and Turn Out the Lights, 12 January 1994, NSWCDD Administrative Vault, Drawer: 3/97, Folder 5000-5999 General Admin. & Management, Jan.-Mar. 1994.

52. Ted Buckley to C/D, 31 May 1994, with enclosed comments by Gene Lutman on Supervisor-Employee Memo, NSWCDD Administrative Vault, Cabinet 3/97, Drawer 4, Folder Emp-Supvr Relationship; e-mail from Michael W. Masters to Dr. Tom Clare, 21 May 1994, NSWCDD Administrative Vault, Cabinet 3/97, Drawer 4, Folder Emp-Supvr Relationship; Thomas C. Pendergraft Memorandum to C/D, Subject: C/D Draft Memo on Supervisor-Employee Relationship, 26 May 1994, NSWCDD Administrative Vault, Cabinet 3/97, Drawer 4, Folder Emp-Supvr Relationship; e-mail from JFerreb to MLindem and CKalivr, 13 May 1994, Subject: TQL Ramblings, NSWCDD Administrative Vault, Cabinet 3/97, Drawer 4, Folder Emp-Supvr Relationship.

53. Commander and Executive Director, NSWCDD, to Commander and Technical Director, NSWC, Subject: Management of High Grade Freeze, 4 December 1992, NSWCDD, Executive Director Signature File, Folder Clare, 1992; e-mail from DTabler to TClare, Subject: Organization Views, 26 May 1994, NSWCDD Administrative Vault, Cabinet 3/97, Drawer 4, Folder Emp-Supvr Relationship; Scott Oral History, 4-5.

54. Scott Oral History, 5-6; Gates Oral History, 42-43.

55. Naval Inspector General to Commander, NAVSEA, Subject: Naval Occupational Safety and Health (NAVOSH) Oversight Inspection, Naval Surface Warfare Center Division (NAVSURFWARCENDIV), Dahlgren, Virginia, NSWCDD "Wine Cellar" Collection, Box Incoming Faxes, 6/93-9/96, Folder 6/21/93-Nov. 93; Commander, Dahlgren Division, Naval Surface Warfare Center, to Commander, Naval Surface Warfare Center, Subject: NAVOSH Oversight Inspection, 11 August 1993, NSWCDD Administrative Vault, Cabinet 2, Drawer 3, Folder NAVOSH Inspection (Aug 93); "Navy Orders Worldwide Safety Check," *The Capital*, 15 November 1989; "Navy Activities Halted to Reexamine Safety," *Washington Post*, 15 November 1989; House of Representatives, "Recent Navy Accidents," *Hearing Before the Military Personnel and Compensation Subcommittee and the Seapower and Strategic and Critical Materials Subcommittee of the Committee on Armed Services*, 101st Cong., 1st sess., 20 November 1989.

56. Captain N. Scott, e-mail to D, C1, C2, D1, G, W, C8, Subject: NAVOSH Oversite [sic], 12 August 1993, NSWCDD Administrative Vault, Cabinet 2, Drawer 3, Folder NAVOSH: Oversight Board; Naval Surface Warfare Center to Commander, Naval Surface Warfare Center, Subject: NAVOSH Oversight Inspection, 11 August 1993, NSWCDD Administrative Vault, Cabinet 2, Drawer 3, Folder NAVOSH Inspection (Aug 93); Vice Admiral Stephen F. Loftus to Vice Admiral Kenneth C. Malley, 28 September 1993, NSWCDD "Wine Cellar" Collection, Box Incoming Faxes, 6/93-9/96, Folder 6/21/93-Nov. 93; Vice Admiral Kenneth C. Malley to Vice Admiral Stephen F. Loftus, Deputy CNO (Logistics), 1 November 1993, NSWCDD Administrative Vault, Cabinet 3/97, Drawer 4, Folder Untitled; Vice Admiral Kenneth C. Malley to Vice Admiral Stephen F. Loftus, Deputy CNO (Logistics), 10 November 1993, NSWCDD "Wine Cellar" Collection, Box Incoming Faxes, 6/93-9/96, Folder 6/21/93-Nov. 93.

57. Commander, Dahlgren Division, Naval Surface Warfare Center, to Commander, Naval Surface Warfare Center, Subject: NAVOSH Oversight Inspection, 24 September 1993, NSWCDD Administrative Vault, Cabinet 2, Drawer 3, Folder NAVOSH Inspection (Aug 93); Commander, Dahlgren Division, Naval Surface Warfare Center, to Commander, Naval Surface Warfare Center, Subject: NAVOSH Oversight Inspection, 15 October 1993, NSWCDD Administrative Vault, Cabinet 2, Drawer 3, Folder NAVOSH Inspection (Aug 93); Malley to Loftus, 1 November 1993, NSWCDD "Wine Cellar" Collection, Box Incoming Faxes, 6/93-9/96, Folder 6/21/93-Nov. 93; Malley to Loftus, 10 November 1993, NSWCDD "Wine Cellar" Collection, Box Incoming Faxes, 6/93-9/96, Folder 6/21/93-Nov. 93; Clare warned SPAWAR as early as 20 June 1991 that a mandated 20 percent employee reduction would result in an "imbalance" between the workforce and funded programs, and that NSWC was already "operating at a level less than that which funded programs would support." Dr. Thomas A. Clare, by direction, Commander, Naval Warfare Center, to Commander, Space and Naval Warfare Systems Command, 20 June 1991, NSWCDD, Executive Director Signature File, Folder Clare, 1991.

58. For the most recent and best scholarly treatment of the Navy's role in Operation DESERT STORM, see Edward J. Marolda and Robert J. Schneller Jr., *Shield and Sword: The United States Navy and the Persian Gulf War* (Annapolis, Md.: Naval Institute Press, 2001).

59. Ibid.

60. Clare to Commander, SPAWAR, 20 June 1991, NSWCDD, Executive Director Signature File, Folder Clare, 1991; *Command History of 1990* (Dahlgren, Va.: NSWC, 1990), 15, 21-24; Clare Oral History, 39-40; Scott Oral History, 17-18; "Naval Surface Warfare Center Brief," 30 September 1991, 11-12, NSWCDD Administrative Vault, Shelf Materials, Folder Loose.

61. "Naval Surface Warfare Center Brief," 11-12.

62. Sheila Young, interview by James Rife, Dahlgren, Va., 5 June 2003, 23-24; Gates Oral History, 41.

63. Ibid.

64. "Chemical Arms Threat Called Small" and "Dahlgren's Role," Fredericksburg *Free Lance-Star*, 10 August 1990.

65. Clare to Commander, SPAWAR, 20 June 1991, NSWCDD, Executive Director Signature File, Folder Clare, 1991; *Command History of 1990*, 22-24; "Naval Surface Warfare Center Brief," 11-12; "Chemical Arms Threat Called Small" and "Dahlgren's Role," Fredericksburg *Free Lance-Star*, 10 August 1990; "Events Spotlight Dahlgren," Fredericksburg *Free Lance-Star*, 21 August 1990; "NSWC Tests Help U.S. Deal with Chemical Warfare Threat," *Westmoreland News*, 30 August 1990.

66. Ibid.

67. Thomas James Yencha, "A Chemical Warfare Naval Simulation Model for Surface Ships," *Technical Digest*, September 1991 (Dahlgren, Va.: NSWC, 1991), 46-53; "Naval Surface Warfare Center Brief," 11-12.

68. "Naval Surface Warfare Center Brief," 11-12.

69. Robert L. Hudson listing in "Senior Scientific and Engineering Personnel," NSWC/DL MP-24/77, December 1977, NSWCDD Administrative Vault, Shelf Materials; Charles E. "Gene" Gallaher, interview by James P. Rife, Dahlgren, Va., 14 May 2004 (hereafter Gallaher/Rife Oral History), 1-13.

70. See Dean C. Allard, "Interservice Differences in the United States, 1945-1950: A Naval Perspective," *Airpower Journal*, Winter 1989, for a brief discussion of early joint relationships among the armed services, for which we draw heavily here; Charles E. "Gene" Gallaher, interview by Rodney Carlisle, Dahlgren, Va., 2 May 2003 (hereafter Gallaher/Carlisle Oral History), 10-13, 23-24. For an overview and planned utilization of U.S. joint experimentation modeling and simulation in the future, see Todd Morgan, Modeling and Simulation Branch (J955), U.S. Joint Forces Command, PowerPoint presentation entitled *Joint Experimentation M&S Requirements: From Today's Concepts to Tomorrow's Capabilities*, undated but post-2000, available at http://www.msiac.dmso.mil/experimentation_documents/ 2JEM_S_Rqmts_Overview.ppt.

71. Gallaher/Carlisle Oral History, 11-12, 26-27; Robert L. Hudson, "Comments Relevant to History Document," 1-2, and "Specific Contribution for Article," NSWCDD Public Affairs Office, 17 June 2005, hereafter cited as "Hudson Comments" and "Hudson Contribution," respectively.

72. Hudson Comments, 1-2.

73. For detailed first-hand accounts of Operation EAGLE CLAW and why it failed, see Colonel Charlie Beckwith, *Delta Force: The Army's Elite Counterterrorist Unit* (New York: Avon Books, 1983, with epilogue written by C. A. Mobley, 2000), and Command Sergeant Eric L. Haney, *Inside Delta Force: The Story of America's*

Elite Counterterrorist Unit (New York: Delacorte Press, 2002). For a concise but thoughtful analysis of Operation EAGLE CLAW, its problems, and aftermath, see Charles G. Cogan, "Desert One and Its Disorders," *The Journal of Military History* 67, no. 1 (January 2003): 201-16. Concerning the C-130/rocket booster escape scheme, Lem Hill later quipped, "God help us that that never happened . . . . We'd have probably killed them all!" Hill Oral History, 50-51; Hudson Contribution; Hudson Comments, 1-2.

74. Beckwith, *Delta Force*, 295-310, 326-30; Haney, *Inside Delta Force*, 185-209; Cogan, "Desert One," 211-14.

75. Cogan, "Desert One," 201-2, 216.

76. Ibid.; Edward N. Luttwak, *The Pentagon and the Art of War: The Question of Military Reform* (New York: Simon and Schuster, 1985). For the best and most thorough history of the Goldwater-Nichols Act, see James R. Locher III, *Victory on the Potomac: The Goldwater-Nichols Act Unifies the Pentagon* (College Station, Tex.: Texas A&M University Press, 2002); Christopher M. Bourne, "Unintended Consequences of the Goldwater-Nichols Act," *Joint Force Quarterly*, Spring 1998, 99-108.

77. Ronald H. Cole, *Operation URGENT FURY: The Planning and Execution of Joint Operations in Grenada, 12 October-2 November 1983* (Washington, D.C.: Joint History Office, Office of the Chairman of the Joint Chiefs of Staff, 1997), 66-68; Ronald H. Cole, Operation JUST CAUSE: The Planning and Execution of Joint *Operations in Panama, February 1988-2 January 1990* (Washington, D.C.: Joint History Office, Office of the Chairman of the Joint Chiefs of Staff, 1995), 71-74.

78. Gallaher/Carlisle Oral History, 12-13, 27-28; Gallaher/Rife Oral History, 1-13; Hudson Comments, 2.

79. Ibid.

80. Ibid.

81. Gallaher/Carlisle Oral History, 1-15; Charles E. Gallaher listing in "Senior Scientific and Engineering Personnel," NSWC MP 82-247, January 1983, NSWCDD Administrative Vault, Dahlgren, Virginia; Captain Lyal Davidson, interview by Rodney Carlisle, Dahlgren, Va., 5 February 2004 (hereafter Davidson Oral History), 13-16.

82. Davidson Oral History, 12-13, 21-24, 27-29, 39-41.

83. Hudson Comments, 3.

84. Scott Oral History, 18; Hudson Comments, 4.

85. Scott Oral History, 18; Gallaher/Carlisle Oral History, 10-11, 21-22, 28; Gallaher/Rife Oral History, 4-8.

86. Scott Oral History, 18; Gallaher/Carlisle Oral History, 10-11, 21-22, 28; Gallaher/ Rife Oral History, 4-8; Memorandum for Commander, Naval Surface Warfare Center, from Vice Admiral Paul David Miller, Ser 745/OU649708, Subject: "Establishment of a Support Organization at NSWC," 13 February 1990, located in the files of Charles E. Gallaher, "J" Department, NSWCDD, Dahlgren, Virginia.

87. Gallaher/Carlisle Oral History, 23-24, 26; Gallaher/Rife Oral History, 22-24; Hudson Contribution. Those same forces that impeded Gallaher's use of the word "Joint" in his department also stopped Clare from pursuing his "vision" of NSWCDD becoming the "DOD Warfare Systems Engineer," since "it became politically unacceptable for a Navy Warfare Center division to have a vision that was a DOD vision. The Navy didn't want to deal with that, and so we changed it to Department of the Navy Warfare Systems Engineer" (Clare Oral History, 25-26). For internal cultural opposition to jointness among the armed services, see Douglas Macgregor, "The Joint Force: A Decade, No Progress," *Joint Force Quarterly*, Winter 2000-01, 18-23.

88. Gallaher/Rife Oral History, 4-12.

89. Hudson Comments, 6.

90. Gallaher/Carlisle Oral History, 20-21, 40; Scott Oral History, 17-18. Some of "J" Department's classified work during DESERT STORM can be inferred from a description of the current Joint Warfare Analysis Center's mission at the U.S. Joint Forces Command's Internet website, http://www.jfcom.mil/about/ com_jwac.htm. The two American pilots involved in the accident were actually indicted by a British court, but the British government soon let the issue quietly drop in the interest of Anglo-American relations and cooperation.

91. Gallaher/Carlisle Oral History, 26; Davidson Oral History, 13-16; Hudson Comments, 7.

92. Gallaher/Rife Oral History, 6-7, 12.

93. Ibid., 9-10; Hudson Comments, 5, 7.

94. Hudson Comments, 7.

95. Hudson Comments, 7-8; Gallaher/Rife Oral History, 10-11, 15-16.

96. Robert L. Hudson, Final Review of Manuscript, 3 October 2005.

97. Ibid., 11, 16.

98. Gallaher/Carlisle Oral History, 28; Osama bin Laden remarks to John Miller of ABC News, 28 May 1998, quoted in Bernard Lewis, *The Crisis of Islam: Holy War and Unholy Terror* (New York: Random House, 2003), 162.

Chapter 8: *Dahlgren Forever, 1995-2003*

1. Biography of Captain John C. Overton, USN, NSWCDD Administrative Vault, Cabinet 2/96, Drawer 3, Folder 30 October-1 November, Wes 21 (Vision 21); Memorandum from M1 (Jim Falter) to D, Subject: Statement of the Short-Term Structural Problem, 21 October 1994; Memorandum from S10 (Bruce Franks) to D (Carlton Duke), Subject: Short Term Restructuring, 27 October 1994; Memorandum from C2 (J. B. Wilkinson) to D', 01, Subject: Near Term Restructuring, 26 October 1994; Memorandum from J (C. E. Gallaher) to D', 01, Subject: Short Term Restructuring Effort, 31 October 1994; E-mail from Al Kidd to Carlton W. Duke, Subject: Fwd: Near Term Restructuring, 15 November 1994; Memorandum from D1, 01 (Ted Buckley and Carlton Duke) to DC, Subject: Short Term Restructuring Team, 14 November 1994; Memorandum from A (Chris Kalivretenos) to Carlton Duke and Ted Buckley, Subject: Short-Term Task, 24 October 1994, all in NSWCDD, Executive Director Signature File, Folder Clare, 1994.

2. Memorandum from Commander and Executive Director, Dahlgren Division, to Division Council, Subject: Long Term Restructuring, 9 March 1995, NSWCDD, Executive Director Signature File, Folder Clare, 1995 (Looseleaf); Rodney P. Carlisle, *Navy RDT&E Planning in an Age of Transition: A Survey Guide to Contemporary Literature* (Washington, D.C.: NL/CCG and Naval Historical Center, 1997), 65-85; John B. Hattendorf, *The Evolution of the U.S. Navy's Maritime Strategy, 1977-1986* (Newport, R.I.: Naval War College Press, 1989); Memorandum from Ira Blatstein to Tom Clare and Dave Skinner, with attached Point Paper, 29 January 1996, NSWCDD "Wine Cellar" Collection, Box Incoming Faxes, 6/93-9/96; Memorandum from Commander and Executive Director to Distribution List F, Subject: Topics for Discussion, 6 November 1995, NSWCDD, Executive Director Signature File, Folder Clare, 1995; T. A. Clare and J. C. Overton, "A View of the Future of the Dahlgren Division of the Naval Surface Warfare Center," October 1995, NSWCDD, Executive Director Signature File, Folder Clare, 1995 (Looseleaf).

3. Memorandum from Commander and Executive Director, Dahlgren Division, to Division Council, Subject: Long Term Restructuring, 9 March 1995; Memorandum from D (Tom Clare) to D', 01, Subject: Flexibility for the Future, 30 May 1995, both in NSWCDD, Executive Director Signature File, Folder Clare, 1995 (Looseleaf).

4. Memorandum from Commander and Executive Director to All Hands, Subject: Financial Condition: Why and What are We Doing?, 15 March 1996, NSWCDD, Executive Director Signature File, Folder Clare, 1996; Robert Gates, interview by James Rife, Dahlgren, Va., 4 June 2003 (hereafter Gates Oral History), 52.

5. Memorandum from Commander and Executive Director to All Hands, Subject: Financial Condition: Why and What are We Doing?, 15 March 1996, NSWCDD, Executive Director Signature File, Folder Clare, 1996.

6. Memorandum from Commander S. R. Pietropaoli to Distribution, Subject: RIF Announcements, 1 February 1996; Draft Memorandum for Interested Members of Congress from Captain Jay M. Cohen, Subject: 1996 Projected Reductions in Force (RIF) at Navy Activities, January 1996, both in NSWCDD "Wine Cellar" Collection, Box Incoming Faxes, 6/93-9/96.

7. Michael L. Marshall, Presentation to NSWCDD Division Executive Board, "External Environmental Scan," NSWCDD Administrative Vault, Cabinet 3/96, Drawer 1, Folder DEB Minutes 1996.

8. Michael L. Marshall, Presentation to NSWCDD Division Executive Board, "External Environmental Scan," NSWCDD Administrative Vault, Cabinet 3/96, Drawer 1, Folder DEB Minutes 1996; John D. Moteff, Congressional Research Service Issue Brief IB10022: "Defense Research: DoD's Research, Development, Test and Evaluation Program" (Washington, D.C.: Congressional Research Service, 13 August 1999); U.S. General Accounting Office, *Defense Acquisition Infrastructure: Changes in RDT&E Laboratories and Centers*, GAO/NSIAD-96-221BR, September 1996, 8-13, 18-19; Gates Oral History, 28-31; Report to the President and Congress, *Vision 21: The Plan for 21ˢᵗ Century Laboratories and Test-and-Evaluation Centers of the Department of Defense*, 30 April 1996, 1-6, NSWCDD Administrative Vault, Cabinet 3/96, Drawer 2, Folder Vision 21, 1996; Memorandum from Rear Admiral R. A. Riddell (Director, Test and Evaluation and Technology Requirements) to Distribution, Subject: Vision 21 Baseline Cost Determination Data Call Demonstration, 5 December 1996, NSWCDD Administrative Vault, Cabinet 3/96, Drawer 2, Folder Vision 21, 1996.

9. Moteff, "Defense Research"; GAO, *Defense Acquisition Infrastructure*, 8-13, 18-19; Gates Oral History, 28-31; Report to the President and Congress, Vision 21, 1-6; Riddell to Distribution, "Vision 21 Baseline Cost Determination."

10. Report to the President and Congress, *Vision 21*, 1-6; Riddell to Distribution, "Vision 21 Baseline Cost Determination"; Gates Oral History, 28-31.

11. Memorandum from D1B to C (Captain John C. Overton), Subject: Vision 21 Staffing Issue, 12 December 1996, with handwritten note from Overton to Bill Cocimano, 13 December 1996, NSWCDD Administrative Vault, Cabinet 3/96, Drawer 2, Folder Vision 21, 1996.

12. Gates Oral History, 27-31.

13. Ibid., 28-31; Moteff, "Defense Research."

14. Gates Oral History, 28-31; Moteff, "Defense Research."

15. Gates Oral History, 28-31.

16. Gates Oral History, 28-31; Moteff, "Defense Research"; Press Release No. 006-04, "Department of Defense Begins Gathering Data for BRAC 2005," 6 January 2004 (Washington, D.C.: Department of Defense, 2004); Efficient

Facilities Initiative (EFI) Factsheet (Revised), 3 August 2001, available online at www.defenselink.mil/news/Aug2001/d20010802efi.pdf.

17. Memorandum from C and D to BOD, Subject: New Organization Stand-up and Issues Resolution, 13 September 1996, NSWCDD, Executive Director Signature File, Folder Clare, 1996.

18. See biographical Information for Thomas C. Pendergraft in NSWCDD Public Relations Packet NSWCDD/MP-01/50, 2003.

19. Gates Oral History, 37-38; Barry Dillon and Joe Francis, interview by James Rife, Dahlgren, Va., 4 June 2003 (hereafter Dillon/Francis Oral History), 3-6; Dr. Tom Clare, interview by Rodney Carlisle, Dahlgren, Va., 1 May 2003 (hereafter Clare Oral History), 8-9; Dr. Joseph Marchese Interview with Dr. Thomas Clare, NSWC, Dahlgren, Va., 4 April 1990, 1-19, NL/CCG Archives, Operational Archives Branch, Naval Historical Center, Collection #0008, "Oral Histories," Box 870, Folder Clare, Dr. Thomas, 4 April 1990.

20. Dillon/Francis Oral History, 3-6; Gates Oral History, 44.

21. Dillon/Francis Oral History, 6-12.

22. Ibid.

23. Ibid.

24. Ibid.

25. Ibid.

26. Ibid.; U.S. *Senate, Statement of Rear Admiral Michael W. Cramer, Director of Naval Intelligence, before the Subcommittee on Seapower of the Senate Armed Services Committee*, 104th Cong., 3rd sess., 8 April 1997.

27. Dillon/Francis Oral History, 6-12; Charles V. Peña, Foreign Policy Briefing No. 60, *From the Sea: National Missile Defense is Neither Cheap Nor Easy*, 6 September 2000 (Washington, D.C.: CATO Institute, 2000), 2-8; Charles V. Peña, Policy Analysis No. 309, *Theater Missile Defense: A Limited Capability is Needed*, 22 June 1998 (Washington, D.C.: CATO Institute, 1998), available at http://www.cato.org/pubs/pas/pa-309.html; Ivan Eland with Daniel Lee, Foreign Policy Briefing No. 65, *The Rogue State Doctrine and National Missile Defense*, 29 March 2001 (Washington, D.C.: CATO Institute, 2001), 2-10; White House Office of the Press Secretary, "National Policy on Ballistic Missile Defense Fact Sheet," 20 May 2003, available at http://www.whitehouse.gov/news/releases/2003/05; National Missile Defense Act of 1999, Public Law 106-38, 106th Cong., 1st sess. (22 July 1999).

28. Dillon/Francis Oral History, 6-12; Peña, From the Sea, 2-8; Peña, *Theater Missile Defense*; Eland, with Lee, *Rogue State Doctrine*, 2-10; White House Office of the Press Secretary, "National Policy on Ballistic Missile Defense Fact Sheet,"

20 May 2003, available at http://www.whitehouse.gov/news/releases/2003/05; National Missile Defense Act of 1999, Public Law 106-38, 106th Cong., 1st sess. (22 July 1999).

29. Commission to Assess the Ballistic Missile Threat to the United States, "Executive Summary of the Report of the Commission to Assess the Ballistic Missile Threat to the United States," Pursuant to Public Law 201, 104th Cong., 15 July 1998 (Washington, D.C.: Government Printing Office, 1998); Dillon/Francis Oral History, 6-12; Peña, *From the Sea*, 2-8; Peña, *Theater Missile Defense*; Eland, with Lee, *Rogue State Doctrine*, 2-10; White House Office of the Press Secretary, "National Policy on Ballistic Missile Defense Fact Sheet," 20 May 2003, available at http://www.whitehouse.gov/news/releases/2003/05; National Missile Defense Act of 1999, Public Law 106-38, 106th Cong., 1st sess. (22 July 1999).

30. Dillon/Francis Oral History, 6-12, 29-33; Peña, *From the Sea*, 2-8; Peña, *Theater Missile Defense*; Eland, with Lee, *Rogue State Doctrine*, 2-10; White House Office of the Press Secretary, "National Policy on Ballistic Missile Defense Fact Sheet," 20 May 2003, available at http://www.whitehouse.gov/news/releases/2003/05.

31. For USACOM's background, mission, and evolution through 1998, see U.S. General Accounting Office, *U.S. Atlantic Command: Challenging Role in the Evolution of Joint Military Capabilities*, GAO/NSIAD-99-39, 17 February 1999 (Washington, D.C.: Government Printing Office, 1999); Charles E. "Gene" Gallaher, interview with James P. Rife, Dahlgren, Va., 14 May 2004 (hereafter Gallaher/Rife Oral History), 15-16.

32. GAO, *U.S. Atlantic Command*; James R. Locher III, "Taking Stock of Goldwater Nichols," *Joint Forces Quarterly*, Autumn 1996, 10-17; Christopher M. Bourne, "Unintended Consequences of the Goldwater-Nichols Act," *Joint Forces Quarterly*, Spring 1998, 99-108; *Joint Vision 2010: America's Military: Preparing for Tomorrow* (Washington, D.C.: Government Printing Office, July 1996); *Joint Vision 2020* (Washington, D.C.: Government Printing Office, June 2000).

33. Douglas A. Macgregor, "The Joint Force: A Decade, No Progress," *Joint Forces Quarterly*, Winter 2000-01, 18-23; Leslie Lewis, Roger Allen Brown, C. Robert Roll, *Service Responses to the Emergence of Joint Decisionmaking*, MR-1438-AF (Santa Monica, Calif.: RAND Corporation, 2001), xiii-xv, 49-60, 81-89; Charles E. "Gene" Gallaher, interview by Rodney Carlisle, Dahlgren, Va., 2 May 2003 (hereafter Gallaher/Carlisle Oral History), 24-26, 29; Gallaher/Rife Oral History, 23-25.

34. Gallaher/Carlisle Oral History, 40-43; for descriptive discussions of "J" Department's various unclassified programs during the 1990s, see the 2003 edition of NSWCDD's Technical Digest, subtitled "Joint and National Needs," NSWCDD/MP-02/113 (Dahlgren, Va.: NSWCDD, 2003); for FAST-CD, or "BUGSPLAT," see Brian K. Wade, Randy D. Wagner, and Erin M. Swartz, "Collateral Damage Estimation," *Technical Digest* (Dahlgren, Va.: NSWCDD, 2003), 208-13.

35. Gallaher/Carlisle Oral History, 40-43; Wade, Wagner, and Swartz, "Collateral Damage Estimation," 211-13.

36. Wade, Wagner, and Swartz, "Collateral Damage Estimation," 211-13; Locher, "Taking Stock of Goldwater Nichols," 10-17; Bourne, "Unintended Consequences of the Goldwater-Nichols Act," 99-108; Gallaher/Carlisle Oral History, 40-43.

37. Gallaher/Carlisle Oral History, 33; "Pentagon Widens Search for New Technologies," *National Defense Magazine*, January 2004, on-line article at http://www.nationaldefensemagazine.org; Captain Lyal Davidson, interview by Rodney Carlisle, Dahlgren, Va., 5 February 2004 (hereafter Davidson Oral History), 13-14; A. Kris Indseth, "Naval Operations Other Than War Technology Center: Incorporating Dahlgren's Legacy in Today's Changing World," *Technical Digest* (Dahlgren, Va.: NSWCDD, 2003), 40-46.

38. Ibid.

39. "Technology Briefs: Non-Lethal Weapons Demonstration, C7F AT/FP Exercise Conducted Aboard U.S.S. *Blue Ridge* (LCC-19)," in *The Dahlgren Leading Edge* 2, no. 4 (October-December 2002): 16-17.

40. Gallaher/Carlisle Oral History, 31-34; Senate Armed Services Committee, *Statement of Admiral Vern Clark, U.S. Navy Chief of Naval Operations on Force Protection*, 107th Cong., 1st sess., 3 May 2001; Hun Kim, "Critical Infrastructure Protection for the Local Installation," *Chips Magazine*, Summer 2002, article on-line at http://www.chips.navy.mil/archives/02_Summer/ authors/index2_files/cip.htm; "Protecting Critical Military Infrastructures," DefenseLINK News, 7 December 2001, article on-line at http://www.defenselink.mil/news/Dec2001/; Steven W. Zehring, "Critical Infrastructure Protection: Supporting Joint and National Needs," Technical Digest (Dahlgren, Va.: NSWCDD, 2003), 48-56; Presidential Decision Directive/NSC-63, 22 May 1998, available on-line at http://www.fas.org/irp/offdocs/pdd/pdd-63.htm.

41. Ibid.

42. Gallaher/Carlisle Oral History, 36; Naval Sea Systems Command, "Naval Surface Warfare Center, Dahlgren Division," NSWCDD/MP-01/50 (Dahlgren, Va.: NSWCDD, 2003), 36.

43. NAVSEA, "Naval Surface Warfare Center, Dahlgren Division," 34; Elizabeth R. D'Andrea, "Leveraging Technologies: Making an Impact Against the Asymmetric Threat," *Technical Digest* (Dahlgren, Va.: NSWCDD, 2003), 106-14; Davidson Oral History, 30-31.

44. Gallaher/Carlisle Oral History, 24-26; Gallaher/Rife Oral History, 27-30; *NSWCDD Strategic Plan for 1998-1999*, NSWCDD/MP-98/92 (Dahlgren, Va.: NSWCDD, 1998).

45. Gates Oral History, 43-44, 46-48.

46. Ibid., 46-48.

47. Clare Oral History, 30-36.

48. James Colvard, "Warriors and Weaponeers: Reflections on the History of Their Association within the Navy" (unpublished paper), NSWCDD, Executive Director Signature File, Folder Clare, 1995 (Looseleaf), 3-6; Lewis, Brown, and Roll, *Service Responses to the Emergence of Joint Decisionmaking*, 49-60.

49. Clare Oral History, 30-36.

50. Ibid.

51. Admiral Vern Clark, USN, "Sea Power 21 Series-Part I: Projecting Decisive Joint Capabilities," in *U.S. Naval Institute Proceedings*, October 2002; Vice Admiral Mike Bucchi and Vice Admiral Mike Mullen, USN, "Sea Power 21 Series-Part II: Sea Shield: Projecting Global Defensive Assurance," in *U.S. Naval Institute Proceedings*, November 2002; Vice Admiral Cutler Dawson and Vice Admiral John Nathman, USN, "Sea Power 21 Series-Part III: Sea Strike: Projecting Persistent, Responsive, and Precise Power," in *U.S. Naval Institute Proceedings*, December 2002; Vice Admiral Charles W. Moore and Lieutenant General Edward Hanlon Jr., USMC, "Sea Power 21 Series-Part IV: Sea Basing: Operational Independence for a New Century," in *U.S. Naval Institute Proceedings*, January 2003; Vice Admiral Richard W. Mayo and Vice Admiral John Nathman, USN, "Sea Power 21 Series-Part V: ForceNet: Turning Information into Power," in *U.S. Naval Institute Proceedings*, February 2003; Vice Admiral Mike Mullen, USN, "Sea Power 21 Series-Part VI: Global Concept of Operations," in *U.S. Naval Institute Proceedings*, April 2003; Vice Admiral Alfred G. Harms Jr., Vice Admiral Gerald L. Hoewing, and Vice Admiral John B. Totushek, USN, "Sea Power 21 Series-Part VII: Sea Warrior: Maximizing Human Capital," in *U.S. Naval Institute Proceedings*, June 2003; Admiral Robert J. Natter, USN, "Sea Power 21 Series-Part VIII: Sea Trial: Enabler for a Transformed Fleet," in *U.S. Naval Institute Proceedings*, November 2003; Admiral Michael G. Mullen, USN, "Sea Power 21 Series-Part IX: Sea Enterprise: Resourcing Tomorrow's Fleet," in *U.S. Naval Institute Proceedings*, January 2004. All of the "Sea Power 21 Series" articles are located online at http://www.usni.org/proceedings/.

52. Naval-Industry R&D Partnership Conference, Panel on Transformational Technologies/Capabilities, *FORCEnet* PowerPoint presentation, available on-line at http://www.onr.navy.mil/about/conferences/rd_partner/confmat/ 2002; Admiral Vern Clark, USN, "Sea Power 21 Series-Part I: Projecting Decisive Joint Capabilities," in *U.S. Naval Institute Proceedings*, October 2002; Vice Admiral Richard W. Mayo and Vice Admiral John Nathman, USN, "Sea Power 21 Series-Part V: ForceNet: Turning Information into Power," in *U.S. Naval Institute Proceedings*, February 2003.

53. Dedication to Dr. Thomas A. Clare, *Technical Digest: Technology Transition and Dual-Use Technology*, 1998 issue (Dahlgren, Va.: NSWCDD, 1998), 177; Thomas C. Pendergraft biography in NSWCDD Public Relations Packet NSWCDD/MP-01/50, 2003; Meeting minutes from Dr. Rodney P. Carlisle and James P. Rife with Mr. Thomas C. Pendergraft and Captain Lyal Davidson, 23 April 2003, NSWCDD, Dahlgren, Virginia; Gates Oral History, 48-49; Davidson Oral History, 7.

54. For NAVSEA Dam Neck's capsule history and mission, see http://www.navseadn.navy.mil/page/whoweare.htm.

55. Dale W. Sisson Jr., "The Development and Application of the Shipboard Collective Protection System (CPS)," in *Technical Digest: Technology Transition and Dual-Use Technology*, 1998 issue, (Dahlgren, Virginia: NSWCDD, 1998), 74-85; "Virginia State Senator John Chichester at Herbert Bateman Chemical and Biological Defense Center Dedication," in *The Dahlgren Leading Edge* 2, no. 4 (October-December 2002): 23; "Technology Briefs," in *The Dahlgren Leading Edge* 3, no. 2 (April-June 2003): 18, 21-22; Naval Sea Systems Command, *On Watch: Supporting the 21st Century Warfighter*, 2003 issue (Washington, D.C.: Office of Congressional and Public Affairs, 2003), 20.

56. Wayne Harman, "Science and Technology: A Dahlgren Cornerstone," Part I, *The Dahlgren Leading Edge* 2, no. 4 (October-December 2002): 5-6; Wayne Harman, "Mathematics: Linchpin of Dahlgren Science," Part III, The Dahlgren Leading Edge 3, no. 2 (April-June 2003): 12.

57. Harman, "Science and Technology," 5-6; Harman, "Mathematics," 12.

58. Ibid.

59. Ibid.

60. "Technology Briefs: Test Flight," *The Dahlgren Leading Edge* 3, no. 2 (April-June 2003): 17.

61. U.S. General Accounting Office, *Evaluation of the Navy's 1999 Naval Surface Fire Support Assessment*, GAO/NSIAD-99-225, 14 September 1999 (Washington, D.C.: Government Printing Office, 1999); Richard A. Frazer and John E. Bibel, "Advanced Technology Demonstration of the Naval Tactical Missile System," Technical Digest: Strategic and Strike Warfare Systems, 1997 issue (Dahlgren, Va.: NSWCDD, 1997), 120-34.

62. Frazer and Bibel, "Naval Tactical Missile System," 124, 133-134; GAO, *Evaluation of the Navy's 1999 Naval Surface Fire Support Assessment*; Frazer and Bibel, "Advanced Technology Demonstration," 120-34; U.S. General Accounting Office, *Naval Surface Fire Support Program Plans and Cost*, GAO/NSIAD-99-91, 11 July 1999 (Washington, D.C.: Government Printing Office, 1999).

63. Tommy Tschirn, interview by James Rife, Dahlgren, Va., 25 June 2003 (hereafter Tschirn Oral History), 5-10; GAO, *Naval Surface Fire Support Program Plans and Cost*; GAO, *Evaluation of the Navy's 1999 Naval Surface Fire Support Assessment.*

64. NAVSEA, "Naval Surface Warfare Center, Dahlgren Division," Publication MP-01/50 (Dahlgren, Va.: NSWCDD, 2003), 12; Kenneth McCollum, *Dahlgren* (Dahlgren, Va.: NSWC, 1977), 143-46; Tschirn Oral History, 4-6, 21-22, 26.

65. Tschirn Oral History, 20-24.

66. Ibid., 4-6, 21-22, 26; GAO, *Naval Surface Fire Support Program Plans and Cost.*

67. NAVSEA, *Year in Review 2001* (Washington, D.C.: Office of Congressional and Public Affairs, 2001), 34; "Technology Briefs: ERGM Conducts Successful Guided Gunfire Flight Test," *The Dahlgren Leading Edge* 2, no. 4 (October-December 2002): 15, 17-18; NAVSEA, On Watch, 12-13.

68. Tschirn Oral History, 6-7.

69. Ibid., 27; GAO, *Naval Surface Fire Support Program Plans and Cost*; GAO, *Evaluation of the Navy's 1999 Naval Surface Fire Support Assessment.*

70. John Bean provides an excellent overview of rail gun technology and its history and capabilities in his article "Technology Spotlight: The Time is Right for Naval Railguns," in *The Dahlgren Leading Edge* 3, no. 2 (April-June 2003): 6-8.

71. "Professional Notes," *U.S. Naval Institute Proceedings* 48, no. 1 (January 1922): 143; Rudolph Lusar, *German Secret Weapons of the Second World War*, trans. R. P. Heller and M. Schindler (New York: Philosophical Library, Inc., 1959), 156, 160; Ian V. Hogg, *German Secret Weapons of the Second World War: The Missiles, Rockets, Weapons and New Technology of the Third Reich* (Mechanicsburg, Pa.: Stackpole Books, 1999), 98-99; W. A. Walls, W. F. Weldon, S. B. Pratap, M. Palmer, and Lt. David Adams, USN, "Application of Electromagnetic Guns to Future Naval Platforms," PR-253, delivered at 9th EML Symposium, Edinburgh, Scotland, May 1998 (Austin, Texas: Center for Electromechanics, 1998), available on-line at http://www.utexas.edu/research/cem/publications/PDF/PR253.pdf; Lt. Cmdr. J. R. Lyman and Cmdr. J. B. Colwell, Experimental Department Memorandum No. 1190-45, "Hypervelocity Guns," 5 November 1945 (Microfilmed), Technical Library, NPG M-1190-45, C.1, NSWCDD, Dahlgren, Virginia.

72. Albert F. Reidl III, "Preliminary Investigation of an Electromagnetic Gun", NWL Technical Not No. TN-E-10/72, July 1972, NSWCDD, Technical Library; Walls et al., "Application of Electromagnetic Guns"; Tschirn Oral History, 7-10, 30; Bean, "Technology Spotlight," 6-8.

73. Bean, "Technology Spotlight," 6-8; Tschirn Oral History, 7-10, 30.

74. Bean, "Technology Spotlight," 6-8.

75. Tschirn Oral History, 7-10, 27, 30.

76. The following detailed information concerning endo- and exoatmospheric interceptors was kindly communicated to us via email by "G" Department's Walter E. Hoye and June Drake, who are both intimately familiar with the technology; Hoye to James Rife, 27 July 2004.

77. For a comprehensive background to the 9/11 attacks and discussions of the complex history of terrorism, the rise of militant Islam, and the struggle to combat Al Qaeda, see Simon Reeve, *The New Jackals: Ramzi Yousef, Osama Bin Laden, and the Future of Terrorism* (Boston, Mass.: Northeastern University Press, 1999); Bernard Lewis, *The Crisis of Islam: Holy War and Unholy Terror* (New York: Random House, 2004); Daniel Pipes, *Militant Islam Reaches America* (New York: W.W. Norton & Company, Inc., 2002); Ahmed Rashid, *Taliban: Militant Islam, Oil, and Fundamentalism in Central Asia* (New Haven, Conn.: Yale University Press, 2000), Ahmed Rashid, *Jihad: The Rise of Militant Islam in Central Asia* (New Haven, Conn.: Yale University Press, 2002); Colonel Russell Howard, USA, and Major Reid Sawyer, USA, *Terrorism and Counterterrorism: Understanding the New Security Environment, Readings, and Interpretations* (New York: McGraw-Hill/Dushkin, 2003); Joshua Sinai, "How to Forecast and Preempt Al-Qaeda's Catastrophic Terrorist Warfare," in *Journal of Homeland Security*, August 2003, available on-line at http://www.homelandsecurity.org/journal/articles.asp.

78. For Bush and Rumsfeld's reaction to 9/11 and the start of the global "War on Terrorism," see Rowan Scarborough, *Rumsfeld's War: The Untold Story of America's Anti-Terrorist Commander* (Lanham, Md.: Regnery Publishing, 2004).

79. NAVSEA, Commander's Message, "NAVSEA Contributes to America's War on Terrorism, Sees Challenges Ahead," in *Year in Review 2001* (Washington, D.C.: Office of Congressional and Public Affairs, 2001), 2.

80. Dillon/Francis Oral History, 30-31.

81. Davidson Oral History, 23-25, 27-28; Gallaher Oral History, 35; PowerPoint Presentation by Dale Galyen, Deputy Division Executive Director, Acting, NSWCDD, on Naval Surface Warfare Center, Dahlgren Division, 30 September 2003; Gallaher/Rife Oral History, 27-37.

82. "Dahlgren Base Gets New Technology for War on Terrorism," Fredericksburg Free Lance-Star, 26 August 2003; "NSWC Dahlgren Opens NITMAC Headquarters," Navy Newsstand, 3 September 2003, available on-line at http://www.news.navy.mil.

83. Associated Press Report from U.S. Central Command, "Missile Strike Kills 200 Iraqi Paramilitary Fighters," 29 March 2003.

84. Gallaher/Carlisle Oral History, 40-41; "Dahlgren Delivers Solutions," *The Dahlgren Leading Edge* 3, no. 2 (April-June 2003): 3; Dillon/Francis Oral History, 31-33.

85. Captain Lyal B. Davidson and Executive Director Thomas Pendergraft, quoted in "Dahlgren Delivers Solutions," *The Dahlgren Leading Edge* 3, no. 2 (April-June 2003): 2-3.

86. Vice Admiral Phillip M. Balisle, quoted in "Dahlgren Delivers Solutions," *The Dahlgren Leading Edge* 3, no. 2 (April-June 2003): 2-3.

87. "Dahlgren Delivers Solutions," *The Dahlgren Leading Edge* 3, no. 2 (April-June 2003): 2-3; Commander's Message, Vice Admiral Phillip M. Balisle, in NAVSEA, On Watch, 2-3; "Aiming High," Fredericksburg *Free Lance-Star*, 21 July 2003.

88. U.S. Department of Defense, Press Release No 006-04, "Department of Defense Begins Gathering Data for BRAC 2005," 6 January 2004; "Aiming High," Fredericksburg *Free Lance-Star*, 21 July 2003.

89. U.S. Department of Defense, Press Release No 006-04, "Department of Defense Begins Gathering Data for BRAC 2005," 6 January 2004; "Aiming High," Fredericksburg *Free Lance-Star*, 21 July 2003; Davidson Oral History, 71-72; Meeting minutes from Dr. Rodney P. Carlisle and James P. Rife with Mr. Thomas C. Pendergraft and Captain Lyal Davidson, 23 April 2003, NSWCDD, Dahlgren, Virginia.

90. "Aiming High," Fredericksburg *Free Lance-Star*, 21 July 2003; Gregory E. Monteith, Jeffrey H. McConnell, Alvin C. Murphy, John P. Chapman, Thomas A. Micek, Christopher M. Thompson, and Russell G. Acree Jr., "Networked Engineering . . . From Concept to Reality— The Navy's Distributed Engineering Plant (DEP)," *Technical Digest* (Dahlgren, Va.: NSWCDD, 2001), 192-204. The attacks of 11 September 2001 led DOD to create the office of Assistant Secretary of Defense for Homeland Defense (ASD/HD). In September 2003 the ASD/HD requested that a Defense Program Office for Mission Assurance (DPO/MA) be established at Dahlgren from the Joint Program Office for Special Technology Countermeasures (JPO/STC) in "J" Department. On 31 October 2003, NSWC commander Rear Admiral Alan B. "Brad" Hicks directed Dahlgren to establish that office. The DPO/MA serves as DOD's critical infrastructure protection analysis and assessment organization supporting the ASD/HD, combatant commanders, and the armed services.

91. Cynthia Rouse interview with Dr. James E. Colvard, 10 December 1976, NSWC, Dahlgren, Va., 17-18.

Postscript: The Way Ahead, 2004-2006

1. Lucia Sanchez, "NAVSEA Poised for Evolution," *South Potomac Pilot*, 13 October 2005.

2. Statement of Admiral Michael G. Mullen, Chief of Naval Operations, Before the House Armed Services Committee, 1 March 2006; CNO's Posture Hearing FY 2007 Budget.

3. Strategic Plan for the Naval Surface Warfare Center, Dahlgren Laboratory, October 2005; CNO Guidance for 2006: Meeting the Challenge of a New Era.

4. Naval Sea Systems Command Corporate Communications, "NSWC Dahlgren and PSNS & IMF Receive the 2005 Nathaniel Stinson EEO Awards," NAVSEA *Newswire*, NOV17-01.

5. Ibid.

6. Policy on Academic Development and Professional Certification Incentives Program (12410, Ser XDPC/511 22 DEC 05), Naval Surface Warfare Center, Dahlgren Division, C1 Files.

7. Department of Defense Base Closure and Realignment Report, vol. I, part 2 of 2: Detailed Recommendations, Section 10, Recommendations-Technical Joint Cross Service Group, May 2005.

8. Jeff Branscome and Emily Battle, "Dahlgren Receives Reprieve from Cuts," Fredericksburg *Free Lance-Star*, 26 August 2005; Defense Base Closure and Realignment Commission, Final Report to the President, 25 August 2005, vol. I, 292, 295-96 (hereafter DBRAC Final Report).

9. "Dahlgren Receives Reprieve"; DBRAC Final Report, 291-92.

10. DBRAC Final Report, 296.

11. "Dahlgren Receives Reprieve."

12. United States Government Accountability Office, *Progress of the DD(X) Destroyer Program*, GAO-05-752R, 14 June 2005. On 23 November 2005, the Undersecretary of Defense for Acquisition finally signed the "destroyer acquisition memorandum" that approved a low initial production of eight ships. On 7 April 2006, the Navy announced that the class and lead ship would carry the designation and hull number DDG 1000 *Zumwalt*. In March 2005, the Government Accountability Office issued a report (*Defense Acquisitions: Assessments of Major Weapon Programs*, GAO-04-248, March 2004, 45, 57) stating that five of ERGM's twenty "critical technologies" were not mature. Because of program delays and cost overruns, the Navy opened the $600 million program to a competition and may order a "shoot-off" in eighteen months to compare technologies between Raytheon and its competitor, Alliant Techsystems Inc., which was awarded a $30 million contract in May 2004 to develop an alternative extended-range guided munition. See "Raytheon Naval Bid Shaky," in the *Arizona Daily Star*, 22 November 2005.

13. John Joyce, "Sen. Warner Calls NSWC Dahlgren 'Crown Jewel,'" *The Waterline* (Naval District of Washington), 20 August 2004.

14. Ibid.

Bibliography

A Note About Sources

Primary sources used during the preparation of this book were collected from a number of different repositories and collections. These included the Dahlgren Laboratory Technical Library; the National Archives and Records Administration facilities in Washington, D.C., College Park, Maryland, and Philadelphia, Pennsylvania; the Library of Congress (for congressional hearing transcripts and records), and the Navy Laboratory/Center Coordinating Group Archives of the Operational Archives at the Naval Historical Center in Washington, D.C. The Navy Department Library at the Washington Navy Yard was particularly useful for its complete run of annual reports of the Bureau of Ordnance and the Secretary of the Navy and its extensive collection of senior officer biographies. Additionally, Dahlgren's Headquarters Administrative Vault and the Executive Directors' Signature Files, both located in the Administration Building, proved to be a treasure trove of documents that are usually unavailable to historians but fortunately were opened for the authors' research use.

Although Dahlgren no longer maintains a museum, Patricia Albert has preserved a large portion of its former historical collection, including vintage station news sheets and periodicals as well as scrapbooks and newspaper clippings, which she helped the authors to access and review throughout the course of the project. In addition, technical operations manager and "K" Department guidance engineer Dr. Robert V. Gates made available his personal collection of unclassified "K" Department materials concerning the histories of Dahlgren's computers and the Navy's Submarine Launched Ballistic Missile program. Dr. Gates's special file collection from the 1980s and 1990s, currently held in the so-called "Wine Cellar," or Room 0008 of Building 183, provided an abundance of primary source materials pertaining to naval RDT&E organizational matters, Base Realignment and Closure (BRAC), and Vision 21. Administrative and technical specialist Karen A. Melichar's collection of administrative and organizational files likewise proved invaluable for analyzing and understanding more recent changes in naval and Department of Defense RDT&E policies as well as in Dahlgren's organizational structure. Finally, the authors were able to identify, locate, and download a wide variety of government documents, including official statements, press releases, reports, and directives that have been published on the World Wide Web for public use.

Wherever possible, the authors direct the reader via endnotes to primary source repository, collection, record group, box, file, date, and/or web address, if appropriate.

Secondary and background sources were mostly accessed and utilized at the Library of Congress and the Navy Department Library. Dahlgren's leadership also provided the authors with a complete run of *Technical Digests*, from 1991 through the present, as well as numerous other internal publications such as *The Dahlgren Leading Edge* and NAVSEA's *On Watch*. Captain Paul L. Anderson, Dahlgren's commanding officer from 1977 to 1981, also kindly allowed the authors to borrow and use his personal bound collection of back issues of *On the Surface* to cover the period from the late 1970s through the early 1980s. Other sources were utilized as necessary from the McKeldin Library of the University of Maryland.

Primary Sources

Beckwith, Charlie A., Colonel. *Delta Force: The Army's Elite Counterterrorist Unit*. With epilogue by C. A. Mobley. New York: Avon Books, 2000.

Brown, David R., Jr. *The Contributions of the U.S. Naval Weapons Laboratory in Geoballistics and Digital Fire Control to the Fleet Ballistic Missile Program*. Dahlgren, Va.: U.S. Naval Weapons Laboratory, 1964.

------. *Establishing the Dahlgren Role in the Fleet Ballistic Missile System, 1956-1965*. Monograph, Naval Surface Warfare Center, Dahlgren, Virginia, 1995.

Bucchi, Mike, Vice Admiral, and Vice Admiral Mike Mullen, USN. "Sea Power 21 Series-Part II: Sea Shield: Projecting Global Defensive Assurance." *U.S. Naval Institute Proceedings*, November 2002.

Bureau of Naval Personnel, Standards and Curriculum Division Training. *Naval Ordnance and Gunnery*. Washington, D.C.: Government Printing Office, 1946.

Clark, Vern, Admiral, USN. "Sea Power 21 Series-Part I: Projecting Decisive Joint Capabilities." *U.S. Naval Institute Proceedings*, October 2002.

Cochran, Thomas B., William M. Arkin, and Milton M. Hoenig. *Nuclear Weapons Databook*. Vol. 1, *U.S. Nuclear Forces and Capabilities*. Cambridge, Mass.: Ballinger Publishing Company, 1984.

Commission to Assess the Ballistic Missile Threat to the United States. "Executive Summary of the Report of the Commission to Assess the Ballistic Missile Threat to the United States." Pursuant to Public Law 201, 104th Congress, 15 July 1998. Washington, D.C.: Government Printing Office, 1998.

Congressional Information Service. *Presidential Executive Orders and Proclamations*, 1918-1919.

Dawson, Cutler, Vice Admiral, and Vice Admiral John Nathman, USN. "Sea Power 21 Series-Part III: Sea Strike: Projecting Persistent, Responsive, and Precise Power." *U.S. Naval Institute Proceedings*, December 2002.

Haney, Eric, Command Sergeant Major. *Inside Delta Force: The Story of America's Elite Counterterrorism Unit.* New York: Delacorte Press, 2002.

Harms, Alfred G., Jr., Vice Admiral, Vice Admiral Gerald L. Hoewing, and Vice Admiral John B. Totushek, USN. "Sea Power 21 Series-Part VII: Sea Warrior: Maximizing Human Capital." *U.S. Naval Institute Proceedings*, June 2003.

Hedrick, David I., Captain. *History, April 1918-December 1945.* Dahlgren, Va.: U.S. Naval Proving Ground, 1945.

Hill, Lemmuel L., and Captain Carl A. Anderson. *A Strategic Perspective on the Future of the Naval Surface Warfare Center: Today's Commitments, Tomorrow's Challenges.* Dahlgren, Va.: Naval Surface Warfare Center, 1988.

Mayo, Richard W., Vice Admiral, and Vice Admiral John Nathman, USN. "Sea Power 21 Series-Part V: ForceNet: Turning Information into Power." *U.S. Naval Institute Proceedings*, February 2003.

Moore, Charles W., Vice Admiral, and Lieutenant General Edward Hanlon, Jr., USMC. "Sea Power 21 Series-Part IV: Sea Basing: Operational Independence for a New Century." *U.S. Naval Institute Proceedings*, January 2003.

Moteff, John D. "Defense Research: DoD's Research, Development, Test and Evaluation Program." Congressional Research Service Issue Brief IB10022, 13 August 1999.

Mullen, Michael, Vice Admiral, USN. "Sea Power 21 Series-Part VI: Global Concept of Operations." *U.S. Naval Institute Proceedings*, April 2003.

Mullen, Michael G., Admiral, USN. "Sea Power 21 Series-Part IX: Sea Enterprise: Resourcing Tomorrow's Fleet." *U.S. Naval Institute Proceedings*, January 2004.

National Archives and Records Administration. Record Group 74, Records of the Bureau of Ordnance.

------. Record Group 330, Records of the Office of the Secretary of Defense.

------. Record Group 402, Records of the Bureau of Naval Weapons.

Natter, Robert J., Admiral, USN. "Sea Power 21 Series-Part VIII: Sea Trial: Enabler for a Transformed Fleet." *U.S. Naval Institute Proceedings*, November 2003.

Naval Historical Center/Operational Archives Branch. Commanding Officer Biographical Files.

------. Senior Commander Biographies.

Naval Research Advisory Committee/Laboratory Advisory Board for Ordnance. Minutes, 1968-1977.

Naval Sea Systems Command. *On Watch: Supporting the 21st Century Warfighter*. Washington, D.C.: NAVSEA Office of Congressional and Public Affairs, 2003.

------. *Year in Review 2001*. Washington, D.C.: NAVSEA Office of Congressional and Public Affairs, 2003.

Naval Surface Warfare Center, Dahlgren Division. *Command Histories*, 1990-1992.

------. *The Dahlgren Leading Edge*, 2001-2004.

------. *Technical Digest*, 1991-Present.

------. *Meeting the Challenge: A 1986 History of the Naval Surface Weapons Center*. Washington, D.C.: Government Printing Office, 1987.

------. *Revolution at Sea Starts Here: A 1987 History of the Naval Surface Warfare Center*. Washington, D.C.: Government Printing Office, 1987.

------. *On the Surface*, 1991-Present.

Naval Surface Weapons Center. *Command Histories*. Dahlgren, Va., 1974-1981.

Officers of the United States Navy. *Naval Ordnance: A Textbook Prepared for the Use of the Midshipmen of the United States Naval Academy*. Annapolis, Md.: United States Naval Institute, 1921.

Smith, Bernard. *Looking Ahead from Way Back: An Autobiography*. Richmond, Ind.: Prinit Press, 1999.

U.S. Congress. *Congressional Record.* House. *Proceedings and Debates of the 1st Session of the 67th Congress of the United States of America.* Vol. 61, pt. 6, "Public Bills, Resolutions, and Memorials," 24 August 1921.

------. House. *Proceedings and Debates of the 2nd Session of the 67th Congress of the United States of America.* Vol. 62, pt. 6, "Naval Appropriations," 18 April 1922.

------. Senate. *Proceedings and Debates of the 2nd Session of the 67th Congress of the United States of America.* Vol. 62, pt. 9, "Increase of the Navy," 16 June 1922.

U.S. Congress. *United States Statutes at Large.* 65th Cong., 1917-1919. Vol. 40, pt. 1, *H.R. 10783: An Act to Authorize the Secretary of the Navy to Increase the Facilities for the Proof and Test of Ordnance Material, and for Other Purposes,* 26 April 1918.

------. House. Committee on Armed Services. *Report on the Staff Investigation into the Iraqi Attack on the USS Stark.* 100th Cong., 1st sess., June 1987.

------. House. Investigations Sub-Committee and the Defense Policy Panel of House Committee on Armed Services. *The July 3, 1988 Attack by the* Vincennes *on an Iranian Aircraft.* 102nd Cong., 2nd sess., 21 July 1992.

------. House. Committee on Naval Affairs. *Hearings on Estimates Submitted by the Secretary of the Navy, 1918.* 65th Cong., 18 January 1918.

------. Committee on Naval Affairs. *Hearings on Estimates Submitted by the Secretary of the Navy, 1919.* 66th Cong., 2 and 3 June 1919.

------. Committee on Naval Affairs. *Hearings on Estimates Submitted by the Secretary of the Navy, 1920.* 66th Cong., 27 January 1920.

------. Committee on Naval Affairs. *Hearings on Sundry Legislation Affecting the Naval Establishment, 1921.* 67th Cong., 1st sess., 25-26 July 1921 and 27 October 1921.

------. Committee on Naval Affairs. *Hearings before the Committee on Naval Affairs of the House of Representatives on Sundry Legislation Affecting the Naval Establishment, 1922-1923.* 67th Cong., 2nd, 3rd, and 4th sess.

------. Sub-Committee of House Committee on Appropriations. *First Deficiency Appropriation Bill, 1919.* 65th Cong., 2nd sess., 4 October 1918.

------. Sub-Committee of House Committee on Appropriations. *Hearing on Navy Department Appropriation Bill, 1924.* 67th Cong., 4th sess., 1922.

U.S. Congress. Senate. Sub-Committee of Senate Committee on Appropriations. *Navy Department Appropriation Bill, 1923: Hearings on H.R. 11228.* 67th Cong., 2nd sess., 31 May 1922.

------. Committee on Armed Services. *Investigation into the Downing of an Iranian Airliner by the U.S.S. "Vincennes."* 100th Cong., 2nd sess., 8 September 1988.

------. Committee on Armed Services. *Statement of Admiral Vern Clark, U.S. Navy Chief of Naval Operations on Force Protection.* 107th Cong., 1st sess., 3 May 2001.

U.S. Department of Defense. *Joint Vision 2010: America's Military: Preparing for Tomorrow.* Washington, D.C.: Government Printing Office, July 1996.

------. *Joint Vision 2020.* Washington, D.C.: Government Printing Office, June 2000.

U.S. Department of the Navy. *Annual Report of the Chief of the Bureau of Ordnance to the Secretary of the Navy.* Washington, D.C.: Government Printing Office, 1890-1940.

------. *Annual Report of the Secretary of the Navy.* Washington, D.C.: Government Printing Office, 1890-1947.

------. *DOD Base Closure and Realignment Report to the Commission: Analyses and Recommendations.* Vol. 4. Washington, D.C.: Department of Defense, March 1995.

------. *FBM Facts/Chronology: Polaris-Poseidon-Trident.* Washington, D.C.: Strategic Systems Programs, 1990.

------. Bureau of Ordnance. *NAVORD Report 4866: Naval Ordnance Computation Center, U.S. Naval Proving Ground, Dahlgren, VA.* Washington, D.C.: Bureau of Ordnance, 1955.

U.S. General Accounting Office. *Defense Acquisition Infrastructure: Changes in RDT&E Laboratories and Centers.* GAO/NSIAD-96-221BR (September 1996). Washington, D.C.: Government Printing Office, 1996.

------. *Evaluation of the Navy's 1999 Naval Surface Fire Support Assessment.* GAO/NSIAD-99-225 (14 September 1999). Washington, D.C.: Government Printing Office, 1999.

------. *Naval Surface Fire Support Program Plans and Cost.* GAO/NSIAD-99-91 (11 July 1999). Washington, D.C.: Government Printing Office, 1999.

------. *U.S. Atlantic Command: Challenging Role in the Evolution of Joint Military Capabilities.* GAO/NSIAD-99-39 (17 February 1999). Washington, D.C.: Government Printing Office, 1999.

U.S. Naval Proving Ground, Dahlgren, Virginia. Advisory Council Minutes, 1953-1959.

------. *Lab Log*, 1955-1978.

------. *News Sheets*, 1939-1955.

------. *Technical Reports*, 1919-2003.

U.S. Naval Weapons Laboratory, Dahlgren, Virginia. Advisory Council Minutes, 1959-1974.

------. *Command Histories*, 1960 and 1963.

------. *SPASUR Automatic System, Mark I: Operating System Description.* Dahlgren, Va.: U.S. Naval Weapons Laboratory, 1962.

White House Office of the Press Secretary. "National Policy on Ballistic Missile Defense Fact Sheet," 20 May 2003.

Wilson, Desmond P., and Cdr. Nicholas Brown, USN. *Warfare at Sea: Threat of the Seventies.* Professional Paper No. 79, 4 November 1971. Arlington, Va.: Center for Naval Analyses, 1971.

Secondary Sources

Adams, James. *Bull's Eye: The Assassination and Life of Supergun Inventor Gerald Bull.* New York: Times Books, 1992.

Alden, John D. *The American Steel Navy.* Annapolis, Md.: Naval Institute Press, 1972.

Allard, Dean C. "Interservice Differences in the United States, 1945-1950: A Naval Perspective." *Airpower Journal*, Winter 1989. Available on-line at http://www.airpower.au.af.mil/airchronicles/apj/apj89/allard.html.

Anspacher, William B., Betty H. Gay, Donald E. Marlowe, Paul B. Morgan, Samuel J. Raff. *The Legacy of the White Oak Laboratory*. Dahlgren, Va.: Naval Surface Warfare Center, Dahlgren Division, and Department of the Navy, 2000.

Baer, George W. *One Hundred Years of Sea Power: The U.S. Navy, 1890-1990*. Stanford, Calif.: Stanford University Press, 1994.

Baile, Kenneth C. *Historical Perspective of NAVSWC/Dahlgren's Organizational Culture*. Dahlgren, Va.: Naval Surface Warfare Center, 1991.

Baldwin, Ralph B. *The Deadly Fuze: The Secret Weapon of World War II*. San Rafael, Calif.: Presidio Press, 1980.

Barlow, Jeffrey G. *Revolt of the Admirals: The Fight for Naval Aviation, 1945-1950*. Washington, D.C.: Naval Historical Center, 1994.

Bashe, Charles J., Lyle R. Johnson, John H. Palmer, and Emerson W. Pugh. *IBM's Early Computers*. Cambridge, Mass.: MIT Press, 1986.

Borland, Stuart A., LCDR. "BOGHAMMER–Make My Day! Phalanx Enters the Surface Warfare Arena." *Technical Digest: Ship Defense Technology*. Dahlgren, Va.: Naval Surface Warfare Center, Dahlgren Division, 1994.

Bourne, Christopher M. "Unintended Consequences of the Goldwater-Nichols Act." *Joint Force Quarterly*, Spring 1998, 99-108.

Boyd, William B., and Buford Rowland. *U.S. Navy Bureau of Ordnance in World War II*. Washington, D.C.: Government Printing Office, 1954.

Breck, Edward. *The United States Naval Railway Batteries in France*. Washington, D.C.: Government Printing Office, 1922.

Brown, David R. *The Royal Navy and the Falklands War*. Annapolis, Md.: Naval Institute Press, 1987.

Brown, David R., Jr. *The Navy's High Road to Geoballistics: Part 1, The Founding Heritage: Exterior Ballistics*. Dahlgren, Va.: Naval Surface Warfare Center, 1986.

Brown, Robert Vance. *The Navy's Mark 15 (Norden) Bomb Sight: Its Development and Procurement, 1920-1945.* Washington, D.C.: Department of the Navy, Office of the General Counsel, 1946.

Bull, Gerald V., and Charles H. Murphy. *Paris Kanonen - The Paris Guns (Wilhelmgeschutze) and Project HARP: The Application of Major Calibre Guns to Atmospheric and Space Research.* Herford, Germany: E.S. Mittler, ca. 1988.

Burden, Rodney A., Michael I. Draper, Douglas A. Rough, Colin R. Smith, and David L. Wilton. *Falklands: The Air War.* British Aviation Research Group, 1986.

Burrows, William E. *Deep Black: Space Espionage and National Security.* New York: Random House, 1986.

Bush, Vannevar. *Modern Arms and Free Men: A Discussion of the Role of Science in Preserving Democracy.* New York: Simon and Schuster, 1949.

Carlisle, Rodney P. *Management of the U.S. Navy Research and Development Centers During the Cold War: A Survey Guide to Reports.* Washington, D.C.: Navy Laboratory/Center Coordinating Group and the Naval Historical Center, 1996.

------. *Navy RDT&E Planning in an Age of Transition: A Survey Guide to Contemporary Literature.* Washington, D.C.: Navy Laboratory/Center Coordinating Group and the Naval Historical Center, 1997.

------. *Powder and Propellants: Energetic Materials at Indian Head, Maryland, 1890-2001.* 2nd ed. Denton, Tex.: University of North Texas Press, 2002.

------. *Where the Fleet Begins: A History of the David Taylor Research Center.* Washington, D.C.: Naval Historical Center and Department of the Navy, 1998.

Carlisle, Rodney P., and William M. Ellsworth. *Shaping the Invisible: Development of the Advanced Enclosed Mast/Sensor System, 1992-1999.* West Bethesda, Md.: Naval Surface Warfare Center, Carderock Division, 2000.

Christman, Albert B. *Sailors, Scientists, and Rockets.* Washington, D.C.: Naval History Division, 1971.

------. *Target Hiroshima: Deak Parsons and the Creation of the Atomic Bomb.* Annapolis, Md.: Naval Institute Press, 1998.

Cogan, Charles G. "Desert One and Its Disorders." *Journal of Military History* 67, no. 1 (January 2003).

Cohen, I. Bernard. *Howard Aiken: Portrait of a Computer Pioneer.* Cambridge, Mass.: MIT Press, 1999.

Cohen, I. Bernard, and Gregory W. Welch, eds. *Makin' Numbers: Howard Aiken and the Computer.* With the cooperation of Robert V. D. Campbell. Cambridge, Mass.: MIT Press, 1999.

Cole, Ronald H. *Operation JUST CAUSE: The Planning and Execution of Joint Operations in Panama, February 1988-2 January 1990.* Washington, D.C.: Joint History Office, Office of the Chairman of the Joint Chiefs of Staff, 1995.

------. *Operation URGENT FURY: The Planning and Execution of Joint Operations in Grenada, 12 October-2 November 1983.* Washington, D.C.: Joint History Office, Office of the Chairman of the Joint Chiefs of Staff, 1997.

Coletta, Paolo. *Admiral Bradley A. Fiske and the American Navy.* Lawrence: University of Kansas, 1979.

Colvard, James. "Warriors and Weaponeers: Reflections on the History of their Association within the Navy." White Paper, NSWCDD, Dahlgren, Va., 1995.

Connery, Robert H. *The Navy and Industrial Mobilization in World War II.* Washington, D.C.: Government Printing Office, n.d.

Cooling, Benjamin F. *Grey Steel and Blue Water Navy.* Hamden, Conn.: Archon, 1979.

D'Andrea, Elizabeth R. "Leveraging Technologies: Making an Impact Against the Asymmetric Threat" *Technical Digest: Joint and National Needs.* Dahlgren, Va.: Naval Surface Warfare Center, Dahlgren Division, 2003.

Defense Threat Reduction Agency. *Defense's Nuclear Agency, 1947-1997.* Washington, D.C.: Department of Defense, 2002.

Driscoll, Daniel C., and Diane M. LaMoy. "Joint Programs in Chemical and Biological Detection." *Technical Digest: Joint and National Needs.* Dahlgren, Va.: Naval Surface Warfare Center, Dahlgren Division, 2003.

Earle, Ralph, ed. *Navy Ordnance Activities: World War, 1917-1918.* Washington, D.C.: Government Printing Office, 1920.

Eckert, Wallace J., and Rebecca Jones. *Faster, Faster: A Simple Description of a Giant Electronic Calculator and the Problems It Solves.* New York: International Business Machines Corporation, 1955.

Edwards, Donald T., and Doreen H. Daniels. *The SLBM Program: Brief History of NSWC's Involvement in the FBM/SLBM Fire Control Systems, 1956-1984.* Dahlgren, Va.: Naval Surface Warfare Center, 1984.

Eland, Ivan. "Protecting the Homeland: The Best Defense Is to Give No Offense." Policy Analysis No. 306, 5 May 1998. Washington, D.C.: CATO Institute, 1998.

Eland, Ivan. "The Rogue State Doctrine and National Missile Defense." With Daniel Lee. Foreign Policy Briefing No. 65, 29 March 2001. Washington, D.C.: CATO Institute, 2001.

Fahrney, Delmer S. *The History of Pilotless Aircraft and Guided Missiles* (Microfilm A-171). Washington, D.C.: Navy Department Library, n.d. but probably 1949 through 1958.

Farrow, Edward S. *American Guns in the War with Germany.* New York: E. P. Dutton and Company, 1920.

Fiske, Bradley A. *From Midshipman to Rear Admiral.* New York: Century Company, 1919.

Ford, Brian J. *German Secret Weapons: Blueprint for Mars.* New York: Ballantine Books, Inc., 1969.

Frazer, Richard A., and John E. Bibel. "Advanced Technology Demonstration of the Naval Tactical Missile System (NATACMS)." *Technical Digest: Strategic & Strike Warfare Weapons Systems.* Dahlgren, Va.: Naval Surface Warfare Center, Dahlgren Division, 1997.

Friedman, Norman. *Seapower and Space: From the Dawn of the Missile Age to Net-Centric Warfare*. London: Chatham Publishing, 2000.

------. *U.S. Naval Weapons: Every Gun, Missile, Mine, and Torpedo Used by the U.S. Navy from 1883 to the Present Day*. Annapolis, Md.: Naval Institute Press, 1983.

Fuhrman, Robert A. "The Fleet Ballistic Missile System: Polaris to Trident." *Journal of Spacecraft* 15, no. 5 (September-October 1978).

Furer, Julius Augustus. "Research in the Navy." *Journal of Applied Physics* 15 (March 1944).

Gates, Robert V. "Strategic Systems Fire Control." *Technical Digest: Mathematics, Computing, and Information Science*. Dahlgren, Va.: Naval Surface Warfare Center, Dahlgren Division, 1995.

Haggart, James A., Lieutenant Commander, U.S. Navy. "The Falkland Islands Conflict: Air Defense of the Fleet." Quantico, Va.: Marine Corps Command and Staff College, 1984.

Hattendorf, John B. *The Evolution of the U.S. Navy's Maritime Strategy, 1977-1986*. Newport, R.I.: Naval War College Press, 1989.

Hedrick, David I. "Research and Experimental Activities of the U.S. Naval Proving Ground." *Journal of Applied Physics* 15 (March 1944).

Heffernan, Natalie F. "JCC(X)-The Future of Joint Command at Sea." *Technical Digest: Joint and National Needs*. Dahlgren, Va.: Naval Surface Warfare Center, Dahlgren Division, 2003.

Herrick, Robert Waring. *Soviet Naval Doctrine and Policy, 1956-1986*. Book 3, *Studies in Russian History*, vol. 8c. Lewiston, N.Y.: The Edwin Mellen Press, 2003.

Hoag, David G. *Ballistic Missile Guidance*. Symposium on New Technology and the Arms Race, 26-30 June 1970. Cambridge, Mass.: Charles Stark Draper Laboratory of the Massachusetts Institute of Technology, 1970.

Hobson, Chris, and Andrew Noble. *Falklands: Air War*. Aerofax Midland Pub. Ltd., 2002.

Hodges, Peter. *The Big Gun: Battleship Main Armament, 1860-1945.* Annapolis, Md.: Naval Institute Press, 1980.

Hogg, Ian V. *German Secret Weapons of the Second World War: The Missiles, Rockets, Weapons and New Technology of the Third Reich.* Mechanicsburg, Pa.: Stackpole Books, 1999.

Hovgaard, William. *Modern History of Warships.* Annapolis, Md.: United States Naval Institute, 1971.

Howard, Russell, Colonel, USA, and Major Reid Sawyer, USA. *Terrorism and Counterterrorism: Understanding the New Security Environment, Readings, and Interpretations.* New York: McGraw-Hill/Dushkin, 2003.

Howeth, Linwood S. *History of Communications-Electronics in the United States Navy.* Washington, D.C.: Bureau of Ships and Office of Naval History, 1963.

Hughey, Raymond H., Jr. "History of Mathematics and Computing Technology at the Dahlgren Laboratory." *Technical Digest: Mathematics, Computing, and Information Science.* Dahlgren, Va.: Naval Surface Warfare Center, Dahlgren Division, 1995.

Indseth, A. Kris. "Naval Operations Other Than War Technology Center: Incorporating Dahlgren's Legacy in Today's Changing World." *Technical Digest: Joint and National Needs.* Dahlgren, Va.: Naval Surface Warfare Center, Dahlgren Division, 2003.

Isenberg, Michael T. *Shield of the Republic: The United States Navy in an Era of Cold War and Violent Peace, 1945-1962.* New York: St. Martin's Press, 1993.

Johnson, Leland. *Sandia National Laboratories: A History of Exceptional Service in the National Interest.* Albuquerque, N.Mex., and Livermore, Calif.: Sandia National Laboratories, 1997.

Jones, Vincent C. *United States Army in World War II: Special Studies, Manhattan: The Army and the Atomic Bomb.* Washington, D.C.: Center of Military History, 1985.

Katrancha, Mark F. "Validation of Synthetic Time Domain Response of Mode-Stirred Cavities." *Technical Digest: Joint and National Needs*. Dahlgren, Va.: Naval Surface Warfare Center, Dahlgren Division, 2003.

Kim, Hun. "Critical Infrastructure Protection for the Local Installation." *Chips Magazine*, Summer 2002. On-line article at http://www.chips.navy.mil/archives/02_Summer/authors/index2_files/cip.htm.

Leary, H. F. "Military Characteristics and Ordnance Design." *United States Naval Institute Proceedings* 48, no. 7 (July 1922): 1125-37.

Leith, Charles H. *Charles Candy Middlebrook, Esq. (1896-1965) and His Part in the Norden Bombsight of World War II*. Fredericksburg, Va.: C. H. Leith, 1988.

LeVine, Susan D. "Assessing Joint and National Needs for Nonlethal Capabilities." *Technical Digest: Joint and National Needs*. Dahlgren, Va.: Naval Surface Warfare Center, Dahlgren Division, 2003.

Levinson, Jeffrey L., and Randy L. Edwards. *Missile Inbound: The Attack on the Stark in the Persian Gulf*. Annapolis, Md.: Naval Institute Press, 1997.

Lewis, Bernard. *The Crisis of Islam: Holy War and Unholy Terror*. New York: Random House, 2003.

Lewis, Leslie, Roger Allen Brown, and C. Robert Roll. *Service Responses to the Emergence of Joint Decision Making*. Publication MR-1438-AF. Santa Monica, Calif.: RAND Corporation, 2001. Available on-line at http://www.rand.org/publications/MR/MR1438/.

Locher, James R., III. "Taking Stock of Goldwater-Nichols." *Joint Force Quarterly*, Autumn 1996, 10-17.

------. *Victory on the Potomac: The Goldwater-Nichols Act Unifies the Pentagon*. College Station, Tex.: Texas A&M University Press, 2002.

Lord, Clifford L. *History of United States Naval Aviation*. New Haven: Yale University Press, 1980.

Lusar, Rudolph. *German Secret Weapons of the Second World War*. Translated by R. P. Heller and M. Schindler. New York: Philosophical Library, Inc., 1959.

Luttwak, Edward N. *The Pentagon and the Art of War: The Question of Military Reform*. New York: Simon and Schuster, 1984.

Macfadzean, Robert H. "Army Combat Operations and Navy Land Attack." *Technical Digest: Joint and National Needs*. Dahlgren, Va.: Naval Surface Warfare Center, Dahlgren Division, 2003.

Macgregor, Douglas A. "The Joint Force: A Decade, No Progress." *Joint Force Quarterly*, Winter 2000-01.

Macknight, Nigel. *TOMAHAWK Cruise Missile*. Osceola, Wis.: Motorbooks International, 1995.

Marolda, Edward J., and Robert J. Schneller, Jr. *Shield and Sword: The United States Navy and the Persian Gulf War*. Annapolis, Md.: Naval Institute Press, 2001.

Mathis, Dan. "The Concept for DoD's Critical Infrastructure Protection Program: Critical Asset Assurance—A System-of-Systems Approach." *Technical Digest: Joint and National Needs*. Dahlgren, Va.: Naval Surface Warfare Center, Dahlgren Division, 2003.

McBride, William M. *Technological Change and the United States Navy, 1865-1945*. Baltimore, Md.: Johns Hopkins University Press, 2000.

McCollum, Kenneth G., ed. *Dahlgren*. Dahlgren, Va.: Naval Surface Weapons Center, 1977.

Meyer, Victor A. "A Vision of Naval Surface Force Structure in 2030." *Technical Digest*. Dahlgren, Va.: Naval Surface Warfare Center, Dahlgren Division, 1991.

Miller, David, and Chris Miller. *Modern Naval Combat*. New York: Crescent Books, 1986.

Miller, B. Larry, Bruce R. Hermann, Everett R. Swift, Robert W. Hill, Alan G. Evans, James P. Cunningham, Jeffrey N. Blanton. "Contributions to the NAVSTAR Global Positioning System (GPS) by the Naval Surface Warfare Center, Dahlgren Division." *Technical Digest: Technology Transition and Dual-Use Technology*. Dahlgren, Va.: Naval Surface Warfare Center, Dahlgren Division, 1998.

Monteith, Gregory E., Jeffrey H. McConnell, Alvin C. Murphy, John P. Chapman, Thomas A. Micek, Christopher M. Thompson, and Russell G. Acree, Jr. "Networked Engineering . . . From Concept to Reality— The Navy's Distributed Engineering Plant (DEP)." *Technical Digest: Theater Air Defense*. Dahlgren, Va.: Naval Surface Warfare Center, Dahlgren Division, 2000-2001.

Moritz, Elan, Rafael R. Rodríguez, David P. DeMartino, Helmut H. Portmann, Jr., Delbert C. Summey. "Systems Engineering of Unmanned Systems for Naval Readiness Augmentation." *Technical Digest: Joint and National Needs*. Dahlgren, Va.: Naval Surface Warfare Center, Dahlgren Division, 2003.

Moses, William M. "Research in the Bureau of Ordnance." *Journal of Applied Physics* 15 (March 1944).

Muir, Malcolm, Jr. *Black Shoes and Blue Water: Surface Warfare in the United States Navy, 1945-1975*. Washington, D.C.: Naval Historical Center, 1996.

Naval Ordnance Laboratory. *History of the Naval Ordnance Laboratory, 1918-1945: Administrative History*. Washington, D.C.: United States Navy Yard, 1946.

------. *History of the Naval Ordnance Laboratory, 1918-1945: Scientific History*. Washington, D.C.: United States Navy Yard, 1946.

Naval Ordnance Station. "A Narrative History of the Naval Ordnance Station, Indian Head, and of the Gun Systems Division." Typescript. Indian Head, Md.: Naval Ordnance Station, ca. 1975.

Niemann, Ralph A. *Dahlgren's Participation in the Development of Computer Technology*. Dahlgren, Va.: Naval Surface Weapons Center, 1982.

Nitze, Paul H., Leonard Sullivan, Jr., and the Atlantic Council Working Group on Securing the Seas. *Securing the Seas: The Soviet Naval Challenge and Western Alliance Options*. Bouler, Colo.: Westview Press, 1979.

Okun, Nathan. "Major Historical Naval Armor Penetration Formulae." Available on-line at http://www.warships1.com/index_nathan/Hstfrmla.htm (26 November 2001).

Padfield, Peter. *Guns at Sea*. London: Hugh Evelyn, 1973.

Palmer, Michael A. *Origins of the Maritime Strategy: American Naval Strategy in the First Postwar Decade*. Washington, D.C.: Naval Historical Center, 1988.

Pardini, Albert L. *The Legendary Secret Norden Bombsight*. Atglen, Pa: Schiffer Pub., 1999.

Peña, Charles V. "From the Sea: National Missile Defense is Neither Cheap Nor Easy." Foreign Policy Briefing No. 60, 6 September 2000. Washington, D.C.: CATO Institute, 2000.

------. "Theater Missile Defense: A Limited Capability is Needed." Policy Analysis No. 309, 22 June 1998. Washington, D.C.: CATO Institute, 1998. Available on-line at http://www.cato.org/pubs/pas/pa-309.html.

Podlesny, Robert E. "NSWCDD Innovative Technology and Problem Solving for Homeland Defense." *Technical Digest: Joint and National Needs*. Dahlgren, Va.: Naval Surface Warfare Center, Dahlgren Division, 2003.

Poff, Ray E., Doyle B. Green, and Nelson B. Mills. "Joint Warfighting Counterfire System (JWCS)-NetFires-Organic Fire Support for Littoral and Expeditionary Operations." *Technical Digest: Joint and National Needs*. Dahlgren, Va.: Naval Surface Warfare Center, Dahlgren Division, 2003.

Pollard, James R., and Bernard G. Duren. "An Overview of Total Ship System Engineering—A 'Gang of Six' Initiative." *Technical Digest: Mathematics, Computing, and Information Science*. Dahlgren, Va.: Naval Surface Warfare Center, Dahlgren Division, 1995.

Pompeii, Michael A. "Ship ACADA: A New Capability in Chemical Warfare Defense." *Technical Digest: Joint and National Needs*. Dahlgren, Va.: Naval Surface Warfare Center, Dahlgren Division, 2003.

Rashid, Ahmed. *Jihad: The Rise of Militant Islam in Central Asia*. New Haven, Conn.: Yale University Press, 2002.

------. *Taliban: Militant Islam, Oil, and Fundamentalism in Central Asia*. New Haven, Conn.: Yale University Press, 2000.

Reardon, Steven L. *History of the Office of the Secretary of Defense.* Vol. 1, *The Formative Years, 1947-1950.* Washington, D.C.: Office of the Secretary of Defense, 1984.

Reeve, Simon. *The New Jackals: Ramzi Yousef, Osama Bin Laden, and the Future of Terrorism.* Boston, Mass.: Northeastern University Press, 1999.

Reynolds, Clark G. *Famous American Admirals.* New York: Van Nostrand Reinhold, 1978.

Ritchie, David. *The Computer Pioneers: The Making of the Modern Computer.* New York: Simon & Schuster, 1986.

Robison, Samuel S. *A History of Naval Tactics from 1530 to 1930: The Evolution of Tactical Maxims.* Annapolis, Md.: United States Naval Institute, 1942.

Rojas, Raúl, and Ulf Hashagen, eds. *The First Computers: History and Architectures.* Cambridge, Mass.: MIT Press, 2000.

Roman, Elmer L. "The Challenges of Naval Force Protection." *Technical Digest: Joint and National Needs.* Dahlgren, Va.: Naval Surface Warfare Center, Dahlgren Division, 2003.

Rosenberg, David A. "American Postwar Air Doctrine and Organization: The Navy Experience." *Air Power and Warfare.* Edited by Alfred F. Hurley and Robert C. Ehrhart. Washington, D.C.: Office of Air Force History, 1979.

Scarborough, Rowan. *Rumsfeld's War: The Untold Story of America's Anti-Terrorist Commander.* Lanham, Md.: Regnery Publishing, 2004.

Schneider, Mark Bernard. *Nuclear Weapons and American Strategy, 1945-1953.* Ph.D. diss., University of Southern California, June 1974.

Schneller, Robert John. *A Quest for Glory: A Biography of Rear Admiral John A. Dahlgren.* Annapolis, Md.: Naval Institute Press, 1996.

Schwoebel, Richard L. *Explosion Aboard the Iowa.* Annapolis, Md.: Naval Institute Press, 1999.

Shelton, Frank H. *Reflections of a Nuclear Weaponeer.* Colorado Springs, Colo.: Shelton Enterprise Inc., 1988.

Simonoff, Adam J. "Collaborative Planning and Replanning." *Technical Digest: Joint and National Needs*. Dahlgren, Va.: Naval Surface Warfare Center, Dahlgren Division, 2003.

Sinai, Joshua. "How to Forecast and Preempt Al-Qaeda's Catastrophic Terrorist Warfare." *Journal of Homeland Security*, August 2003. Available on-line at http://www.homelandsecurity.org/journal/articles.

Sisson, Dale W., Jr. "The Development and Application of the Shipboard Collective Protection System (CPS)." *Technical Digest: Technology Transition and Dual-Use Technology*. Dahlgren, Va.: Naval Surface Warfare Center, Dahlgren Division, 1998.

Sojdehei, John J. "Magneto-Inductive Communications for Littoral MIW Theater." *Technical Digest: Joint and National Needs*. Dahlgren, Va.: Naval Surface Warfare Center, Dahlgren Division, 2003.

Solka, Jeffery L., David J. Marchette, and Michelle L. Adams. "Applications of Statistical Visualization to Computer Security." *Technical Digest: Joint and National Needs*. Dahlgren, Va.: Naval Surface Warfare Center, Dahlgren Division, 2003.

Spinardi, Graham. *From Polaris to Trident: The Development of U.S. Fleet Ballistic Missile Technology*. Cambridge, U.K.: Cambridge University Press, 1994.

Stares, Paul B. *The Militarization of Space: U.S. policy, 1945-1984*. Ithaca, N.Y.: Cornell University Press, 1985.

Thomas, William H., Michael J. North, Charles M. Macal, and James P. Peerenboom. "From Physics to Finances: Complex Adaptive Systems Representation of Infrastructure Interdependencies." *Technical Digest: Joint and National Needs*. Dahlgren, Va.: Naval Surface Warfare Center, Dahlgren Division, 2003.

Tropp, Henry. "Computer Report VII: The Effervescent Years: A Retrospective." *Spectrum* (Institute for Electrical and Electronic Engineers) 11, no. 2 (1 February 1974).

Turnbull, Archibald D., and Clifford L. Lord. *History of United States Naval Aviation*. New Haven: Yale University Press, 1949.

U.S. Naval Weapons Laboratory. *The NWL Story (1918-1971): Proving Ground to Weapons Laboratory*. Dahlgren, Va.: U.S. Naval Weapons Laboratory, 1971.

Van Auken, Wilbur Rice. *Notes on a Half Century of United States Naval Ordnance, 1880-1930*. Washington, D.C.: George Banta Publishing Company, 1939.

Wade, Brian K., Randy D. Wagner, and Erin M. Swartz. "Collateral damage Estimation." *Technical Digest: Joint and National Needs*. Dahlgren, Va.: Naval Surface Warfare Center, Dahlgren Division, 2003.

Walls, W. A., W. F. Weldon, S. B. Pratap, M. Palmer, and Lt. David Adams, USN. "Application of Electromagnetic Guns to Future Naval Platforms." PR-253. Paper delivered at 9th EML Symposium, Edinburgh, Scotland, May 1998. Austin, Tex: Center for Electromechanics, 1998. Available on-line at http://www.utexas.edu/research/cem/publications/PDF/PR253.pdf.

Westrum, Ron. *Sidewinder: Creative Missile Development at China Lake*. Annapolis, Md.: Naval Institute Press, 1999.

Williams, Michael R. *A History of Computing Technology*. 2nd ed. Los Alamitos, Calif.: IEEE Computer Society Press, 1997.

Yencha, Thomas James. "A Chemical Warfare Naval Simulation Model for Surface Ships." *Technical Digest*. Dahlgren, Va.: Naval Surface Warfare Center, Dahlgren Division, 1991.

Zehring, Steven W. "Critical Infrastructure Protection: Supporting Joint and National Needs." *Technical Digest: Joint and National Needs*. Dahlgren, Va.: Naval Surface Warfare Center, Dahlgren Division, 2003.

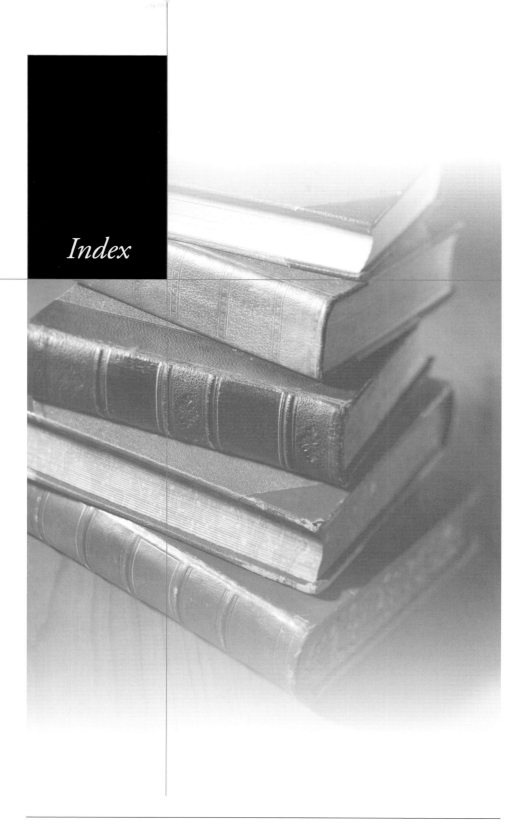

Index

A

B

B-36 *Peacemaker*, 84
Baer, George, 159
Baldwin Locomotive Works, 20
Balisle, Phillip M., 256, 259, 260
Ballantine, John J., 39, 44, 45
ballistic missile technology, 82, 230, 243–244
Ballistiches Institut, 35
Barth, Theodore Harold, 60
base housing at Dahlgren, 24, 27–28, 29, 31, 70, 120
Base Realignment and Closure. *see* BRAC
Bateman, Herbert H., 242
"Bats" (ASM-N-2), 175, 176
Battle of the Bulge, 75
Battle of Jutland, 17, 33, 36, 37
Battle of Tarawa, 70
Battle of Tsushima Strait, 312
Beckwith, Charlie, 211
Beirut, Lebanon, 211
Bennett, Ralph D., 95, 100, 151–152
Benton, Thomas Hart, 9
Berlin Wall, 190
Bernard, Charles "Chuck", *B–4*, 124, 148
Bernard Smith Award, 143
Bigelow, 173
bin Laden, Osama, 220, 252
biological and chemical warfare (BW/CW)
       agents, 138, 207–208, 263–264
Blackistone Island, 18
Blandy, William H. P. "Spike", *A–9*, 67, 72, 74
Blitz, 72
Bloch, Claude C., 33–34
blue-collar employees, 52–54. *see also* civilian
       scientists and workforce at
       Dahlgren
Blue Ridge, 234–235
Blyn, Nadine, *B–6*
BMD (Ballistic Missile Defense), 230
Bock's Car, 77
Boettcher, A. H., 59
BOGHAMMAR, 174
bombsights, *A–7*, 45, 58–62, 71–72
Boomtown, 70, 120
Bowen, Harold G., 65, 81
BRAC (Base Realignment and Closure)
       1991, 194–198
       1993, 199–200
       1995, 224–225
       2005, 226–227, 256, 262–264

Base Structures Committee, 194
       data calls, 33
       environmental issues at Dahlgren and, 201
       recovery from effects of, 237
       warfare centers moved by, 239
       White Oak closure as result of, 198–200
Bradbury, Norris E., 77, 129
Bramble, Charles, 63, 89–90, 92, 94–102, 129
Braun, Boynton, 62
Breslow, Arthur, 76
Brindel, Glenn, 174
Brown, David R., Jr., *B–1*, 105, 107, 108,
       115–116
Bryce, James Wares, 91
BUAER (Bureau of Aeronautics), 56
BUENG (Bureau of Engineering), 43
BUG SPLAT (Fast Assessment Strike Tool-
       Collateral Damage), 233
Building 492, 86
Bull, Gerald, 139–140, 245
BULLPUP missiles, *B–1*
Bunker Hill, *B–7*
BUORD, 81–82
Burke, William, *A–11*, 94
Burroughs, Sherman E., 56
Bush, George H. W., 190, 192–193, 205
Bush, George W., 226, 231–232, 252–255, 256,
       264
Bush, Vannevar, 74, 75, 81, 88–89, 90
BUSHIPS, 92
Butler, Thomas S., 24, 25–26, 27, 28
"buzz bombs", 176

C

C4ISR programs, 262–263, 264
CADs (Cartridge Actuated Devices), 114
calculators, *A–10*, *A–11*, 91–94, 96
Caldwell, W.H., 18
Campbell, Robert, 92
Cann, Gerald A., 193, 194
Carderock, 149, 150
Carl L. Norden, Inc., 60, 61, 62, 102
Carter administration, 160, 161, 209, 211, 348
Cartridge Actuated Devices (CADs), 114
CDC's "Cyber" series, 114
CENTCOM (U.S. Central Command), 254
chemical warfare, 138, 207–208, 263–264
Cheney, Richard B. "Dick", 193, 195
Chief of Naval Operations (OPNAV), 115
China, 169, 230

H

K

L

M

M712 COPPERHEAD 155-mm laser-guided
projectile, 138–139
Machodoc Creek site, 16–19, 21
Macy, Archer M., Jr., 260
Maddox, 132
Mahaffey, Vaughn E., 238
Mahan, Alfred Thayer, 159
Main Battery, *A–5, B–5*, 52, 69
management structure at Dahlgren
"Dahlgren Way", 2–5, 65, 122, 181–182
diversity initiatives, 262
junior management, 188
lines of authority, 34
reorganization under Lyddane, 109
retreats under Anderson, 155
rotation system, *B–3*, 124–125, 146, 148,
164, 181, 188
under Smith, 123–126
systems philosophy, 5, 142–143, 181–182,
195–197, 209
Theater Warfare Systems, 227–232
under Thompson, 3–4
Total Quality Leadership (TQL)
management, 202–203
of White Oak vs., 150
Manhattan Project, 75–77
Marietta, Martin, 138
Maritime Strategy, 222
Mason, John Y., 10
Mason, Newton E., 13
Massachusetts, 71
Masters, Michael W., 203
Maxcy, Virgil, 9
Mayflower, 14
Mayo, Richard W., 240
McCormick, Benjamin B., 40–42
McCullum, Ken, *B–6*
McDonald, David L., 115
McGettigan, Joseph L., 260, 261–262, 265
McGinley, Skip, 204
McGlade, Joe, *B–6*
McNamara, Robert S., 114, 122, 126–127, 147
McVay, Charles B., Jr., 21, 24, 25, 26–27, 29,
31–32, 42–43
MDL (Mine Defense Laboratory). *see* Panama
City, Florida
Measured Response Options (MRO)
program, 237
Merrill, 177

metallurgical research, 64–67, 82, 115
Meyer, Wayne, 164–165, 166, 168–169
Meyers, Wesley W., 86
microfilm, 105–106, 116
microwave weapon efforts, 135
Middendorf, J. William, II, 151
Middlebrook, Charles, 61
Mikulski, Barbara A., 199
Miller, John, 220
Miller, Paul D., 215, 217
Milligan, Richard, 171
Mills, James H., Jr., *B–4*, 115, 136
Mine Defense Laboratory (MDL). *see* Panama
City, Florida
mine warfare, 152
minorities at Dahlgren, 262
Mirick, Carlo B., 43–44
Missile Defense Agency (MDA), 235–236, 251
MIT, 89, 90, 92, 94, 175
Mk 148 computer, 162
Mk 160 Gun Computer System (GCS), *B–9*
Mk 45 Mod 4, *B–8, B–9*, 245
Mk 46 torpedoes, 167
Mk 75 gun mounts, 162
Mk 8 Light Case, *A–11*, 84–88
Mk 88 Fire Control System, 117
MLRS (MULTIPLE LAUNCH ROCKET
SYSTEM), 244
modernization of Dahlgren, 119–120
Monger, Albert, 152
Moody, William H., 12
Moore, Robert W., 28
Morella, Constance A., 197
Morse, Robert, 127
Moulton, Ray, 35–36
MRO (Measured Response Options
program), 237
Mudd, Sydney E., 23, 24, 26, 27, 28, 29, 30
Mullen, Mike, 261, 262
Multiple Independently Targetable Reentry
Vehicle (MIRV) warheads, 117
Multiple Reentry Vehicles (MRVs), 116–117
Mundy, Carl E., Jr., 222–223
Muroc Field, 76

N

Nagasaki mission, 75, 77
naming of and name changes of Dahlgren,
21–22, 110, 172, 184–185, 298–299
Nanos, G. P., Jr., 253

T

W

Y

Z

THE SOUND OF
FREEDOM

1918 - 2006